The Art and Philosophy of the Garden

The Art and Philosophy of the Garden

DAVID FENNER AND ETHAN FENNER

OXFORD
UNIVERSITY PRESS

OXFORD
UNIVERSITY PRESS

Oxford University Press is a department of the University of Oxford. It furthers
the University's objective of excellence in research, scholarship, and education
by publishing worldwide. Oxford is a registered trade mark of Oxford University
Press in the UK and certain other countries.

Published in the United States of America by Oxford University Press
198 Madison Avenue, New York, NY 10016, United States of America.

Library of Congress Cataloging-in-Publication Data
Names: Fenner, David E. W., author.
Title: The art and philosophy of the garden / David Fenner and Ethan Fenner.
Description: New York, NY : Oxford University Press, [2024] |
Includes bibliographical references and index.
Identifiers: LCCN 2023045775 (print) | LCCN 2023045776 (ebook) |
ISBN 9780197753590 (hardback) | ISBN 9780197753613 (epub)
Subjects: LCSH: Gardens—Philosophy. | Gardening—Philosophy. |
Aesthetics. | Form (Aesthetics)
Classification: LCC SB454.3.P45 F46 2024 (print) | LCC SB454.3.P45 (ebook) |
DDC 635.01—dc23/eng/20240110
LC record available at https://lccn.loc.gov/2023045775
LC ebook record available at https://lccn.loc.gov/2023045776

DOI: 10.1093/oso/9780197753590.001.0001

Printed by Sheridan Books, Inc., United States of America

Contents

Acknowledgments

David's Acknowledgments

When Ethan and I began to talk about writing this book, we had both read David Cooper's *A Philosophy of Gardens* but little else on the topic. The next book I read was Mara Miller's *The Garden as an Art*, and I was so impressed by it that, after consulting a friend who knew her, I wrote to Mara to say so. While this book has a literature to react to, Mara's book, being the first of its kind since the eighteenth century, largely had to invent the discussions that Ethan and I now join. Since that time, Mara has been singularly helpful in moving our project forward. Her encouragement has been invaluable, and my greatest debt of gratitude is owed to her.

I was introduced to gardens by my grandmother, Roberta Pearl Fenner. Her shade garden in Florida's St. Petersburg Beach, dominated by a large Jacaranda tree frequently covered with purple flowers, was the first garden I understood to be a garden. It had a clear design and was beautifully balanced. Later, when I moved to my own house in Jacksonville, it was my friend Father Neil Gray whose camellias became the parents of more than sixty camellias that now populate my garden. Neil and my grandmother are the spiritual foundations of my interest in gardens and gardening.

I thank Chuck Hubbuch, the head of horticulture at my university for many years, for his encyclopedic knowledge of plants and his boundless patience in guiding me as I planned the occupants for my garden. I thank Larry Weaver and his crew for partnering with me in the maintenance of my garden—Larry takes care of the routine gardening and "heavy lifting" so I can do all the fun bits. I thank Pam Kimball and Sam Kimball, Shelby Miller and David Courtwright, and Scott Hochwald and Joanne Hochwald—colleagues and good friends—for listening to me talk about nothing except this book for months on end. I thank Yuriko Saito for her encouragement as I submitted my first papers on gardens to *Contemporary Aesthetics*—and all the anonymous and generous reviewers of this book and those papers—and I thank Peter Ohlin for his kindness and support in making this book a reality.

Ethan's Acknowledgments

I came to a love of gardens on a meandering pathway, with subtle direction coming from more friends, family members, teachers, mentors, and coworkers than I can list here or even recall. I thank my mother, Mynette Fenner, and my father, David Fenner, whose home garden in Jacksonville, Florida, I helped develop over many weekends growing up; Father Neil Gray, whose back garden provided the setting for an annual playing of Berlioz's Requiem and many more hours of formative garden appreciation; my high school biology teacher Mrs. Norah Betancourt, who moved me in the direction of studying botany; my academic advisor and thesis sponsor at New College of Florida, Dr. Amy Clore, whose tireless devotion to her students ensured my undergraduate education was the best it could possibly be.

During my college years I began to appreciate the work of botanic gardens and the direct impact they can have on the conservation and study of the world's threatened plant species. I thank Dr. Bruce Holst at Marie Selby Botanical Gardens for early guidance and Cheryl Peterson at Bok Tower Gardens for patiently assisting research and conservation of Florida natives; Louisa Hall, Rebecca Hilgenhoff, and everyone else I met during my internship at the Tropical Nursery of the Royal Botanic Garden, Kew, who inspired me to become a professional horticulturist; and all the many teachers, instructors, volunteers, and classmates with whom I worked at the School of Professional Horticulture at the New York Botanical Garden. Special thanks to Kristen Schlieter and Yukie Kurashina, with whom I worked at NYBG, and George Longoria, with whom I worked at the San Francisco Botanical Garden, who reminded me of the human aspect of the work we do in the garden.

I thank Marta McDowell and Toshi Yano for their important feedback on the first chapters of this book, and the many friends who donated pictures used between chapters: Adam Dooling, Marta McDowell, Toshi Yano, Samantha Bachert, Eric Hupperts, and Sarah Snow. Many thanks to my talented coworkers at the University of California Botanical Garden, who teach me something new every day, to our volunteers and visitors, and to the horticulturists who came before me in the Southern African Collection. Thanks to my friends Sean Cameron, Gideon Dollarhide, Clare Loughran, and John Ruzel, who over the past couple years have listened and provided feedback to stray points brought up during the course of writing, and thanks to Mr. Mendoza and Cornerstone in Berkeley for providing a place to write in the evenings after work.

Cover: Knot Garden, Filoli, Woodside CA. Initially designed and installed by Woodside-Atherton Garden Club, 1976. Photo by Ethan Fenner, Spring 2018.

Introduction

Gardens, as a kind, are one of the most ubiquitous and popular of all aesthetic forms.

Aesthetic forms are all around us. The vast majority of objects and events[1] we encounter on a daily basis have distinctive aesthetic characters that motivate us to seek them out, acquire them, and make them part of our lives. Where and in what we choose to live, what cars we drive, how we attire ourselves, what we consume, and what we use to prepare ourselves to face a day clean and fragrant—all of these have pronounced aesthetic characters that strongly stand out, in many cases as their most central feature. While many of the aesthetic forms with which we choose to surround ourselves and imbue our lives are created by us, just as many come directly from nature. Throughout history and across cultures we have memorialized the aesthetic forms of plants, making them the subject of countless works of art and the default choice for adornment, from the Acanth leaves at the tops of Corinthian columns to the botanical flourishes that create the border of an Ottoman rug. Think of the place in our lives of cut flowers, one of the most prevalent of aesthetic forms. These days we have dozens of options of cut flowers from which to choose, with a range of shapes, colors, and scents. We may choose some to say "congratulations," others to say "condolences," but their ultimate purpose is the same: to beautify our living spaces for a week or two. The profitability of the cut flower industry speaks to our affinity toward the most anthropocentrically "useless" and ephemeral part of the plant, which neither feeds nor clothes nor shelters us.

The plants that occupy the pots on our back patio and the boxes on our windowsills require an investment of our time and attention to keep them healthy and beautiful. A whole front and back yard, which may exceed the square footage of a house, calls for specialty equipment, some amount of know-how, and hours of often strenuous work to keep these spaces up to par. If we fall behind on this work, we risk incurring the judgments of our neighbors and the wrath of a Home-Owners Association. In urban settings, where green space is at a premium, new

---

[1] From here on, "objects" will stand for both "objects" and "events."

White Garden, Sissinghurst Castle Garden, Cranbrook, UK
Vita Sackville-West, 1950
Photo by Adam Dooling, Summer 2016

The Art and Philosophy of the Garden. David Fenner and Ethan Fenner, Oxford University Press.
© Oxford University Press 2024. DOI: 10.1093/oso/9780197753590.001.0001

developments require that at least some land be devoted to plants—a border of shrubs around an apartment building, tree pits on the sidewalk, and plants to fill spaces in medians and parking lots. These too require time, effort, and money to keep alive, and perhaps our cities and suburbs would be more efficient without them. But we have decided that green spaces are worth it—we prefer the streets with flower baskets, the yards with trees, and the neighborhoods with parks.

While we may decorate the things we make and the spaces we occupy with plants, the quintessential expression of the intersection between our aesthetic sensibilities and the natural world is a garden. Whether we create and maintain our own gardens, tour destination gardens, pass along or through gardens on our way to work or as we return home, gardens typically present themselves as places of intimate connection between nature and human endeavor, of respite and relaxation, of renewed perspective, but perhaps most of all as places of beauty. At their most developed, gardens occupy the foreground rather than the background of our time and attention. Most are the result of many generations of daily care and planning, involving the work of specialists in many trades working in concert to create a cohesive whole. We visit gardens because we seek positive experiences, and those able to come away with such experiences represent an enormously wide swath of the human population.

Gardens, like the many plants they typically contain, are "aesthetic-forward." So it is surprising that not more has been written about them as aesthetic objects. There are of course reasons for this, and we will examine those, but considering the incredible popularity of gardens and gardening, they are not given much attention in the academic literature. The project of examining the aesthetic character of gardens is motivated by the same attitudes and values that motivate examining the aesthetic character of anything, especially of those things that are "aesthetic-forward." We seek to understand a basic human inclination to experience and appreciate an object in terms of our sensory relationship with that object and in terms of everything relevant to that experience and appreciation (which can be for many aestheticians a larger project than the straightforwardly sensory one). But beyond this—and perhaps more basically—we seek to enhance the value of such experiences. We are motivated to seek out aesthetic opportunities and turn our eyes (ears, nose, and so forth) to objects in ways we commonly describe as aesthetic. This motivation is fueled by something of value—it might be simple felt pleasure; a satisfaction that comes with order and purposiveness; a satisfaction that comes with working out the logic embedded in a cognitive puzzle; to feel a new feeling or feel more intensely or in a new context; to establish a relationship with another human being who has expressed themselves[2] in revelatory ways—what motivations prompt a human being to invest

---

[2] Throughout this book we will use the plural "they" or "their," even with singular sentence subjects, to avoid gendered singular pronouns.

the time and effort to engage with an object in ways we commonly think of as aesthetic may be quite plural, but they are all based on some value held by the person so motivated. The enhancement of that value by attempting to understand it carefully, systematically, and deeply through disciplined analysis and critical reflection is a goal of this book.

We have four tasks to accomplish in this introduction. First, we wish to explain what it is to undertake a consideration of the aesthetics of gardens and gardening. Second, we want to offer a scan of the main literature devoted to garden aesthetics and garden theory that is focused to a large degree on garden aesthetics. This scan is not meant to be comprehensive or constitute a history; rather it is meant to provide a baseline in terms of contemporary work done on garden aesthetics and the philosophy of gardens so that this book is appropriately nested in the recent work of others. That scan will focus on how those theorists understood the projects they undertook, and some reasons they took to be why more work of a theoretical nature about gardens has not been undertaken. Third, we want to ask and answer the question "why now?" Why is this the right time to take up a project in garden aesthetics? There are a few answers to this question, and at least one will refer to the garden aesthetics and philosophy literature, conceptualizing this book as a continuation of a conversation others took up over the last three decades. Fourth, we want to lay out some terms and their meanings. This will be very brief. As in all academic projects, it is crucial that how we use particular words in this book, what we take them precisely to mean, be entirely transparent. There is not a great deal of vocabulary to survey—and that again speaks to the popularity of gardens and how the language we apply to gardens and our experiences of them is broadly transparent already—but this will be the last task of this chapter before moving to a conversation about what gardens are, or perhaps what The Garden is, and what characteristics they all commonly share (if any).

The Project

Of all the aesthetic matters to which to direct academic attention, why this one? Why focus on the aesthetics of gardens and of gardening? Before attempting an answer, it should be noted that this question is already "up two levels" from more basic questions: first, why direct philosophical attention to gardens? And, second and more basically, why direct any academic attention at all to gardens?

It is simplistic merely to say that gardens are worthy of academic attention because of their ubiquity and popularity. Many things are very present and very popular—breakfast cereal, laundry detergent, mailboxes, stop signs—but on the whole we do not invest academic attention in these things. The reason gardens are appropriately the subject of academic attention—whether that attention is biological, anthropological, historical, art-historical, philosophical, or from the perspective of a practical or professional discipline like one focused on architecture or agriculture—is because they possess aspects that allow such attention to gain weight and gravity.

Gardens have been with us for an extremely long time. They may have been with us—as a focus of our deliberate planning and our conscious efforts—from almost the time we moved from being nomadic hunters and gatherers to when we started putting down our first agrarian roots. As we spread and cultures formed that grew distinct from one another—and as having enough of a character to be judged distinct from one another—gardens took on that character and became cultural expressions that, viewed from our vantage point in the twenty-first century, not only are outward symbols of those cultures but are part of the fabric of what gives those cultures their very identities.

Three excellent examples come from (1) Chinese imperial gardens that bring the natural and supernatural worlds into the orbit of the emperor, (2) Japanese tea gardens and Japanese (or Japanese-style) *karesansui* gardens (dry gardens of the sort commonly associated with Buddhist temples in Kyoto), and (3) what we might broadly call "Islamic gardens"—*chahar bagh* gardens and Mughal gardens—which reinstantiate or represent the form of paradise or the Garden of Eden.[3] All of these garden styles not only express the cultures of which they are part; they reinforce and reestablish their cultures, providing a means of cultural identification that spans the life of the garden both as a physical garden and as part of a cultural history. They not only have the style they have because of the culture of which they are part; they assist in creating that culture.[4]

No human lives within a vacuum; we all have social and cultural contexts, even hermits and recluses. To understand other humans—or of perhaps greater importance, to understand ourselves as human—we must understand these contexts. Academic attention directed at gardens advances that goal, and it does so in ways that are not merely ancillary but definitive of our identities. What gardens we create, how we create them, whether we create them individually or collectively, whether an individual can create a garden—all these issues speak not only to our physical contexts in obvious ways but to our cultural contexts and, for many of us within a particular culture or society, to our individual identities.

---

[3] "Walpole had argued that every culture has 'designed' the garden of Eden in its own image." John Dixon Hunt, *Greater Perfections: The Practice of Garden Theory* (London, UK: Thames and Hudson, 2000), p. 208. But it is in *chahar bagh* gardens that this is most obvious.

[4] While it is our intention, as will become plain in successive chapters, to speak to the character of all gardens, we will return to many of the same garden examples throughout the book to illustrate our points because (1) they are easily identifiable with a style, a gardenist, a place, or a time, and they stand out as icons; (2) they occupy a particular place in the history of and literature on gardens; and/ or (3) they are gardens of which readers can easily find photographs and histories on the internet or in reference books.

Biological (botanical, horticultural, agricultural, agronomical) attention may be self-evidently justified, and the argument above may be enough to justify attention directed at gardens that is social-scientific—attention from anthropology and history, perhaps—and attention from the perspective of professions like architecture—but does it justify attention from, say, the humanities?

For The Garden to justify attention from the humanities, it must engage with either or both of two things. First is an engagement with meaning. Whether or how gardens can be interpreted and understood as bearers or conveyers of meaning is something we will examine in Chapters 7 and 8 of this book. But there is perhaps a more fundamental way of thinking about the relationship of meaning and gardens—or, perhaps better stated, there is a way to think about gardens and "meaningfulness" or the garden's possession of meaning for and in the lives of those who attend to them. This is the focus of David E. Cooper's book *A Philosophy of Gardens*.[5]

> What explains the immense significance that human beings locate in the making and experiencing of gardens? . . . The kind of significance that concerns me—the kind germane to the "fundamental question" of "Why garden?"—is the significance gardens have *for* people: for Sir Francis Bacon, say, when judging the garden to be "the purest of human pleasures"[6]

His "fundamental question" is not only the organizing principle of his study; it informs the content of everything in the book. Cooper is interested in what meaning—or again, what meaningfulness—The Garden holds for humans. His work is philosophy written by a philosopher, but the psychological tone of the question and of the answer he works toward is evident. In one sense, Cooper's study is an exercise in garden interpretation, not in terms of interpreting any particular garden but rather interpreting the meaning of The Garden. As we will see later, Cooper discusses garden interpretation in the particularized and sited way we will, but the interpretation of any particular garden is not Cooper's general focus or what motivates his study as a whole. He instead is interested in exploring his "fundamental question" in terms that focus on The Garden and on the value(s) that motivate one to engage in thinking about gardens and engage in gardening practices. That "Why Garden?" is a question that strikes us not only as legitimate but as one that might well lead, if asked and answered well, to insights and illuminations about human values speaks to the worth of considering the question in the first place. And if Cooper's "fundamental question" may lead to

---

[5] David E. Cooper, *A Philosophy of Gardens* (Oxford, UK: Oxford University Press, 2006).
[6] Cooper, p. 3.

illumination, it easily may be expected that other such theoretical treatments of related questions will, too.

If gardens are not only meaningful as manifestations of the values of humans translated into actions, but they are also bearers of meaning—separately as individual gardens or collectively as The Garden—it suggests that the work of garden interpretation is worthwhile as meaning is discovered and revealed. This takes the argument in support of paying attention to gardens in abstractly theoretical ways beyond merely philosophic treatments and shares the general project with the rest of the humanities, from those invested in projects of reading texts and textual meaning to those who develop portfolios of historical and contextual information about gardens as aesthetic (and perhaps art) objects and so on. The particularized nature of much of the practice of focusing critical attention on particular gardens requires that the philosopher share with their colleagues in other humanities disciplines the responsibility of thinking about gardens in these abstract ways.

In *A Philosophy of Gardens*, Cooper writes:

> While I have urged that a philosophy of gardens should not be restricted to the aesthetics of gardens, it is nevertheless within the domain of aesthetics that one might above all have expected discussion of the garden to have been more prominent than it has been.[7]

He follows this up with four reasons for why this expected discussion has not materialized in the way one might expect—and we will look at these four reasons in a moment—but the point of our quoting him here is to move from an argument for thinking about gardens at all, through an argument for thinking about them in philosophical terms, to at least an expectation that thinking about gardens aesthetically should be central to what it means to think about gardens philosophically.

There are some theorists who believe an aesthetic approach to thinking about gardens is too limited and too limiting.[8] And we would agree, as the discussion

---

[7] Cooper, p. 8.

[8] Mara Miller, *The Garden as an Art* (New York, NY: State University of New York Press, 1993). On page 5, Miller writes, "This makes all the more important a second burning preoccupation behind this study, namely, the question whether aesthetic theory as we know it is adequate to art and to its description and to the recognition and analysis of its effect? The example of the garden strongly suggests that it is not." And on page 32 she writes, "But if aesthetic value is inherent in gardens of all types, it is also true that gardens have a large number of other functions as well." We do not disagree with Miller at all, but we believe, first, that an aesthetic consideration of gardens can transcend simply talking about aesthetic value per se, and, second, that while a full consideration of The Garden in philosophical terms likely will transcend merely the aesthetic, the aesthetic is a good place to begin. We take her central question about whether The Garden is an artform to be a question within traditional aesthetic discourse.

above should illustrate. But too often some reduce the notion of adopting an aesthetic approach to merely an engagement with the perceptually based or the formal aesthetic features of a garden and nothing more. Such a view is not in keeping with the practice of philosophical aesthetics as illustrated in its literature. A wider view may be the case over the entire historical course of the discipline of philosophical aesthetics, but it is certainly the case in the nineteenth, twentieth, and twenty-first centuries as aesthetics, in relating to the movements and offerings of the art world—thought of in its broadest terms—has had to focus on matters external to the perceptual or formal properties of works of art. Indeed, as Pablo Picasso led the movement to dispense with aesthetic virtues—such as beauty—as the sine qua non of art, aesthetics as a discipline has been forced to bifurcate discussions into foci on aesthetics per se and, on the other hand, on art and the philosophy of art. Apparently not all art objects are aesthetic objects, at least post-Picasso, and yet philosophizing about art has been the province of disciplinary aesthetics since at least that time.

A similar case can be made when thinking about the role of subjectivity and the place of the audience member—the person having an aesthetic experience as they focus attention on an aesthetic object, with that focus commonly today characterized not only by attention directed toward the formal aesthetic properties of the object but also to all manner of apparently relevant externalities or contextualities having to do with the origins of the object; with the object's relations to other objects and to all manner of relevant themes and motifs; and with associations, identifications, tastes, and relevant experiential contextualities that characterize the relationship created as the subject considers the object, especially what the subject brings to that relationship. These matters transcend the aesthetic conceived narrowly as about perceptually based aesthetic properties of objects. And yet these matters to one degree or another have been focal in the professional aesthetic literature. The point is that taking an aesthetic approach to gardens should not be thought of as limited to a pre-nineteenth-century notion of aesthetics as only dealing with what is available to the senses in an unmediated way or what is based on or closely related to that, and as we will see, the broader spectrum of aesthetic interaction is crucial to a full understanding of gardens as aesthetic objects.

If an aesthetic treatment of gardens is not limited to merely the sensory (or perhaps we should say "formal" to ensure inclusion of the artform "literature" and representational and expressive features of aesthetic objects), what then might such a treatment cover? Questions we intend to take up in this book include the following.

Classificatory and Ontological

- What is "a garden"? When does a garden come into being?
- What is it "to garden"? What are gardening practices?
- Is a garden "an object"? If so, what kind?
- When does a garden cease being a garden or change into a different garden?
- What is it for a garden to be an aesthetic object?
- What is it to be an aesthetic property (feature, aspect, etc.) of a garden?
- What is it to have an aesthetic experience of a garden? How does this relate to other aesthetic experiences? To experiences commonly?
- What is it to be an aesthetically relevant contextual property of a garden or the experience of a garden?
- Can a garden be a work of art? Is The Garden an artform? A craft? A subsidiary or subset of another artform? If The Garden is an artform unto itself, how does it relate to other artforms or art kinds?
- If The Garden is an artform, what is it to be a "garden artist"?
- What is it to be a member of a garden's audience? What is it to be an appreciator of a garden?
- Can a garden be "faked" or "forged"?

Epistemological and Hermeneutical

- How do/can we come to know (see, hear, sense) a particular garden as an object?
- Can a garden be a conveyor of knowledge or meaning? If so, how?
- How can a garden be "read" or interpreted?
- Can The Garden possess a meaning? If so, how do we come to know it?

Axiological and Normative

- Should gardens exist? If so, under what conditions or constraints (if any)? Is there any kind of garden that should not exist?
- Are there particular ways within which gardens should be regarded, valued, or engaged? Is there any special perspective one should take to value a garden correctly?
- Can one garden be better than another? Can a garden be evaluated? Aesthetically evaluated? Artistically evaluated?
- How do we manage the subjective in our appreciation or evaluation of a garden? Should personal associations and identifications play a role?

- What is the role of taste in the evaluation of gardens? The role of cultural context? The role of function?
- How are the values associated with gardens related to other values? Virtues? Vices? Religious values?
- Can or should gardens be thought about in moral/ethical terms?

At one point or another, and to one degree of centrality or another, all these kinds of topics have figured into professional aesthetic discourse. And so they will figure into our consideration of gardens and will fall under the heading of "the aesthetics of gardens and gardening."

The goal of this project, and the means of judging its theoretical significance, is to enhance our understanding of what it means to consider The Garden or any particular garden from an aesthetic point of view. This goal, if it is reached, will be reached through considering answers to the various questions posed directly above and to questions related to them. If the goal is reached, then this will in turn lead to the achievement of another result: the enhancement of garden appreciation in general. Just as so many aesthetic experiences and art experiences are made deeper and richer through the introduction of information, through critical reflection, and through the application of theories that address ways to frame, to inform, and to guide those experiences, so the same might be expected when these same efforts are directed toward our considerations of gardens, aesthetic forms—to repeat—that are perennially ubiquitous and popular.

The Literature

This brief review will focus not on the breadth of the views of the theorists we survey but rather on their motivations for uptake and participation in the conversation about garden theory. Many of these theorists offer justifications for why attention should be focused on garden theory and explanations for why more attention has not focused this way.

One could think of garden literature beginning with the first mentions of garden design. This apparently occurred in the Sumerian Epic of Gilgamesh, originally written around 2000 BCE[9] (the Garden of Eden would have been described in the *Torah*, originally written—or compiled—as early as 500 BCE[10]).

---

[9] Stephanie Dalley (ed.), *Myths from Mesopotamia: Creation, the Flood, Gilgamesh, and Others* (Oxford, UK: Oxford University Press, 2000).

[10] Biblical scholars believe Genesis was written as a compilation of three or four original sources and the final compilation was written sometime after the Babylonian exile (starting shortly after 600 BCE), during the Persian rule over Judah, which began in 539 BCE and continued until the Greeks entered the picture in 331/332 BCE. A date of 500 BCE puts the final compilation toward the start of the Persian period.

The oldest writings on gardening practice are found in Book X of *De Re Rustica* by Lucius Junius Moderatus Columella, written sometime before 70 CE (the year in which he died).[11] Japan's oldest garden design manual, likely the oldest instructional manual on garden design in the world, is the *Sakuteiki*, dating from the eleventh century CE. But our interest is in theoretical—philosophically aesthetic—literature focused on The Garden, and the origins of that literature are in eighteenth century Europe.

In Britain, Horace Walpole (Fourth Earl of Orford), one of the leading garden writers of the (mid-) eighteenth century, described The Garden as one of the three sisters of the arts, alongside painting and poetry.[12] William Kent, of about same time period and occupation, is said to have written that "all gardening is landscape painting." Walpole and Kent were working at the time (early to mid-1700s) when Lancelot "Capability" Brown was transforming the estates of the English wealthy by removing their French-influenced formal and Baroque gardens in favor of naturalistic and bucolic settings that stretched for the hundreds of acres that typically surrounded their great houses. Walpole was joined in his discussion of the aesthetic merits of The Garden by William Gilpin, Richard Payne Knight, John Loudon, Alexander Pope, Uvedale Price, Humphry Repton, and Thomas Whately, and by members of the so-called British taste theorists: Anthony Cooper, Third Earl of Shaftesbury, working in the late 1600s and early 1700s; Joseph Addison, writing for the *Spectator* in the early 1700s, and Archibald Alison, who, in 1790, wrote,

> [T]he great superiority of its [gardening's] productions to the original scenes in nature, consists in the purity and harmony of its composition . . . to awaken an emotion more full, more simple, and more harmonious than any we can receive from the scenes of nature itself.[13]

The Garden had become central to the exercise of taste of those presumed to have taste, and so it was incumbent upon the thinkers of the day to engage garden theory and speak to it in aesthetic terms as well as incorporate it into their more general aesthetic theories. This may be seen as a case of theory following the practice of the arts, and by and large the respect for garden design by those who

---

[11] Some scholars may say that Vitruvius, in his work on architecture in the first century BCE, offered insights on gardens, but we would argue his contributions are more appropriately thought to be about agriculture.

[12] Walpole's assertion forms a major thread in Stephanie Ross's *What Gardens Mean* (Chicago, IL: University of Chicago Press, 1998).

[13] Archibald Alison, *Essays on the Nature and Principles of Taste* (Edinburgh, 1790), p. 85. Despite Alison's celebration of The Garden in this quote, he goes on, a few pages after, to categorize the arts, and he does so in virtue of the control that an artist has over their medium. Gardening falls low on the scale as so much of the medium is presented by nature in a way unaltered by artifactual design. This limits the depth of the artist's expression or expressibility.

were practicing it and benefiting from its practice is reflected in the theory. The Garden was thought entirely appropriate for aesthetic discussion.

On the Continent in the eighteenth and early nineteenth centuries, Immanuel Kant and Georg Hegel continued discussion of The Garden. Their comments were fairly brief, and while neither explicitly denied the "received" view that The Garden was an artform, both discussed The Garden—as art and as aesthetic object—in less than purely positive terms. We will take up discussion of their views in the chapter concerning whether The Garden is a bona fide artform. Kant and Hegel were joined by Karl Heinrich Heydenreich and Friedrich Schiller, whose contributions to garden theory we will review briefly in a moment. Heydenreich and Schiller were more friendly toward The Garden than were Kant or Hegel.

While he does not take himself to be doing philosophy per se, certainly one of the central and earliest voices to this kind of literature is that of John Dixon Hunt. His book-length contributions to this area commence in the late 1970s. Hunt's 2000 book *Greater Perfections: The Practice of Garden Theory* is in the vein of a call to increased focus on garden theory. In fact, it represents not merely a call to increased focus but a detailed map of how that literature should be guided and informed. Hunt believes what we need is not a history per se, and certainly not a philosophy, but rather a kind of anthropology. He believes understanding gardens in terms of their particularity and how they are contextualized in cultures and times is essential to understanding the attitudes and values that frame attention to them, whether that attention is theoretic (concerning their design) or concrete in terms of appreciating gardens and garden practices.

> The subject of landscape architecture has no clear intellectual tradition of its own, either as a history, a theory, or even a practice. . . . [14] Lest we develop too firm a commitment to any one perspective—literary, art historical, philosophical, horticultural, and so on—the garden theorist must invent the subject anew. . . . [W]e must discover within the activities of garden art and landscape architecture themselves the ground of an adequate theory.[15]

Hunt offers a reason why more work in aesthetics and philosophy in general has not been as common as we might expect; he does not offer this reason as encouragement for more philosophic work, though. Actually the opposite—but his view gives us insight on the matter. Michael G. Lee, in writing about Hunt, illustrates Hunt's point clearly:

---

[14]   Hunt, *Greater Perfections*, p. 6.
[15]   Hunt, p. 8.

> [U]nlike recent architectural theory, where philosophy has played a role, doubts about its usefulness as a potential tool for garden theory continue to be raised by some of the field's most prominent scholars. John Dixon Hunt, for example, has recently argued . . . that philosophy is too mired in verbal hair-splitting to be of much use in understanding a phenomenon as physically and sensually saturated as a garden[16]

Lee suggests that the aims of philosophy are different from the aims a historian such as Hunt might embrace, and as their goals are different, so too will be their methodologies and their tolerance for the level of detail that Lee describes as hair-splitting. And indeed, Lee's book, from which this quote comes, is in a series of philosophy books. But for all this we should not miss the larger point. The Garden is indeed sensually saturated, and the level of abstraction that characterizes philosophy may seem too removed to be of use in understanding it. Philosophers involved in, say, the detailed work of definition might seem engaged in projects that cast no real light on understanding The Garden even at the level of abstraction of theoretical garden design, the area in which Hunt works, much less at any level of lesser abstraction. Those who take up gardening practice, either as professionals or amateurs, see the most progress made when hands are in direct contact with the soil. If Hunt's view is standard, this could explain why more philosophical work has not been done about The Garden—it might not be seen as useful.

There are at least two arguments to offer in response. First, the success of the contributions that philosophers have made to garden literature demonstrate that philosophizing about gardens can shed illuminating light on The Garden, not only in ways that satisfy philosophers but in ways that help anyone with passing academic interest in The Garden understand the character and value of both the garden and gardening practice. Philosophy is adept at getting to the most basic issues. To understand a thing thoroughly is to start at ground level, and philosophy is especially good at that through uncovering assumptions and questioning axioms. But in addition to this, philosophy can deal with matters of value and normativity in ways that other disciplines typically cannot or do not. To understand why we invest so much attention in our gardens as we dig in the dirt or in The Garden as a focus for thought is to understand primarily why we value gardens and gardening.

Second, to pick up on Lee's point that gardens are sensually saturated, the greatest usefulness that philosophy may exhibit as it focuses on The Garden is through that area devoted to exploration of how we connect to the world in ways

---

[16] Michael G. Lee, *The German "Mittelweg": Garden Theory and Philosophy in the Time of Kant* (New York, NY: Routledge, 2013), pp. 7–8.

that are many times, in terms of sensory connection, immediate. That of course is the area of aesthetics. We connect to gardens in ways that are exceptionally complex; this is the case when it comes to our "unmediated" sensory connection more so perhaps than with any other object to which we take an aesthetic point of view. Gardens typically envelop us; we typically move through them rather than simply look at them. This makes an "aesthetics of gardens" unique (or close to unique) in terms of how it explains and captures these phenomena, and with that level of uniqueness comes opportunity for revelation that might not come with exploration of aesthetic forms that are more common and significantly less complex.

Does the level of complexity of garden interaction render insights too particularized—too individualized to particular gardens, particular cultures, particular times—to be of use in fashioning ways to understand gardens at large or The Garden, as we have been using the term? This does not have to be the case, but the devil will be in the detail. Insights that can apply broadly to how we connect to gardens aesthetically must be advanced and critiqued to know whether they have promise; to say that a thing is exceptionally complex is not a good reason not to try to understand that thing. Neuroscience would have stopped years ago if we demurred from investigating the complex. Hunt's complaint should not be taken simply as an explanation for why philosophy may not have engaged with The Garden as much as we might expect, but it should also be taken as a challenge to see if useful insights are possible by applying philosophy—and, in particular, aesthetic discipline—to the study of gardens.

Mara Miller's 1993 book *The Garden as an Art* is the first book-length project to reinvigorate contemporary formal philosophical discussion centered on the garden.[17] Her book is an exploration of what it means for The Garden to be classed alongside other canonical artforms as a bona fide art kind. The case she offers is characterized by deep and thorough consideration of what it means for The Garden to be an artform, and her insights will inform many chapters of this book. Miller believes, in part, The Garden is not given the attention other artforms enjoy because of the relatively recent separation of aesthetics from art, where aesthetic virtues and artistic virtues no longer coincide.

Miller's treatment of the topic is not limited simply to the case for The Garden as an artform; her book delves into the nature of language as a signifying system and how language as the preeminent version of such a system is like and unlike The Garden as a signifying system. While this topic may be seen to connect

---

[17] Miller, *The Garden as an Art*. There are earlier articles devoted to garden philosophy—particularly a set in the *British Journal of Aesthetics* in the early to mid-1980s—but Miller's book is the first book to focus on it.

to the semiotic work of Nelson Goodman as he explores the nature of art,[18] Miller's treatment goes beyond this into the formal realm of philosophy of language, thereby taking her work beyond the confines of aesthetics, even writ large enough to include all her work on The Garden as an artform. The significance of this, and of mentioning this, is to illustrate that philosophizing about gardens should indeed transcend the boundaries of thinking about gardens merely in aesthetic ways (no matter how broadly we interpret "aesthetic ways.")

Stephanie Ross offers a central contribution to the philosophical literature focused on interpreting the meaning of gardens in her 1998 book *What Gardens Mean*.[19] Ross's central focus is on whether and in which ways a garden can bear and convey meaning, but her study is framed historically by considering the place of The Garden in aesthetically focused discourse principally in the eighteenth and nineteenth centuries in Europe. She does a deep dive into this context and so the book fulfills to what seems a significant degree what Hunt is looking for in the developing literature. In addition, Ross engages with other nature-based aesthetic forms related to The Garden and in doing so develops insights about The Garden itself.

It is in 2006 that Cooper published his *A Philosophy of Gardens*, mentioned above.

In 2013, Lee published *The German "Mittelweg": Garden Theory and Philosophy in the Time of Kant*. In this book, Lee takes up the task of reviewing how two German philosophers—Heydenreich and Schiller—developed their theoretical views on The Garden and on how garden design should proceed. The book examines the influence of Kantian philosophy—primarily his work in the *Critique of Judgment*—on Heydenreich and Schiller, and through insights gleaned from that examination, Lee explores a particular casting of how Kant may be conceived to have regarded gardens from an aesthetic point of view. To say that Lee's book exhibits depth of scholarship or novelty of insight is to offer a decided understatement.

Heydenreich and Schiller were working (in the late 1700s) not only under the strong influence of Kant's work (aesthetic and otherwise) that had dominated various parts of the German Academy during this time, but they were also working under the influence of garden design of that time in France and in England. The icon of French garden design was Versailles and its designer (in the mid to late 1600s) was André Le Nôtre. The highly formalized style Le Nôtre championed held for many years, and today we continue to think of Le Nôtre's work as quintessential French garden design. England's extensive and expansive

---

[18] Nelson Goodman, *Languages of Art: An Approach to a Theory of Symbols* (Indianapolis, IN: Hackett Publishing Company, 1976).

[19] Ross, *What Gardens Mean*.

work of iconic landscape architecture was pursued by Capability Brown (in the mid-1700s). Brown's style was formed through creation of sweeping landscape vistas surrounding great houses—inclusive usually of lakes that looked like serpentine rivers, substantial architectural bridges, and groupings (or "clumpings") of trees—that mimicked an ideal of a perfect and perfectly natural setting, a vista that looked like it might have been in place for hundreds of years, the natural elements shaped only by evolving nature. These styles were well known, and their iconological characters were established by the time Heydenreich and Schiller took up work on garden design theory. Lee demonstrates how both contributed to the development of a German garden style that took aspects of both French formalism and English "naturalism" in fashioning a middle path—in German, a "*mittelweg*"—between the two styles. While neither Heydenreich nor Schiller are known actually to have designed a garden or created concrete procedures for such design, according to Lee their work contributed to the creation of a German style of garden design.

Lee's book, however, is not merely about this style and its creation. It is, especially in its latter chapters, a book about Kant and how Kantian aesthetic insights and principles could be thought of as applicable to garden design theory. The novelty of applying Kant—whom many take as being one who helped relegate garden aesthetics to the sidelines (we will discuss this later)—in this way is striking. It encourages one to give Kant a second thought, particularly with regard to ferreting out the most mature implications of his aesthetic theory for The Garden. And in this way, perhaps Lee's project is also contributory to the apologetic case for garden aesthetics insofar as he shows in compelling ways that "second thoughts" about theories concerning The Garden are worth having.

The Timing

Why now? Why take up discussion of garden aesthetics now?

First, there has been, as mentioned above, a resurgence of aesthetic interest in The Garden beginning in 1993 with Miller's book *The Garden as an Art*. One could argue that this literature began with the work of Hunt stretching all the way back to the late 1970s, but Hunt, as mentioned earlier, does not take himself to be doing philosophy or philosophical aesthetics (despite the fact it is difficult not to see his theoretical work as philosophical in nature).

Many on earth today—if only among those nations that possess wealth—are increasingly conscious of the aesthetic dimensions and aspects of those objects that populate their lives. Gardens are such objects; as we said, gardens are "aesthetic-forward." So not to engage with their popularity and their aesthetic characters is in some sense to view with a deliberate blind eye. As the breadth and depth of the disciplinary aesthetics literature has grown significantly—as evidenced in part by the number of recent new journals devoted to the area— so it is natural to find that an aesthetic form as popular and present as gardens would make its way back into focus.

Second, within the last century and a half, the world of art has experienced a seismic shift. Art audiences of the nineteenth century would have no basis upon which even to recognize as art some of what we today take to be settled and commonplace works of art—works by Duchamp, Rauschenberg, Warhol, Cage, and many others. This has forced art theory to keep pace, and that has meant that definitions of art that turn on objective features that all works of art share in common have come under heavy scrutiny. Today, aesthetic definitions— definitions that hold that the aesthetic characters of objects and events properly called art are necessary to their identity as works of art—require a good deal of justification in the face of so many works that seem quite distant from the aesthetic. It is more common today to find in ascendancy theories about defining art that turn on social-scientific and historical considerations, on how today's works of art relate to yesterday's and how today's art world relates to yesterday's. As definitions of art evolve to keep pace with the actual world of art, it is natural to ask if these evolutions are indeed staying up to the mark of incorporating every data point they should. Relevant data points include works of art in all recognized or canonical artforms. But the conversation does not end there. We must also ask if we are recognizing all forms that should properly be recognized as artforms or art kinds. And this is where The Garden and where Miller's book come in.

The Garden was once recognized as a "fine art." Should The Garden retake its place as a recognized artform? This question is one that has had theoretical forces raised against it, but many of those arguments fail to endure, and without such opposition, it is natural to include deliberation about artforms in our consideration about how to understand the boundaries of what can be art—and then to consider whether The Garden should be one of these. If the answer is a yes—or even a tentative yes—then the next set of questions to ask will be those that attempt to understand what including The Garden as an artform entails and how it might necessarily alter current trajectories in defining art.

Whether and how The Garden is an artform—and what that means and entails—is a much larger question than simply asking for a thumbs-up or thumbs-down on its proverbial application to be a member of the club. And so thorough treatments of aesthetic consideration of The Garden are necessary, and not merely one. Aesthetic discourse on The Garden must grow (or perhaps regrow) into a literature of its own, where diverse opinions on all manner of aesthetic questions are advanced and debated, where theories are advanced and theories retreat or are repaired as concerted critical scrutiny is offered in response. If such a literature does not manifest, then The Garden is still on the edges of the conversation and, for all the good intentions of those writing on the aesthetics of The Garden, it remains an aesthetic novelty. Miller, Hunt, Ross, Cooper, Lee, and others have offered contributions to creating that literature. And, it is fervently to be hoped, there will be more.

Third, as global warming/climate change becomes each year a more tangible reality, the concern for natural preservation grows. Botanical collections, seed banks, and gardens that host pollinators, migratory birds, soil microbes, and other native wildlife are at least partly the results of efforts that flow from those concerns, but heightened concerns with climate change have given a greater urgency to thought directed toward these concerns and toward concrete efforts to address preservation, or at least well-informed planning. For those who are serious about such concerns, and who wish to invest their resources and efforts in ways that promise to be maximally efficacious, they will need to turn to theory—centrally natural-scientific but also social-scientific—to inform their thinking and their decisions. The framing of science, including normative discussions about the values that inform and guide science, are the purview of the humanities. As thoughts about gardens may be a subset of the thoughts about natural processes and preservation in general, so theory about gardens must be developed in ways that are as sophisticated as the science they inform and are informed by.

Finally, gardens have become increasingly important as extensions of how many of us choose to live. Certainly this can be limited in terms of the space one has (gardens are much easier to create and maintain in suburban settings rather than urban ones, for example), the resources one has (a typical suburban home garden can add hundreds if not thousands of dollars to the cost of homeownership), and the extent of one's knowledge. Planting roses under thick trees and dealing with the frustration that they do not look like the ones on the websites (and in the hundreds of "garden-appreciation picture-books" that populate the shelves of local bookshops, another sign of the popularity of gardens and the aesthetic appreciation of them) is an enterprise easily avoided with a little education. Despite these potential limitations, gardening increases in popularity as a pastime. Part of the explanation of this may be that one wishes to make good impressions on their neighbors and passers-by through the landscaping of their own front yard, but the more satisfying answer seems to lie with the attraction that gardening and gardens holds as an activity and a "result-of-activity" unto itself.

Gardens have experienced a resurgence in popularity over the past decade, a trend that shows no signs of slowing. A new generation of "plant-fluencers" has taken to social media, particularly Instagram, showing off well-lit apartments filled with houseplants and giving us crafty ideas for living walls and dishes filled with succulents. Beginning gardeners and naturalists download apps to identify plants and give care advice with the click of a button. At the same time, the increased attention we devote to the screens of our cellphones has led some to advocate a return to nature as a means of "detoxification" from the evils of the modern world. Similar claims have been made at least since the Industrial Revolution, and led to the development of, for example, Central Park in New York City. Botanical gardens, which are often located near high-population areas, bill themselves as "urban oases" from which to escape fast-paced life. Plant nurseries saw increased sales during the start of the 2020 coronavirus pandemic as people sought to populate their windows with something colorful, to occupy their time with something productive, or perhaps to control a plot of land in a violent and unpredictable world. As we write this, restaurants and venues are finding new ways of accommodating customers outdoors and office work is becoming decentralized. With this there will almost certainly be increased attention to the design and care of our outside spaces.

Designing, creating, and maintaining a garden are activities that give satisfaction on their own. The garden is satisfying as the creation usually of an aesthetically rich and evocative place, one appreciated as one strolls through it, sits within it, or looks at it through a window. That it is a result of one's own labors, creative and manual, makes it all the more satisfying. But the resulting garden is only a part of the complete equation. Gardening practices, even ones that make one sweat and make one's muscles ache, have an attraction as well. We garden not only to have the resulting garden; we garden for the sake of gardening. The very activities of garden practice are valuable to us and worth our endurance of sweat and aches.

It has been common for the last few centuries—following disinterestedness theory as expressed by the British taste theorists, Kant, Schopenhauer, and others, and following art-for-art's-sake movements as expressed by artists like Oscar Wilde and James Whistler—to conceive of aesthetic attention to be centrally or essentially characterized by thorough disregard for any instrumental purposes to which the object (or activity, in our case) may be put, disregard for influence exerted by anything external to the object, and exclusive focus on the properties that may be known through straightforward sensory connection with the object. These movements may be over and done—and we will argue as much later—but they nonetheless give us insight into the nature of the value that gardens and gardening hold for many people. We value them on their own, without reference to anything beyond their own properties—both the garden and the activities of gardening. Such attention would have been recognized readily by the theorists mentioned above as clearly and canonically aesthetic.

The list of reasons offered move from the more particular to the more general. They illustrate, by taking up the matter in different circumferences of orbits, the timeliness of the development of more discussion—both "lay" and academic—devoted to garden theory, especially to garden aesthetics. As Cooper says explicitly, and as our review of the literature strongly suggests if not states outright, an aesthetic approach to The Garden seems the place to start.

The Nomenclature

There are only a few terms that need to be defined before moving on to more substantive work. Defining them will help eliminate confusion and may guide debate about the substance of what is being discussed in ways that are more productive and profitable; too many theoretical discussions get unnecessarily mired in terminological debates that not only cannot see the forest for the trees but cannot see the forest for the blades of grass in the field before it. It is intuitively easy to sympathize with those garden theorists, like Hunt, who lament philosophy's fascination with the "blades" rather than the "forest."

Through this book, we will use the terms "garden" and "landscape" interchangeably. As garden theory has been the purview of both the discipline of horticulture—which grows from the natural sciences and is scaffolded by them—and the discipline of landscape architecture—which grows from architecture per se, it seems overly fussy to attempt a distinction between them for the sake of this project. Certainly, horticulture and landscape design have different spheres in which they operate and how they are commonly and professionally thought of and regarded, but those distinctions are likely not useful to a treatment of gardens through the lens of philosophical aesthetics. Given that, we will use the terms interchangeably.[20]

We will use the term "gardenist"—a term introduced by Walpole[21]—to refer to (1) the creator or designer of a garden, (2) the person who directs the initial installation of a garden (if they do not do it themselves), (3) the person who directs any significant or character-altering changes to the plan of the garden, and (4) the person who in some circles is otherwise called the "landscape designer" or "landscape architect."[22] If The Garden turns out to be a bona fide artform, then the "gardenist" is prima facie equivalent to the garden's artist.

We will use the term "gardener" to refer to the person who maintains and tends the garden, keeping it and its components alive (to the extent to which the garden in question has living elements) and in good functional order; performing both

---

[20] Hunt discusses the term "landscape architecture," a term introduced in 1840, and its relation as a subset of architecture on the first page of *Greater Perfections* and again on page 217.

[21] David E. Cooper, "In Praise of Gardens," *British Journal of Aesthetics* 43:2 (2003), p. 102.

[22] The term "landscape architect" was first introduced in 1862 by Frederick Law Olmsted and Calvert Vaux, the designers of, among many other civic gardens, New York City's Central Park.

routine and special tasks to keep the garden in an order befitting either its purpose, its original design, or an intentional revision; and undertaking limited gardenist-style alterations to the garden as natural changes are accommodated (for instance, moving plants that require sunlight from underneath a maturing tree) in line with the garden's (original or revised) design or purpose. It may be the case, especially in modest home gardens, that the "gardenist" and the "gardener" are the same person, but this is infrequent in larger-scale projects.

The details of "gardening practice" will be examined in a later chapter, but we will use the term to refer to any and all practices of gardenists and gardeners that are directed toward the garden in purposeful ways.

A point likely already made: we use the capitalized term "The Garden" to refer to the category of any and all bona fide gardens.

As will be evident in the next chapter, it is important to us to adopt an approach in which theory follows practice. All terms should be descriptive in the sense of being used in the manner that people who use the term frequently mean by the term. Terms that exclude reference arbitrarily are suspect. This includes, first and foremost, the definition of "garden" or "a garden," and that is why an entire chapter is devoted to defining that term. As is already evident, the terms above all turn on the definition of "garden." Without that first and primary definition, nothing proceeds.

1

What Is a "Garden"?

As we discussed in the book's introduction, gardens are everywhere. They surround us, in both senses of the word: we encounter gardens on our way to work, on our way to the store, on our way home, and sometimes we go well out of our way to visit a particularly compelling garden. And as we enter a garden, that garden fully envelops us, providing rich opportunities of engagement for every one of our senses and sometimes of our emotional and cognitive faculties as well. Every garden presents an occasion for an aesthetic experience—or so we will argue in the course of this book. This is the case not only for those gardens designed to delight our senses but even in those cases where the focus of the garden is utilitarian, say for food or medicine or as a staging area for some other purpose.

Every garden is unique; to use the philosophic term, every garden is autographic. It exists as a thing that cannot be replicated. Just as true, gardens as a kind are special insofar as they represent best among all things the intersection and interplay between humans and nature. Certainly, we can experience this when we hike through "untouched" nature or interact with nature through agriculture, but gardens are special because of the intimacy of the intersection. They exist as artistic co-creations of nature and humans, a series of back-and-forth actions and reactions, where both parties have something to add. Humans, through deliberation and design, express themselves in the creation of a place that cannot be said to be only nature or only artifice; these places become something new as vision leads to expression, expression to cooperation, and cooperation to instantiation.

There are many different kinds of gardens, and in this chapter we attempt to provide as comprehensive an inventory of garden types and styles as we think history will support. It is inevitable that a goal of a fully comprehensive list is not one that can be met. Gardens began thousands of years ago, and given their ubiquity through human history, and given that many gardens will have been the

Kushak, Bagh-e Fin, Kashan, Iran
Unknown gardenists starting 1590
Photo by Marta McDowell, Fall 2015

The Art and Philosophy of the Garden. David Fenner and Ethan Fenner, Oxford University Press.
© Oxford University Press 2024. DOI: 10.1093/oso/9780197753590.003.0001

modest undertakings of names lost to history, even some garden types and styles may have disappeared in a way that is no longer recoverable.

It seems clear to us that the first step in talking about gardens is to determine what we are talking about. We will use the term "garden" in two ways: "garden" means any particular garden or any token of the type; "The Garden" means the type, abstract kind, or ontological category that includes only and all gardens. Used in this second way, "The Garden" is the name of a set of objects.

There are other ways to use the term, and we want to ensure that these are represented in this chapter.

- First, "a garden" can be a subset or a species of a larger "garden." When one visits the gardens at Sissinghurst Castle in Kent, one can tour the White Garden, the Rose Garden, or the Yew Walk. When one visits Hampton Court Palace, one can tour the Knot Garden, the Privy Garden, the Tiltyard, or the Maze. So a "garden" may fit within another "garden," like a set of nested boxes. One way to distinguish between the larger and the smaller is to think of the smaller as a "garden room," and sometimes we may wish to think of the larger complex as a "park" rather than a garden—or, as we did just above, refer to the "gardens" (plural) of Sissinghurst.
- Second, "a garden" may name a garden style or a garden type that is smaller than "only and all gardens." When one speaks of "the French garden," they generally mean a garden style typified by the formal Versailles gardens whose designer was André Le Nôtre. Or when one talks of "a pleasure garden," they mean a garden like Longwood Gardens in Pennsylvania or Butchart Gardens in British Columbia. When one talks of a "dry landscape garden" (known as *karesansui*), they mean the sort of stone-and-gravel gardens found at Ryōan-ji, Daisen-in, and Saihō-ji in Kyoto. For the sake of organization, in this book we will refer to this classification of garden as either a "garden style" or a "garden genre," depending on what seems appropriate for the context.
- Third, "a garden" can be, for someone from Britain, equivalent to "a yard" for an American from the United States. In the United Kingdom and Ireland, "a garden" can simply name the area that surrounds one's house, an area that may be richly planted or modestly so. Such gardens seem always to have some plantings, if only some grass, but a "back garden" in this sense of the word need not be any more than what Americans mean by a "backyard" and vice versa. In complement, what an American may mean by "a yard" may be simply a lawn, with "garden" reserved to refer to those areas of the property that are occupied with flowers, vegetables, or even trees; additionally, "yard" for an American may refer to the property in general terms—lawn, "gardens," and patios included.
- Fourth, it is entirely possible to use the term "garden" in a metaphorical sense, such as in the phrase "a garden of ideas" to mean "a bounty of various

and diverse ideas." There is no practical way to circumscribe all the ways the term can be used metaphorically, of course. Within this chapter, we mean to stick closely to literal uses of the term.

Some scholars believe that defining "garden" in strict terms is impossible.[1] Some invoke Ludwig Wittgenstein's notion of "family resemblance" as the connecting principle that pulls all those things we call "gardens" together into a single kind. The claim is that while many gardens share in common with other gardens characteristics that seem relevant to their identities as gardens, no one characteristic will be present in each and every instance of what we correctly call "a garden." Wittgenstein's observation is a species of "antiessentialism," the view that no single essence can be identified in all instances of the tokens of some type, an essence that once identified can be used to circumscribe the set of those instances in ways that present and prescribe strict boundaries to the correct use of the term.

We offer two responses. First, while Wittgenstein's observation is worthy as an articulation of a modest and intuitively plausible version of antiessentialism, it is merely theoretic. That is, there is no argument for its application in any particular case, nor is there an articulated mechanism by which the observation can be applied to any particular case. Wittgenstein's observation only has normative import if it can be demonstrated that there is absolutely no kind that has among its members some defining shared characteristic. If this can be done, it would be done philosophically, but Wittgenstein does not present an argument that antiessentialism is true of all ontological classifications. His argument is particularist in the sense that he observes "family resemblance" in some instances where essentialist definitions are sought. If the reverse were true—if he advanced an antiessentialist thesis that applied (essentially) to all ontological classifications—then no project of definition could ever be successful. Wittgenstein instead offers a more useful tool to us: a way to understand how it is that we may in some instances class objects (or events) together under a single umbrella term even when all those objects fail to share in some common characteristic. Taken this way, Wittgenstein's offering is helpful, but it does not offer an argument for why defining characteristics of The Garden should not be pursued and such characteristics sought out. On the contrary, the particularist nature of his observation invites the work of critical examination and discovery of whether the members of a set do or do not have essentialist features.

Our second response is that those scholars who adopt Wittgenstein's observation regarding defining the term "garden" do not advance the conversation. Even

---

[1] Stephanie Ross, *What Gardens Mean* (Chicago, IL: University of Chicago Press, 1998), pp. 6–19. Ross does not explicitly reject a single definition of a garden, but she does invoke Wittgenstein in this context.

if the family-resemblance argument happens to apply in the case of "garden," the search itself would prove valuable in exploring and revealing the nature of the various traits that some gardens do in fact share with other gardens. Whether Wittgenstein's observation is correctly applicable in the case of gardens does not stand as a reason not to look for commonalities; instead, it stands precisely as a challenge to do the exploratory work to uncover commonalities, even if these commonalities fail to apply in every garden case.

This work is exactly what we propose to do in this chapter. The value of searching for a satisfying definition of The Garden is easy to understand. It is the lesson in Plato's *Meno*—before we can talk about a thing meaningfully, we must understand what it is that we are talking about. Our approach will begin with an attempt to create a comprehensive inventory of all garden types. Such an inventory is open to both claims that it is not comprehensive (or sufficiently comprehensive) or that it is too broad and includes items that conflict with the views of some that "this so-called garden" or "that so-called garden" is not truly a garden. This cannot be helped; it is an unavoidable danger in doing the work that we meant to justify tackling above.

The first danger—that it may prove insufficient—is one the response to which must be: we have to start somewhere. Even if our inventory is incomplete, at least it will stand as a jumping off point. Yet we expect that our inventory will be comprehensive enough to be valuable to the project of exploring the aesthetic aspects of gardens; if we leave out a garden style or two, the aesthetic project can still proceed.

The second danger is unavoidable. In any definitional project, one must rely at least in part on the ways in which people who speak a certain language commonly refer. And commonly, reference need only go so far as is necessary for functional communication. To know what precise color "chartreuse" refers likely is unnecessary to use the term perfectly competently. In the same way, one's intuitions about what counts as a garden may differ, perhaps significantly, from those of another. So while we will do our best and likely err on the side of inclusion, we are fully conscious that the chances we will convince all readers of the necessity of the inclusion of everything we include are slim. But again, achieving such agreement is not crucial to our aesthetic project.

Below we attempt to explain the mechanism by which we assemble this list, and we will also discuss a variety of ways such an inventory can be approached and why our particular approach seems the best one for the purposes of this book. That is, inventories must follow organizational structures of some sort; we look at the field, pick one, and justify it as the best one for our purposes. However, before proceeding with our project, we think it useful to spend a few minutes surveying the modest field of scholarship in this area to see what others think.

Earlier Definitions

We begin our brief survey of how others defined The Garden with David Cooper, author of the 2006 book *A Philosophy of Gardens*:

> Pressed to say what I mean by the term [garden], my response would be "The same as you who are pressing me mean by it—so you already know what I mean."[2]

Relying on commonly shared intuition to test reference may be key, and this seems the implication in what Cooper is saying. When pressed, Cooper admits he is sympathetic to the working definition that Mara Miller offers—that we will see below—but he adds:

> But such definitions, far from having to be set out before discussion can continue, are of a kind that the whole discussion should be seen as groping towards.[3]

We begin with Cooper because instead of advancing a definition per se, his contribution focuses on the conditions for creating a definition. More importantly for our upcoming inventory project, Cooper defends what we see as highly relevant to our project: that the definition should follow the discussion, or, to put it in different terms, that theory should follow practice, that definitions should only be reached after accounting for the actual use of the term by those who commonly use it.

Miller, author of the 1993 book *The Garden as an Art*, offers the following definition of the term "garden":

> A garden is any purposeful arrangement of natural objects (such as sand, water, plants, rocks, etc.) with exposure to the sky or open air, in which the form is not fully accounted for by purely practical considerations such as convenience . . . an arrangement that is like a garden but composed of purely artificial objects could be a garden only in a metaphorical sense . . . a true garden must have exposure to the open air or sky; enclosed arrangements of plants become a different category—greenhouses, orangeries . . . in my opinion this can only be an imitation of a garden, not the real thing.[4]

---

[2] David E. Cooper, *A Philosophy of Gardens* (Oxford, UK: Oxford University Press, 2006), p. 13.
[3] Cooper, p. 15.
[4] Mara Miller, *The Garden as an Art* (New York, NY: State University of New York Press, 1993), p. 15.

This definition includes four signature elements:

1. A garden must be a purposeful (purposed and deliberate) arrangement of objects.
2. The form of this arrangement is not accountable for through "purely practical considerations."
3. The incorporated objects are natural objects.
4. The garden must have "exposure to the open air or sky."

The first three items on this list—purposeful arrangement, an arrangement that goes beyond simple instrumentality, and a composition of natural objects—are all matters we will discuss in depth in this chapter and in the next.

The criterion "exposed to the open air or sky" is an especially interesting one.[5] The inclusion of this criterion seems laudatory if we believe that contiguity (of whatever degree) of the garden in question with the larger ecosystem of which it is a part is an important definitional feature. The locational context of the garden—its "sitedness"—seems important to its identity. This includes what natural and artificial areas lie just outside its borders; what sort of sun, rain, seasonal influences it experiences; what pollen and seeds are brought into it; and so forth. So the inclusion among Miller's criteria of "exposed to the open air" is certainly understandable. On the other hand, while this excludes greenhouses, glasshouses, orangeries, and conservatories, it is unclear that if the motivation is to secure a full appreciation of the "sitedness" and natural contiguity of a garden, "exposure to the open air" should be a necessary condition.

First, greenhouses, orangeries and the rest participate in a "borrowed ecosystem" in the sense that within these enclosures is a microcosmic ecosystem that while it may represent important aspects of an ecosystem in a different part of the world still contains all the elements of an ecosystem. They must contain all these elements—water, sunlight, heat, humidity, biotic and abiotic components of soil—or their living components would not survive. A "borrowed ecosystem" is still an ecosystem, so while contiguity with nature may be important, these enclosures are indeed contiguous with nature—just a nature that may not be physically or regionally immediate.

Second, all these sites—greenhouses, orangeries, and the rest—do indeed participate in some of the conditions of ecosystem contiguity. Trees that cast shade

---

[5] "Exposed to the open air or sky," as well as other characteristics, are echoed in Isis Brook's brief definition: "The definition of a garden I will be using is an enclosed or demarcated outside space with living plants." Isis Brook, "The Virtues of Gardening," in Dan O'Brien (ed.), *Gardening—Philosophy for Everyone: Cultivating Wisdom* (Chichester, UK: Wiley-Blackwell, 2010), p. 13.

on bedding plants also cast shade on the plants within the glasshouse, and an overcast day outside is an overcast day inside. The soil, even if amended, will still be subject to similar structure and drainage as the surrounding areas. Glasshouse design involves careful consideration of the site to produce the desired effect within its walls—depending on the environment. Heat may be above ground, below, or both; ridge vents may open in a way that allows rain to enter or prevents it from entering; the type of cooling will depend on ambient humidity; the roof design depends on the amount of snowfall; even the orientation of the greenhouse's axis will depend on its proximity to the equator.

Greenhouses may be excluded from Miller's definition based on other criteria (if the arrangement of its plants are purely for convenience, for example), but the purposefully arranged conservatories of botanic gardens are better thought of as "semipermeable" to the wider ecosystem, where some aspects are allowed to enter passively, and other aspects are actively kept out or removed. This is no different from how we look at garden rooms, or even the garden itself. Although many of these ecological modifications are becoming more automated, what happens outside necessarily affects what happens inside, and it is the job of the gardener to monitor the day-to-day environmental conditions and adjust the greenhouse's climate by shading, venting, misting, and so forth.

Third, many conservatories, glasshouses, and orangeries exist on the sites of larger gardens. In this way, we may think of them as "garden rooms" that contribute to the overall experience that one may have while visiting a garden. The feeling of stepping into the hot and humid Palm House at Kew Gardens on a cold and rainy London day is a memorable sensory experience, and it could be argued that a visitor to Kew has not experienced the whole garden just by wandering the grounds. Just as one experiences the contrast moving from a wooded garden to a desert garden, or from moving from an effluent perennial border to a calm slope of grass, glasshouses as subsections of gardens complement their surroundings and work as parts of integrated wholes. In our inventory of gardens, we will encounter other enclosed or semienclosed spaces, beyond glasshouses and conservatories, that are necessary components of the gardens in which they are placed. Seen as a part of an integrated whole garden, glasshouses lend themselves to garden aesthetics not only by their interior contents and design but also by their outward appearance.

Finally, let us consider John Dixon Hunt's definition of a garden. Hunt is the author of many books in garden theory, stretching back to the 1970s, but perhaps his most central contribution to what we might call garden philosophy is his 2000 book *Greater Perfections: The Practice of Garden Theory.*

I would provisionally define landscape architecture as exterior place-making; at that simplest level, place-making is to landscape architecture what building is to architecture.[6] . . . Gardens are privileged, then, because they are concentrated or perfected forms of place-making.[7]

The expression "place-maker," Hunt says, was used self-referentially by Capability Brown, who designed gardens usually on the scale of dozens or hundreds of acres. Hunt qualifies gardens as a particular kind of place-making—a "concentrated or perfected" form of "exterior" place-making. These two are necessary qualifications for two reasons. First, the term "place-making" might easily be seen to apply in the realm of architecture in a way that is even more intuitive than its application to gardens. While a garden is certainly a designed space, so of course are all buildings. So the qualification of "exterior" is important. Second, Hunt's conception of the history of gardens relies heavily on an evolutionary model he adopts from two theorists working in the mid-1500s, Bartolomeo Taegio and Jacopo Bonfadio, following an earlier distinction offered by Cicero.[8] "The Three Natures" is the name of this model: "First Nature" is the wild; "Second Nature" is agriculture (and, for Hunt, urban developments[9]); and "Third Nature" is The Garden.[10] Hunt writes:

> Gardens now take their place as a third nature in a scale or hierarchy of human intervention into the physical world: gardens become more sophisticated, more deliberate, and more complex in their mixture of culture and nature than agricultural land, which is a large part of Cicero's "second nature." By implication, the first nature becomes for Bonfadio the territory of unmediated nature, what today we might (provisionally and awkwardly) call wilderness.[11]

The distinction between The Garden and agriculture expressed in the difference between Second Nature and Third Nature is borne on considerations of sophistication and variety in design. Agriculture as Second Nature, inclusive of urban developments—where Hunt says we largely live our lives[12]—is the best match to an intuitive sense of place-making. It is when we set down roots and leave our nomadic existence for a settled one—when we leave First Nature for Second Nature—that we do our most obvious and fundamental place-making. The Garden as place-making then must be set apart from that first move; Third Nature must be qualitatively different from Second Nature. Hunt describes this

---

[6] John Dixon Hunt, *Greater Perfections: The Practice of Garden Theory* (London, UK: Thames and Hudson, 2000), p. 1.

[7] Hunt, p. 9.

[8] Hunt, p. 32.

[9] Hunt, p. 58.

[10] Although Hunt does not, we will capitalize "First Nature," etc., each time we use the expression.

[11] Hunt, p. 34.

[12] Hunt, p. 59.

difference as one of degree of design: "more sophisticated, more deliberate, and more complex in their mixture of culture and nature."

Hunt's full definition of a garden is this:

> A garden will normally be out-of-doors, a relatively small space of ground (relative, usually, to accompanying buildings or topographical surroundings). The specific area of the garden will be deliberately related through various means to the locality in which it is set: by the invocation of indigenous plant materials, by various modes of representation or other forms of reference (including association) to that larger territory, and by drawing out the character of the site (the *genius loci*). The garden will thus be distinguished in various ways from the adjacent territories in which it is set. Either it will have some precise boundary, or it will be set apart by the greater extent, scope, and variety of its design and internal organization; more usually, both will serve to designate its space and its actual or implied enclosure. A combination of inorganic and organic materials are strategically invoked for a variety of usually interrelated reasons—practical, social, spiritual, aesthetic—all of which will be explicit or implicit expressions or performances of their local culture. The garden will therefore take different forms and be subject to different uses in a variety of times and places. To the extent that gardens depend on natural materials, they are at best ever-changing (even with human care and attention that they require above all other forms of landscape), but at worst they are destined for dilapidation and ruin from their very inception. Given this fundamental contribution of time to the being of a garden, it not only exists in but also takes its special character from four dimensions. In its combination of natural and cultural materials, the garden occupies a unique place among the arts, and it has been held in high esteem by all the great civilizations of which it has been a privileged form of expression.[13]

Hunt does not take himself to be doing philosophy, so his definition is not subject to the same sort of critique that normally would be applied to the style of philosophical definition characterized by necessary-and-sufficient conditions. Nonetheless, there are many discernable elements of his definition that might be taken, individually, to constitute criteria for a formal definition:

1. "A garden will normally be out-of-doors."
2. "The specific area of the garden will be deliberately related through various means to the locality in which it is set."
3. "The garden will . . . be distinguished in various ways from the adjacent territories in which it is set."

---

[13] Hunt, pp. 14–15.

4. "The garden will . . . take different forms . . . "
5. " . . . and be subject to different uses in a variety of times and places."
6. "To the extent that gardens depend on natural materials, they are . . . ever-changing."

The first of these criteria we discussed above, and Hunt includes the modifier "normally." The second distinguishing mark is about a garden's "sitedness"—we began discussion of this topic above and will continue it, as promised, as the book unfolds. The third distinguishing mark of a garden is about the "boundedness" of a garden; the sixth is about the ever-changing nature of The Garden—these two topics we will review later. They are both important aspects of gardens and deserve serious consideration. Items 4 and 5 on the list above, while important, are not particularly unique to gardens, so they only modestly advance the project of defining The Garden in ways that separate it from all other things.

What may be of greatest importance about Hunt's definition is that he, like Miller, offers one. The extent to which Hunt's—or Miller's—definitions are serious attempts, ones that enjoy at least intuitive appeal and some degree of success as definitions, provide reason for the definitional project to continue and to garner respect. Their legitimacy as definitions—whether they fully succeed or not—speak to the optimism that one is entitled to feel in approaching the possibility that a definition of The Garden is possible.

How do we start to define the term "garden"? There are a plurality of approaches we could adopt. Ultimately we proceed by creating an inventory of gardens and looking for common traits among the types of gardens listed there. This follows one approach, but there are others, so let's take a moment to consider the possibilities.

Defining "The Garden" Etymologically

We might begin by turning to the etymology of the word "garden." The earliest usages of the European words that were the ancestors of our word "garden" all made explicit reference to "enclosure."[14] The English "yard" and the Latin *hortus* both derive from the same Proto-Indo-European root as "garden," and their earliest usages would have described enclosed spaces. Francois Berthier, in an essay titled "Reading Zen in the Rocks: The Japanese Dry Landscape Garden," says that the first Japanese use of the term "garden"—during the 700s CE—was signified by the word *shima*, meaning "island."[15] These two uses, European and

---

[14] Hunt, p. 19.
[15] Francois Berthier, *Reading Zen in the Rocks: The Japanese Dry Landscape Garden*, ed. and trans. Graham Parkes (Chicago, IL: University of Chicago Press, 2000), p. 12.

Japanese, echo one another. A garden is seen to be markedly divided from the larger environment in which it is placed.

"Paradise" is another word whose etymological origins provide a clue to the nature of the earliest gardens. The Greek *paradeisos*, an enclosed park or pleasure ground, was used at various times to describe the Garden of Eden and heaven itself. This in turn came from an Iranian word *paridaiza*, which described enclosed, stocked hunting grounds for the rulers of the Achaemenid Empire (around 550 BCE). The prefix para/pari/pairi/peri- ("around") directly relates to the idea of enclosure, while the suffix dise/dis/daeza/daiza relates to heaping up, building, kneading, or shaping (the English "dough" shares an early ancestor). The *paridaiza* collected various species of plants and animals as the Achaemenid Empire expanded, including fruit trees and flowers, which were protected from the desert by a sophisticated irrigation system.

The garden as we know it probably developed from walled areas to protect delicate crops from the burning heat and winds of the desert. Gardens took on the dual task of providing food as well as pleasant areas where one could rest under the shade of date palms. The earliest record of a garden layout—again emphasizing enclosure—comes from an Egyptian tomb of around 2000 BCE.[16] It describes a walled area with a central rectangular fishpond with flowering lotus, surrounded by fig trees and edged with flowerbeds. All is symmetrical and there are tall shade trees. The basic pattern of the earliest Egyptian tomb gardens—walled, open air spaces with mostly symmetrical arrangements of plants—crosses many cultures and many centuries. The Persian *paridaiza* developed into the *chahar bagh* (or *charbagh*), an enclosed garden divided into four parts by central rills of water. In Europe, we see the enclosed *peristylium* gardens in Roman houses, and later, the *hortus conclusus* and cloister gardens of medieval castles and monasteries.

Later, we will discuss the "bounded" nature of The Garden. A garden being "bounded" is not, however, the same as a garden being walled or enclosed in the way expressed above. In antiquity, "enclosed" meant "walled" or something comparable, such as "fenced." Today this seems no longer the case. There are many examples of gardens that do not have walls or fences; there are many examples of gardens that are so large as to make this sort of enclosure an expensive if not impractical suggestion.[17]

This is a case of theory following practice. While ancient gardens may have all shared in the common characteristic of being enclosed—usually walled—spaces,

---

[16] Penelope Hobhouse, *Plants in Garden History* (London, UK: Pavilion, 1992), p. 11.

[17] Landscape Historian Marta McDowell says, "Most expansive gardens have some enclosed areas. Frederick Law Olmsted's 'clearings,' such as at Central Park and Prospect Park, open out from visual bottlenecks—gateways of a sort—to heighten their impact." We offer our heartfelt thanks to Professor McDowell for her assistance with this chapter.

many postantiquity gardens do not. As definitions of art must adapt to introductions of works of art and forms of art that transcend early mimetically focused definitions, so a definition of The Garden must keep pace to include all those spaces, all those objects (if a garden can be called "an object"), we commonly take to be gardens.

In complement to the presence of counterexamples rendering an "enclosure definition" problematic, there are many spaces that are walled or fenced—prisons, forts, and castles come to mind—that are not gardens. So the etymological route to defining The Garden, while interesting and inspiring of ideas we will discuss later, is a nonstarter in terms of providing us with a usable and inclusive definition on which to base any further discussion.

Defining "The Garden" Historically

We might next turn to history. Stephanie Ross, one of the earliest contemporary garden philosophers and author of the 1998 book *What Gardens Mean*, writes:

> Historians today trace gardens to two ultimate sources: (1) sacred groves and nymphaea dedicated to particular pagan deities, and (2) utilitarian kitchen and medicinal gardens.[18]

Archeologist Jo Day writes that "the earliest [written] reference to gardens [is] found in the third millennium BCE Sumerian Epic of Gilgamesh."[19]

> There was the garden of the gods; all round him stood bushes bearing gems. Seeing it he went down at once, for there was fruit of carnelian with the vine hanging from it, beautiful to look at; lapis lazuli leaves hung thick with fruit, sweet to see. For thorns and thistles there were haematite and rare stones, agate, and pearls from out of the sea.[20]

The earliest pictorial depictions—that is, not descriptions but rather pictures—of gardens appeared as early as 1400 BCE in Egyptian tomb painting. The funerary chapel of Sennefer, mayor of Thebes during the reign of Amenhotep II, includes a painting of a garden plan.[21] Both the *Illiad* and the *Odyssey*—written between

---

[18] Ross, *What Gardens Mean*, p. 1.

[19] Jo Day, "Plants, Prayers, and Power: The Story of the First Mediterranean Gardens," in Dan O'Brien (ed.), *Gardening—Philosophy for Everyone: Cultivating Wisdom* (Chichester, UK: Wiley-Blackwell, 2010), p. 66.

[20] *The Epic of Gilgamesh*, translated by N. K. Sanders (Assyrian International News Agency).

[21] Tom Turner, *Garden History: Philosophy and Design 2000 BC to 2000 AD* (New York, NY: Spon Press, 2005), p. 43.

800 and 700 BCE—contain a number of references to gardens.[22] The Persian *paridaiza* we looked at above traces back to about 550 BCE. As mentioned in the introduction, the account of the Garden of Eden was recorded as early as 500 BCE in the Hebrew scripture in the book we know commonly as Genesis.

> Chapter Two: 8 Now the Lord God had planted a garden in the east, in Eden; and there he put the man he had formed. 9 The Lord God made all kinds of trees grow out of the ground—trees that were pleasing to the eye and good for food. In the middle of the garden were the tree of life and the tree of the knowledge of good and evil. 10 A river watering the garden flowed from Eden; from there it was separated into four headwaters. 11 The name of the first is the Pishon . . . . 13 The name of the second river is the Gihon; it winds through the entire land of Cush. 14 The name of the third river is the Tigris; it runs along the east side of Ashur. And the fourth river is the Euphrates. 15 The Lord God took the man and put him in the Garden of Eden to work it and take care of it.

The Garden of Eden has also been described, albeit briefly, in the Islamic scripture, the *Quran*:

> Second Surah: 21. O people! Worship your Lord who created you and those before you, that you may attain piety. 22. He who made the earth a habitat for you, and the sky a structure, and sends water down from the sky, and brings out fruits thereby, as a sustenance for you . . . . 25. And give good news to those who believe and do righteous deeds; that they will have gardens beneath which rivers flow. Whenever they are provided with fruit therefrom as sustenance, they will say, "This is what we were provided with before," and they will be given the like of it. And they will have pure spouses therein, and they will abide therein forever . . . . 35. We said, "O Adam, inhabit the Garden, you and your spouse, and eat from it freely as you please, but do not approach this tree, lest you become wrongdoers."[23]

Styles of gardens following in the model—literally, tacitly, or as a matter of genealogy—of the Garden of Eden as described in the Abrahamic religious literature we will refer to in this book as participating in "Edenic Culture." The clearest examples of Edenic Culture gardens occur in the Middle East and in areas of Asia where Islamic culture flourished. The Islamic view of the Garden of Eden differed a bit from the Judeo-Christian (or Jewish, and so Christian) version. The Islamic Garden of Eden was envisioned as illustrative of Paradise (capital "P")

---

[22] As we see, for instance, in Books 8 and 21 of the *Iliad* and Book 7 of the *Odyssey*.

[23] Translated by Talal Itani, https://www.clearquran.com/.

and so may have been thought of as more closely associated with the Christian heaven instead of being an earthly realm. The association is apparently a matter of some interpretation; the ambiguity lies in the fact that traditional Islamic gardens are laid out in highly structured and formal ways, each containing four channels or small canals of water, each of these eventually symbolizing the four rivers described in the Hebrew account. These rivers of course are earthly. A solution to the ambiguity is to take the earthly Garden of Eden to be a representation of the heavenly Paradise. (To read too much literalism into the Islamic scripture or traditions is likely a mistake; as is also likely the case with the Hebrew scripture). Roger Paden demonstrates the connection between the Garden of Eden and Paradise.

> [T]he *Hortus conclusus* [was] a garden style common during the late Middle Ages with high walls surrounding a relatively small area. This garden was the direct descendent of earlier, more practical gardens, such as kitchen and "physic" gardens. "*Hortus conclusus*" means "enclosed garden," but since both its component terms refer to enclosed spaces, the name seems a bit redundant. This redundancy, however, only serves to emphasize the connection between this type of garden and the idea of "paradise," a term derived from an old Persian word, "pairidaeza," a word formed from two roots, "paire" (around) and "diz" (to form) and usually referring to walled royal gardens. Through this notion of "enclosure," *Horti conclusi* are twice linked to the idea of paradise, the earliest earthly example of which was the Garden of Eden.[24]

The botanic or botanical garden (the difference being largely a matter of British versus American English), as a garden style, has been closely associated with Edenic Culture,[25] as the aim of the botanical garden has been to form collections of plant species. The proverbial ultimate collection of all plant species is, of course, Eden.

This brief review of the earliest history of gardens, including the Egyptian reference above, connects closely to the etymologically based idea of "enclosure" as definitive of The Garden. "Enclosure" as a defining characteristic—or as *the* defining characteristic—not only can be met with the counterexamples mentioned above but it must be understood also as culturally relative. The strictness of adherence to a prescriptive garden style differs culture to culture. While some cultures, like the Iranian, adhere strictly to the formula and have gardens that are all structurally very

---

[24] Roger Paden, "The Ethical Function of Landscape Architecture," *Environmental Philosophy* 15:2 (2018), p. 149.

[25] Hunt, *Greater Perfections*, p. 92.

similar,[26] and while some cultures have expansive variation on the theme but still stick close to the structural formula, like Japanese tea gardens, many cultures either lack a single formula or lack the requirement to stick to a formula or formulas that may be popular in that culture.

US gardens are a good example. While the prairie style garden exemplified in the design work of Jens Jensen (1860–1951) may arguably be the closest thing we have to a true US style of garden design (excepting perhaps the civic work of Frederick Law Olmsted and Calvert Vaux), there are not very many gardens in America that illustrate this prairie style (or, for that matter, the urban style developed by Olmsted and Vaux). Most Americans, if asked, would not be able to identify a quintessential American garden style though some (who have any degree of love of gardens) may be able to describe the French style (the André Le Nôtre/Versailles formal style) or the English cottage garden style. Obligations per se for French gardens to adhere to the Le Nôtre style or for English home gardens to adhere to the standard English cottage garden style do not exist, but the culture of preferring participation in these styles is old and established. Not so in the United States. As cultures are different, so too are not only their garden styles but the degree to which gardens in a particular culture not only do but must adhere to a particular style.

Defining "The Garden" via Gardening Practice

A third way we might approach defining what makes something a garden focuses on the practices involved in gardening. Francois Berthier, quoting Musō Soseki, father of the Zen garden, writes, "He who distinguishes between the garden and the practice cannot be said to have found the true Way."[27] The relationship of these two elements is as close as any characterization of artist and artwork, of creator and created. If there is promise in defining art in terms of the practices of artists, then there is promise in defining gardens in terms of the practices of gardenists and gardeners. Thomas Fuller, oft-quoted seventeenth-century cleric and historian, is thought to have said, "As is the gardener, so is the garden."

What are gardening practices? Brook writes, "Inherent in the idea of a garden is some kind of care or attention beyond the initial design. The actions by a person to nurture plants, to shape and develop, or just to encourage what grows, we call 'gardening.'"[28] Cooper writes, "Gardening, in my view, is just about any

[26] Mohammadsharif Shadidi, Mohamad Reza Bemanian, Nina Almasifar, and Hanie Okhovat, "A Study on Cultural and Environmental Basics at Formal Elements of Persian Gardens (before & after Islam)," *Asian Culture and History* 2:2 (2010).

[27] Berthier, *Reading Zen in the Rocks*, p. 52.

[28] Brook, "The Virtues of Gardening," p. 14.

activity geared to the design, cultivation, and care of the garden."[29] Such practices commonly include:

1. Designing
2. Preparing
3. Planting, Placing (as in stones), Building, and Ornamenting
4. Irrigation
5. Tending to the Ongoing Health of the Living Elements (including finding and treating for pests, staking, cabling and bracing trees, air spading, adding compost and fertilizer, and so forth)
6. Propagating and Grafting
7. Weeding and Pressure Washing
8. Pruning, Cut-Backs, Shaping, and Deadheading
9. Editing and Curating (including transplanting and removing desirable plants that do not belong or are too thick)
10. Replacing and Resurfacing
11. Redesigning

Cooper puts a special emphasis on gardening and gardening practices, an aspect of gardens he believes is easily or frequently overlooked.

> One wonders why so many people who could afford it do not automatically pay others, or invest in "hi-tech" equipment, to do the job for them. . . . It is my impression that . . . the garden has become increasingly appreciated in recent years as a theatre of garden practices, and less as an aesthetic spectacle.[30]

While there is nothing substantively wrong with approaching the question of what counts as a garden through understanding the garden as what results from gardening practices, this approach puts the emphasis centrally on gardenists and gardeners and less on the products of their efforts, the garden itself. It seems intuitively more plausible, if one is defining a thing, to place emphasis on the thing itself rather than the efficient or even formal causes of how it was brought into being. While the inventory approach might be seen as putting too much emphasis on the material and final causes—to continue to invoke Aristotle's schema—these causes are more focused on the object itself. As understanding the garden is subject to the same problems associated with the "intentional fallacy" in the interpretation of literature, focus on the object seems generally the safer course to understanding it.

---

[29] Cooper, *A Philosophy of Gardens*, p. 68.
[30] Cooper, pp. 63 and 65.

There is a second and more telling reason for preferring a focus on The Garden rather than on gardening practice: the object-focused approach provides the platform and scaffolding necessary to ground the project of exploring features common to all gardens (or families of features) in an obvious and obviously potentially fruitful way. Focus on the gardenist and gardener does not.

Defining "The Garden" Linguistically

Let us turn now to a fourth way to approach defining The Garden, a way focused on linguistics and how contemporary users of a language use the terms of that language. Hunt writes that "the experience of a new kind of place-making elicited, even required, fresh definitions of the garden."[31] While it may be common to think that translation of a term from one language to another requires us to check our translation against the intentions of the original speaker to capture their meaning or their reference, the reality is that in the vast majority of cases where we seek to understand the meaning or reference of a term, we simply turn to the rules of the languages involved. If one wishes to understand the Spanish word "gato," the Japanese word "neko," or the French word "chat," one normally will consult some translation device or a lexicon, mechanisms that function through following linguistic rules that govern reference and word-meaning. These mechanisms are not based in a priori considerations but rather on the way that nonidiosyncratic speakers of a language commonly use a word.

What does it mean to apply such an approach to defining The Garden? The Oxford English Dictionary, arguably the most authoritative lexicon of English, defines a garden as "an enclosed piece of ground devoted to the cultivation of flowers, fruit or vegetables." (Again, we see "enclosed.") This definition may work for the purposes of communication, and so also for the purposes of translation, but it does not prove satisfactory in our case since there are many spaces or objects that English-speakers refer to as gardens—and apparently are correct in their reference—that do not fit the OED definition. But while the actual mechanism of consulting a dictionary may not work for us, the theory behind such consultation might. What we want to know is how speakers of English use the word "garden." So the best way for us to know this, following the lexicological approach, is for us to attempt to capture all references to "garden" that have any claim whatsoever to commonality or any claim whatsoever to being intuitively plausible—any reference to "garden" that facilitates communication—and then examine them to find what is common, in terms of meaning and characteristic,

---

[31] Hunt, *Greater Perfections*, p. 16.

to all. This of course is the project described at the top of this chapter and the project for which we argued in the discussion on antiessentialism.

As we proceed with building a taxonomical inventory, following a general linguistically focused approach, we may glean some insight from an alternative linguistic approach—examining where the word "garden" appears in our metaphorical lexicon. "A garden of ideas" and "a garden of earthly delights" come to mind. These relate more to collections of objects, as do usages of "garden" for nonhuman animals. Some leaf-cutting ants in Central and South America make "gardens" of leaves on which an edible fungus grows—this relates less to collection and more to cultivation. Male bowerbirds, in their creation of intricate nests, will collect colorful objects, including berries, as decorations. Some of these berries germinate, resulting in a kind of accidental cultivation. An "octopus's garden" is a collection of stones and shells around an octopus's undersea den; in Ringo Starr's imagining, such a garden was a safe and hidden outdoor space with a comfortable balance of warmth and shade.

Defining "The Garden" via Inventory and Taxonomy

Constructing an inventory means determining a way to approach categorizing items into classifications designed to capture all instances of kinds of gardens. This starts with determining a taxonomical approach. There are several ways to address that task.

1. *Historically or chronologically.* One could attempt to list all gardens within a historical structure that emphasizes "garden eras." Such an approach would likely start in Egypt (of four thousand years ago) and work forward, probably next by visiting China, era by era. This approach would render fascinating results, likely moving from ancient gardens, to classical, to Asian/Islamic, to medieval, renaissance, baroque, and then into the eighteenth century and beyond.[32]

2. *Anthropologically.*[33] Gardens began, following the "evolutionary" model favored by Hunt, at the point where Second Nature turned into Third, and that likely happened as soon as those who left nomadic life to develop agricultural practice decided to "concentrate and perfect" the forms present in agriculture. Agriculture turns into horticulture, and the garden is born. This likely began well before the Sumerians and the Egyptians, but the Egyptians were the first to make a record of it. In theory, cultural

---

[32] This list follows the matter described by Tom Turner in his book *Garden History.*
[33] This is likely the approach that would be favored by Hunt (see p. 16 of *Greater Perfections*).

anthropologists should be able to trace the influence of Egyptian garden-making from Egypt as it spread. Likely the same project would begin in China or perhaps India, and the influences of their garden practices could be traced as they were shared, adopted, and adapted by other cultures. The results of this study probably would be even more fascinating than those of an historical approach.

3. *Wild to Formal.* A taxonomy of gardens could be created through considering the spectrum of all gardens along a continuum of gardens that begin with those that are closest in form to First Nature—that is closest to wilderness, "natural" or "virginal" land—and progress in greater and greater degrees of formality and structure, eventually culminating in a point represented by the most formal garden.[34]

4. *Greatest to Least Great.* One could organize gardens in terms of their greatness—with "greatness" expressed as the result of the application of some evaluative formula or other.

5. *Greatest Gardenist to Least Great.*[35] The same approach as the one directly above could be employed, but rather than looking at gardens per se, we would either look at garden designs (which, unlike gardens, are static) or at the reputations of garden designers. This approach, and the one directly above, are the least preferable of the lot; classifying by some application of normative standard involves all manner of opportunity for challenge and contest.

The taxonomic approach we will use is teleological or purpose-driven. It is teleological in two interrelated ways. First, every garden has a design—a necessary condition of being a garden (or so we will argue). So a teleological approach should in theory capture representation of every garden type that exists. Second, as every design is necessarily artifactually deliberate—that is, every design is the product of deliberation and decision in the creation of some object that bears the mark of being artifactual in some degree—a teleological approach allows us to ask the questions: "why was this garden created?" "what is the purpose of this garden?" Uncovering the purpose of a garden's existence and its design will help us understand its aesthetic features as discovery of the interrelational fit of the aesthetic properties of any object helps to understand not only its worth as an aesthetic object but also its meaning as one. (While this project may sound as if the gardenist's intentions are necessary to understanding the garden's properties

---

[34] This likely would be Hunt's least favorite option (see p. 213).

[35] As defined in this book's introduction, a "gardenist" is a garden designer or garden artist; the term is used to contrast with a "gardener" who maintains the garden.

and purposes, this is not necessarily the case. We will return to this topic when we discuss the meaning and interpretation of a garden.)

Adopting a teleological approach is not question-begging in favor of the artifactual nature of gardens. We will not presume every garden is designed but instead argue for that claim later. But adopting such an approach now allows us both to cast an appropriately wide net while at the same time setting up the resulting inventory to be useful for the purpose of discovering what features gardens share in common, or what series of family resemblances are shared among what series of gardens, or if there are no features that seem general enough for us to find definitionally interesting.

It must be noted before we begin that many gardens have, or are even created to have, multiple and various purposes. Identifying a garden as falling within one garden type is not to suggest that it cannot also fall under others. We will try to point that out along the way, but we do not promise to be comprehensive in that endeavor as purposes gardens possess can sometimes be relative to persons or groups of persons—that is, a garden's purpose is not always known through simply reading its objective features. In addition, a garden's purpose can change. A garden originally constructed as a kitchen garden might one day develop into a flower garden. A pleasure garden, containing a wide variety of plants collected together by one (with the resources to do such a thing) for the simple purpose of enjoyment, might one day form the basis for a scientifically focused botanical garden, a preservation garden, or an education garden. Additionally, we will encounter examples where gardens evolved out of First or Second Nature as their use and purpose changed over time. A plot of land that may have been set aside for agriculture—a sheep pasture or an apple orchard—could become a garden if its primary purpose becomes something other than the efficient production of sheep or apples.

Gardens have existed for thousands of years, all throughout the world. To give a thorough and accurate survey of every nuanced garden style would require a level of expertise that few if any possess. So, of necessity, our inventory must be incomplete. In addition, as it is an inherent quality of gardens that they must be maintained, and as garden maintenance requires resources, it must be pointed out that the majority of garden examples from which we have to choose are those owned or sponsored by those with the means to maintain them. This may exclude the many gardens created and used by the working classes that have been lost to time.

We begin our taxonomy with six basic categories of gardens:

1. Identity and Style Gardens
2. Preservation and Collection Gardens
3. Destination Gardens

4. "Function-Forward" Gardens
5. Setting and Framing Gardens
6. Personal Gardens

1. **Identity and Style Gardens.** Gardens that fall under this broad category include those that are focused on communicating a message concerning their ownership or demonstrating aspects about a particular culture. The list of national styles included below is meant merely to represent this type of garden; it includes some of the most identifiable garden styles. A nation may embrace or express many garden styles over the course of its existence; the styles listed below are iconic "nation-identifying" ones that come from a particular time in the life of that nation.

   a. *Wealth and Prestige Gardens*: gardens that are created and possessed by the wealthy principally as a sign of their wealth or status and meant for the enjoyment of their owners and pursuit of the owners' interests. Many gardens associated with royal estates would fit within this category. An excellent extant example of this sort of garden is Oak Spring Garden in Virginia, originally owned and developed by Bunny Mellon. A garden of many hundreds of acres, Oak Spring features apple trees (pruned for aesthetic structure as much as, if not more than, for apple production), several glasshouses, and courtyard gardens kept up to the tastes of the late Mrs. Mellon.

   b. *Central Asian Garden Styles*
      i. Persian. Persian gardens are extremely formal gardens, the iconic style (*chahar bagh* or *charbagh*) typified by division of the space by four straight channels of water into four quarters, with those quarters commonly divided into additional quadrants, all equilateral in shape. Persian gardens are typically planted with trees in the center of each area and plants—flowers and herbs—organized symmetrically around them. Shade and water, given the locale, are very important components. Examples include the Fin Garden and the Eram Garden, both in Iran. As Islam became the dominant religion in the region, the four-quadrant garden became associated with the concept of earthly paradise. An example of the Persian style having moved west to Europe is the Moorish Garden, the most famous example being the Alhambra in Spain.
      ii. Mughal. The Mughal garden is the Persian garden (in particular the *chahar bagh* style) transplanted to India, Afghanistan, and Pakistan. Examples include the Shalimar Gardens of Lahore, the Nishat Bagh of Kashmir, and the Taj Mahal gardens.

c. *East Asian Garden Styles*

    i. Chinese. Chinese gardens historically were built for pleasure, to impress, and for contemplation. They are generally characterized by a microcosmic representation of idealized nature, with significant symbolic representations of both natural and supernatural elements. Their elements, which typically include intricate buildings and bridges, are not all immediately visible but rather are discovered as one walks their paths. This implies an asymmetry uncommon in formal European gardens. Chinese gardens are typically walled and feature rock and water more prominently than plants. They may make use of the "borrowed scenery" of mountains outside the garden's borders. Examples include the gardens at Suzhou (likely the oldest still surviving in China) and the Imperial Gardens in the Forbidden City in Beijing.

    ii. Japanese. Japanese gardens, inspired by early interactions with China, come in a variety of forms, four of the most common being: stroll (or strolling) gardens ("*kaiyū-shiki-teien*"), tea gardens ("*roji*"), the "*karesansui*" or dry landscape gardens or Zen gardens, and small, usually personal urban gardens ("*tsubo-niwa*"). An example of the stroll garden is at the Katsura Imperial Villa in Kyoto, and an example of a tea garden is located at the Shinto shrine complex located at Ise. The term *karesansui* (made by the characters "dry," "mountain," and "water") appears for the first time in the eleventh century in the *Sakuteiki*, the oldest treatise of the art of gardens in Japan.[36] Examples of *karesansui* appear in temple complexes in Kyoto, perhaps the most famous example being at Ryōan-ji, which was founded in the mid-1400s.[37] Other great examples include those found within Daisen-in and Saihō-ji (also known as Kokedera, the Moss Temple). Like Chinese gardens, they are strongly asymmetrical and may contain borrowed scenery, pavilions, structures, and a focus on water and rock. These elements are more muted in color and design. *Karesansui* deserves special attention (we return to it throughout the book) as it is the most pared-down garden style we will encounter.[38]

---

[36] Berthier, *Reading Zen in the Rocks*, p. 19.

[37] Berthier, p. 30.

[38] This section on Japanese gardens only deals with gardens in Japan, but Kendall Brown writes that while there have been gardens in Japan for fifteen centuries, there are now far more "Japanese-style" gardens outside Japan than inside, many of which are in the United States. He writes, "In the 21st century, Japanese gardens may well be considered a universal art ... embraced and adapted so widely and deeply to constitute an expressive language likely meaningful everywhere and available to anyone. Links with their birth culture, once strong, have become weaker as these garden styles accumulate

d. *European Gardens*

    i. Italian. The iconic Italian garden is formal, rectilinear, and axial, incorporating terraces and parterres. Structure is provided by hedges as well as by built elements such as stairs, walls, and sculpture. It is the quintessential expression of the renaissance garden, demonstrating order and symmetry. It was a huge influence on the seventeenth-century French formal garden style and associated English styles. An early example is the Villa Medici at Fiesole, built in the mid-1400s.

    ii. French. The iconic French garden—popular examples being the baroque gardens at Versailles and Vaux-le-Vicomte designed by Andre Le Nôtre in the mid-1600s—is known for being highly formal and symmetrical, including parterres, bosquets, and grottos. Also at Versailles is the quintessential example of the *ferme ornée*, an idealized, ornamental farm: Marie Antoinette's *Hameau de la Reine*.

    iii. English. Prior to 1700, the gardens of England were baroque or typically resembled the highly ordered gardens of France. But the garden designs most commonly associated with England—as a reaction to the manicured style—are the large mid-1700s landscapes of Capability Brown meant to look like idealized countryside, complete with grazing animals, large serpentine lakes made to look like rivers, architecturally significant bridges, "clumps" of trees, views that stretch along lengthy axial arms and frequently end with sights of significant village architecture, and lawns that sweep right up to the castle or manor house in the middle. Brown's designs frequently incorporated "ha-has" which were walls submerged in the ground, excavated on the side opposite the house to prevent animals grazing on the acreage from approaching the house. "Ha-has" provided a barrier, albeit one hidden from view of anyone surveying the landscape from the house. Examples of this style are the landscapes of Stowe, Blenheim Place, and Highclere Castle (where *Downton Abbey* was filmed). (The Picturesque garden style developed in part as a reaction to Brown, and it was influential on American landscape design through the work of Andrew Jackson Downing.) English cottage gardens are another design likely to come to mind

---

identities and functions that may relate to Japan only tangentially. As such, it makes sense to call them Japanese-style gardens, acknowledging gardens based on adaptable values rather than garden make in Japan or about Japan" (Kendall Brown and David Cobb, *Visionary Landscapes: Japanese Garden Design in North American* [Rutland, VT: Tuttle, 2017], p. 6).

when thinking about iconic English gardens. Cottage gardens were originally a mix of vegetable and flower gardens, although more recently they are composed simply of flowers. They are made to look as if the plants have naturally spread to fill the space between them; they are not symmetrical and, despite requiring a good deal of maintenance, are meant to look as if they are the products of benign neglect. They typically are sited against the house. An example would be the gardens bordering the wall of Hampton Court Palace that faces the Thames. English garden designs also include something of a reaction to Brown's design, which we will discuss in the next chapter. These English garden designs were replicated across the globe as British Colonialism spread so that the Colonial British could feel "more at home" in their new surroundings;[39] this transplantation of style was successful in varying degrees depending on the garden designer's sensitivity to the new climate and plants of the region.

iv. German. Michael Lee describes the German style of garden design, arising in the late 1700s, with the word "mittelweg," which means "the middle way" and which signifies that the German style sought to find a middle way between the French (formal) garden style and the English (natural) landscape style. Lee describes this movement as being more academic and inspired by an academic consideration of the ideal of garden design than it was an actual practice.[40]

e. *American Gardens.* The United States does not have an iconic garden style.[41] If we set aside the ubiquitous suburban lawn dotted with shrubbery, there are two garden styles that are connected to the United States: (1) the civic or municipal style of garden creation made famous by Frederick Law Olmsted and his partner Calvert Vaux, perhaps best typified by New York City's Central Park, and (2) the Prairie garden style of Jens Jensen. A characteristic garden style of South America may be that pioneered by Roberto Brule Marx, the Brazilian landscape architect whose use of tropical plants and geometrical modernism is recognized on the promenade of the Copacabana and replicated in landscapes around the world.

---

[39] Isis Brook, "Making Here Like There: Place Attachment, Displacement, and the Urge to Garden," *Ethics, Place, and Environment* 6:3 (2003), pp. 227–236; Michael Moss, "Brussel Sprouts and Empire: Putting Down Roots," in Dan O'Brien (ed.), *Gardening—Philosophy for Everyone: Cultivating Wisdom* (Chichester, UK: Wiley-Blackwell, 2010), pp. 79–92.

[40] Michael G. Lee, *The German "Mittelweg": Garden Theory and Philosophy in the Time of Kant* (New York, NY: Routledge, 2013), pp. 13–14, 42, 44, 61, 137–138.

[41] Michael Pollan, *Second Nature: A Gardener's Education* (New York, NY: Dell, 1991), p. 230.

2. **Preservation and Collection Gardens**
   a. *Conservation Gardens and (some) Nature Preserves.* Whether or not nature preserves are gardens is a question to do with the extent of human participation. Some nature preserves are on the site of larger gardens, maintained by horticulturists, and attempt to present an ideally functioning native ecosystem; others, such as those found in National Forests, are subject to much less management and frequently used for purposes other than aesthetic appreciation. The latter are likely not gardens.
   b. *Scientific Collection Gardens*
      i. Botanical Gardens. Isis Brook writes, "In the early botanical gardens, we see an attempt to recreate the Garden of Eden, which would, of course, have been an ordered world."[42] We often look to *Orto Botanico di Padova* in Padua, Italy, as the first botanical garden, from 1545; this would have originally been a collection of medicinal herbs. Archivist Michael Moss notes that "plant hunting" had become a staple of British horticulture since the 1600s, with the Royal Botanic Garden, Kew, taking center stage in the 1700s.[43] Despite the Eurocentrism we might expect to find at the heart of "plant hunting," anthropologist Susan Toby Evans writes, "We may not realize it, but when we visit or study botanical gardens, we are appreciating a landscape design format that the Aztecs pioneered."[44] Examples of botanical gardens span the globe; examples include RBG Kew in London, NYBG in New York, Kirstenbosch in Cape Town, and the UNAM Botanical Garden in Mexico City.[45]
      ii. Arboretum Gardens, Alpinum Gardens, Palmetum Gardens, Ferneries, Meadows, Moss Gardens, Rose Gardens, and Native Plant Gardens. While most of these gardens represent the systematic collection and placement of various plants, there are a few examples of gardens that are modifications of native land. The Thain Family Forest, a fifty-acre old-growth forest in the center of the New York Botanical Garden, is such an example. Although it has the character of being untouched, it is dutifully maintained to

---

[42] Isis Brook, "Wildness in the English Garden Tradition: A Reassessment of the Picturesque from Environmental Philosophy," *Ethics and the Environment* 13:1 (2008), p. 107.

[43] Moss, "Brussel Sprouts and Empire," p. 89.

[44] Susan Toby Evans, "The Garden of the Aztec Philosopher-King," in Dan O'Brien (ed.), *Gardening—Philosophy for Everyone: Cultivating Wisdom* (Chichester, UK: Wiley-Blackwell, 2010), p. 207. Evans writes, "The botanical gardens laid out by Aztec kings and landscapers were, like their successors in sixteenth-century Europe, compendia of plants that were meant to represent exhaustively a particular region."

[45] For more on botanical gardens, see Thomas Heyd, "Thinking Through Botanic Gardens," *Environmental Values* 15:2 (2006), pp. 197–212.

remove invasive species and reintroduce native trees and shrubs. It should be noted that not all examples of items mentioned here are gardens; as in the case with nature preserves, it depends (at least in part) on the level and character of human involvement.

iii. Species Affinity Gardens: where all the plants share their capacity for living in a particular set of circumstances, like a bog, a dessert, an alpine climate, a tropical climate, and so forth.

iv. Education, Trial, Demonstration, Evolution, and Taxonomic Gardens. Examples include the Evolution Garden at RBG Kew, where plants are grouped by botanical family, and the trial gardens at Mount Cuba Center in Delaware, where cultivars of plants native to the mid-Atlantic are grown for study.

c. *Zoological Gardens and Aquaria.* "Zoos and Aquaria" is an apparent outlier in the list, since the ways in which we approach modern zoos differ considerably from the ways we approach most gardens. However, zoos and their predecessors, menageries, are closely associated with gardens—from the Persian *paridaiza* to Versailles, to some modern estates that have on their acreage collections of exotic animals. We include them in our list, but most instances probably are not gardens.

d. *Wildlife Gardens,* including butterfly gardens and pollinator gardens.

3. **Destination Gardens**

a. *Pleasure Gardens,* including water gardens, glasshouses and orangeries, bosquet, knot gardens, mazes, and children's gardens. Examples include Butchart Gardens in British Columbia and Longwood Gardens in Pennsylvania.

b. *Public/Civic Gardens and Arboreta.* "Public landscapes like New York City's Central Park were conceived as places for reacquainting urban populations with nature's 'charms' and repairing the psychic and spiritual damage caused by everyday life in the modern industrial city."[46]

c. *Botanical Gardens, Zoological Gardens, and Aquaria* (repeated from above).

d. *Art Gardens and (some) Earthworks.* This is a potentially controversial inclusion in this list: do large-scale environmental artworks count as gardens? Many are referred to as such, and if they are not gardens per se, they are closely related. We will include the category, following our intention to cast a wide net. A thorough discussion of these sites is included in Ross's book.[47] Examples include Robert Smithson's

---

[46] Eric MacDonald, "*Hortus Incantans*: Gardening as an Art of Enchantment," in Dan O'Brien (ed.), *Gardening—Philosophy for Everyone: Cultivating Wisdom* (Chichester, UK: Wiley-Blackwell, 2010), p. 107.
[47] Ross, *What Gardens Mean*, pp. 208–224.

2005 *Spiral Jetty* and Alan Sonfist's *Time Landscape*, began in 1965. Both these works appear to share with typical gardens many of the features—perhaps all—that we will review later, although there are other earthworks that apparently do not.

e. *Experimental Gardens*. As the name suggests, these are gardens that push the boundaries of what counts as a garden by being based on or including elements uncommon to typical gardens. Striking examples include the Garden of Cosmic Speculation in Dumfriesshire, Scotland; Little Sparta, also in Scotland; Martha Schwartz's Bagel Garden in Boston; and the High Line in New York City. Each of these examples, we think, are widely and uncontroversially considered gardens.

4. **Function-Forward Gardens**
   a. *Nurseries*: places where plants are grown for sale, for displays, or for collections. While most nurseries would fall under the heading of "Second Nature," since their primary purpose is selling plants, some nurseries create gardens to showcase the aesthetic potential of their product. Annie's Annuals and Perennials in Richmond, California, is an example. In general, only a small subset of nurseries would count as gardens or as inclusive of gardens.

   b. *Consumable Gardens*: food-focused gardens, which would include kitchen, vegetable, and herb gardens, and fruit orchards. Food-focused gardens can incorporate a range of technologies, including raised beds, hydroponics, aeroponics, and aquaponics (where sea creatures are raised in tanks of water with plants growing on top of the water). These gardens skirt the line between Second and Third Nature; the difference is in their intention and use. The Crops of the World garden at the University of California Botanical Garden is meant to highlight the botanical characteristics of various crop plants; The Land exhibit at Disney's Epcot is meant to demonstrate developing food growing technologies. Individual plots for crop growing in a community garden are less about efficiency and profit and more about reaping the benefits of one's own horticultural efforts.

   c. *Health Gardens*
      i. Physic/Medicinal Gardens: gardens designed to grow plants with medicinal properties and harvested for that purpose. Botanical gardens as we know them probably evolved from gardens such as these. The Oxford University Botanic Garden, founded in 1621 as a "physic garden," and the Chelsea Physic Garden in London, founded in 1673, are prime examples.
      ii. Therapeutic Gardens. The field of "ecopsychology" shows correlations between spending time in nature and enhanced mental

attitude and health. Therapeutic gardens exist in this vein. This may include Hospice gardens and prison gardens.

iii. Sensory Gardens. Not all gardens are made for the eyes; some are made for engagement of the olfactory, some to touch, some to hear birds and other wildlife. Gardens in this style are designed to speak directly and immediately to human sensation, usually with the goal of aesthetic engagement and whatever somatic values are served through such engagement.

d. *Tea Gardens, Beer Gardens, Pub Gardens, Entertainment Gardens.* Japanese tea gardens are distinctly different from German beer gardens (or *Biergartens*) or British pub gardens; nonetheless they all focus on contextualizing drinking and (perhaps to a lesser extent) eating. They likely do not belong under "setting and framing" gardens as the garden aspects of these gardens are not ancillary to their functions but rather central to them; this is especially evident in Japanese tea gardens. In addition, from the late 1700s to the late 1800s, Vauxhall Gardens and Ranleigh (or Ranelagh) Gardens, both in London—precursors to the public park movement—were used primarily as venues for public entertainment.

e. *Religious/Philosophical Gardens.* Gardens have served as places for contemplation likely since shortly after their inception. Brook writes, "Our wonderment at this being opens our hearts and in that openness we receive something and are improved by it. To call this experience pleasure, even a higher pleasure, requires that we take away pleasure's hedonic overtones, or perhaps we should just leave pleasure behind and call it grace."[48]

i. Ancient Greek Gardens. Damon Young writes, "[Aristotle's] Lyceum also housed the first botanical garden (probably stocked by the Macedonian Empire), which undoubtedly contributed to his long book *On Plants.* In this, Aristotle was following his teacher Plato, whose Academy was also in a sacred grove. . . . Aristotle's own student and successor, Theophrastus, wrote the first systematic treatise on botany."[49] Epicurus's school was also sited within a garden.[50]

---

[48] Brook, "The Virtues of Gardening," p. 24.

[49] Damon Young, *Philosophy in the Garden* (Minneapolis, MN: Scribe Publications, 2020), pp. 2–3.

[50] Gordon Campbell, "Epicurus, the Garden, and the Golden Age," in Dan O'Brien (ed.), *Gardening—Philosophy for Everyone: Cultivating Wisdom* (Chichester, UK: Wiley-Blackwell, 2010), pp. 220–222.

ii. *Karesansui.* Francois Berthier writes, "[Ryōan-ji in Kyoto] is disconcerting through its not really being a garden: there is no green of trees to be seen, no scent of flowers to be caught, not any birdsong to be heard. . . . [I]t is by seizing the essence of nature that the human being can discover his own 'original nature.' That is why the Zen monks stripped nature bare, retaining only rocks and sand, and a little vegetation."[51] (We will argue that *karesansui* are indeed gardens.)

iii. Religious Institutional Gardens. Examples abound and include Baha'i gardens, temple gardens, church gardens, cloister gardens, monastic gardens, and Mary gardens.

f. *Memory Gardens*

i. Cemeteries and Columbaria. Miller writes, "Richard Etlin has shown how, increasingly since the late eighteenth century, a gardenlike setting has seemed the most fitting environment for our buried dead and the lines between cemeteries and gardens and parks have been blurred."[52] Jo Day writes, "Ineni, an official in the reign of Thutmose I (1504–1492 BCE), not only had his garden painted on his tomb, but the number of each species he planted was listed there. . . . These painted tombs gardens were for the soul of the dead, a place for them to find rest, shade, and refreshment, while the actual real garden surrounding the tomb was for the living relatives and priests to perform funerary ceremonies and rites of remembrance."[53] Greenwood and Woodlawn Cemeteries in Brooklyn and the Bronx provided early New Yorkers respite from the city in the days before Central Park.

ii. History Gardens and Gardens of Remembrance. Many of these exist on the sites of larger gardens or parks—a prime example is the AIDS Memorial Grove in San Francisco's Golden Gate Park. We might even include Maya Lin's Vietnam War Memorial in Washington, DC, and other such outdoor monuments, depending on how they are situated and what surrounds them.

g. *Political Gardens*, such as victory gardens and peace gardens.

h. *Children's Gardens*, such as the Everett Children's Garden at the New York Botanical Garden, where the arrangement of plants is designed to

---

[51] Berthier, *Reading Zen in the Rocks*, pp. vii and 6.
[52] Miller, *The Garden as an Art*, p. 52.
[53] Day, "Plants, Prayers, and Power," p. 72.

be especially appealing to children, complete with tunnels, mazes, miniature dwellings, and floral caterpillars.

i. *Emblem and Representational Gardens.* Some gardens are designed to instantiate, incorporate, or even showcase narrative elements. The most oft-cited example of such a garden is Stourhead in Wiltshire. Stourhead incorporates many monuments surrounding a large lake; a walk around the lake is meant to read as the Trojan hero Aeneas's journey as described in Virgil's *Aeneid*.

5. **Setting and Framing Gardens**
   a. *Residential and Estate Landscape Gardens.* American "yards" and British "gardens" are what we have in mind here when they are designed to frame the house that is the chief focus. Not all residential gardens are gardens, even in Britain; one that is entirely ignored or has no living components we would not think of as a genuine garden.

   b. *Sculpture and Gallery Gardens.* One might think of the sculpture in a sculpture garden as a part of the garden per se, and indeed it is common in all manner of gardens to have statuary and sculptures as a part of the garden. But if the focus is the sculpture in such a way that most any garden setting would accommodate it—that is, when the focus is not the whole garden but rather the sculpture itself—the garden is best understood as a setting. Storm King Art Center in Beacon, New York, is an example.

   c. *School, University, Museum, Hospital and other Public Institutional Gardens.* Most public institutional buildings (that are not in an urban core) have gardens that are like the gardens surrounding our homes; they may be larger and planted in a more fastidious way, but they often do not differ in kind from residential gardens. The gardens within and surrounding the university campuses designed by Frederick Law Olmsted—such as American University, Chicago, Cornell, Mount Holyoke, Stanford, UC Berkeley, Wellesley, and Yale—are examples.

   d. *Workplace, Business, and Commercial Gardens.* The same may be said here as of public institutional gardens with the only additional caveat being that these gardens serve an added purpose. Public institution gardens foster a sense of calm and confidence; workplace gardens do this, but they must also support the profit-making aspect of the business through encouraging confidence in trade—this may reduce to an obligation for these gardens always to look professional and be maintained to a high standard. Examples may include the atria of malls and office buildings, if exposure to the sky or open air is not required.[54]

---

[54] Thomas Leddy, "The Garden as an Art," *International Studies in Philosophy* 28:4 (1996), pp. 126–127.

e. *Sports and Recreation Gardens.* Outdoor sports in particular, including miniature golf and the like, may be played on fields that are bordered with garden areas. Arenas, stadia, and the like commonly have framing gardens. Golf courses are a particularly interesting example—we will discuss golf courses below.

f. *Playgrounds.* As in the case of sculpture gardens, it is possible to see the playground equipment itself—the swings, slide, see-saw—as part of the garden per se. In that case, "playgrounds" should principally fall under "function-forward" gardens. But this would be unusual; generally the garden areas in a playground are only to provide a setting for the play equipment.

g. *Theme Park Gardens.* Walt Disney World makes great use of plants and garden areas, and they have an extensive staff of horticulturists, florists, and interns to accomplish this. But while particular garden areas may be foci unto themselves—as we find in the tea garden in the United Kingdom Pavilion in Epcot, the rose garden in front of the Magic Kingdom's Cinderella Castle, and at the Polynesian Village Resort—and as the extensive topiaries that Disney creates are certainly meant to be admired in their own right, the vast majority of garden areas at Disney are constructed as framing for the rides, attractions, and other "Disney-culture" elements.

h. *Parking Lot and Public Thoroughfare Gardens.* Many parking lots and roadways are bordered with garden areas; they are infrequently extensive as gardens and their maintenance depends almost entirely on available civic budgets. They are included, presumably, to enhance the comfort of our drives.

i. *Pavement Gardens.* Pavement Gardens are a relatively recent phenomenon. They consist in very tiny gardens created in cracks and recesses in sidewalks and other walkways. While they are certainly admired as stand-alone gardens when encountered—and so they may better be classed as some category of pleasure garden—their existence as gardens is dependent on their locations in and around walkways, and they are generally not thought to have any substantive permanence.

j. *Guerilla Gardens.* While some guerilla gardens are intended to produce food, others are the aesthetic correlate to graffiti. In all cases they are installed in spaces not owned or controlled by the gardenist/gardener. They technically involve trespass, and an unsympathetic viewer may say they involve vandalism. A defense of guerilla gardens is offered in "A Deweyan Defense of Guerrilla Gardening."[55] If we keep in mind that

---

[55] Shane Ralston, "A Deweyan Defense of Guerrilla Gardening," *The Pluralist* 7:3 (2012), pp. 57–70.

Banksy is essentially a graffiti artist, we can understand that a guerilla garden may rise to a strong level of quality as a garden, although in practice a so-called guerilla garden may not involve enough human interaction to be a garden. Pavement gardens may count as a species of guerilla gardens, depending on circumstances, and when that is the case, the resulting garden we understand to be a genuine garden.

6. Personal Gardens
   a. *Residential Gardens.* Residential gardens may well be a species of "setting and framing gardens," but we include them here as they are typically owned and maintained by families who live in structures nested within them. Residential gardens typically occupy the spaces adjacent to our homes and commonly extend our living space from within the home into these areas. They can take many forms. They may include lawns, swimming pools, flower gardens, rock gardens, cactus gardens, pollinator gardens, butterfly gardens, xeriscaped gardens, container gardens, window boxes, water gardens, fish ponds, bottle gardens, terraria, flower boxes, orchards—the list is only constrained by imagination.
   b. *Communal Gardens.* These gardens are commonly shared and composed of shared space, such as "squares" within or among housing units, entry areas to and common areas within gated communities, and roof gardens. Care of these gardens is sometimes the responsibility of those who share the space, but it is more common for gardening companies to be hired to attend to the maintenance. We may also look at designed suburbs as an example of communal gardens: Riverside, Illinois, a commuter suburb of Chicago, was designed by Frederick Law Olmsted and features much of what we associate with his other more famous gardens. The walking trails of the North Berkeley Hills also meet the criteria. Originally built as pedestrian shortcuts at a time when few residents owned cars, the steep trails cut through planted side yards, natural waterfalls, and small parks. The Berkeley Path Wanderers Association now maintains and builds upon these paths as informal additions to the city's park system. Perhaps also apropos are the "rights of way" laws that allow for public walking access across many natural but maintained areas in England and in Wales.
   c. *Allotment and Community Gardens.* These kinds of gardens are usually found in urban settings or settings where a residential garden is impossible or impractical. Participants are allotted a small piece of land, usually contiguous with other such allotments, on which they commonly grow vegetables and sometimes flowers. It is unclear whether

an allotment garden is a garden itself or whether the individual plots composing the allotment gardens are the genuine gardens; again, it will depend on the particulars of the situation.

d. *Telegardens.* A "telegarden" is an interesting technical innovation whereby one accesses a small garden remotely and by controlling a robotic arm maintains that small garden. The first of these appeared at the University of Southern California in 1995. While we expect that most gardens will contain animal life of some size and plurality, telegardens may contain the least.

What Does Not Count as a Garden

Lest we be accused of including everything under the sun in our garden inventory, we need to say a few words about what does not count as a garden. First and foremost, following our linguistics-inspired path, anything that no one has ever referred to as a garden is not a garden. Automobiles, banks, lamps—many things are not gardens and no one has ever referred to them as gardens—at least not meaningfully, literally, or for the purpose of communication.

In complement, just because someone refers to something as a garden does not make it a garden. One can use a word incorrectly, and upon correction can realize they did not mean to say "garden"—perhaps as in the phrase "my old watchdog is garden the house." But more to the point, one can use a term metaphorically and, while perfectly meaningful, that usage is not a literal classification of the object of the metaphor as a garden. We discussed this at the top of the preceding chapter.

While the British use "garden" to refer to what Americans use "yard" to refer to, what the British use "yard" to refer to is typically not a garden. A space where items are stored—instrumental items awaiting use or really any item—or a space set aside for such a purpose is not a garden. As an example of a US use, "Boston Yard" as a common contraction of "Boston Navy Yard"—a shipyard—is not a garden.

Anything that is First Nature or Second Nature rather than Third Nature is not a garden. Wilderness or nature untouched, if such a thing still exists, is not a garden. An area devoted to agriculture or some such agrarian purpose is not a garden. Second Nature also typically includes plant nurseries and other outdoor areas designed for profit. A city block composed of buildings, while metaphorically meaningful if referred to as a garden, is not a garden.

Could one create a bona fide garden that includes objects like automobiles, lamps, watchdogs, ideas written on paper, and even ships? Certainly. But what may populate a garden is not unto itself a garden. From the inventory above, we see plants, rocks, water, buildings, and animals populating gardens, but these things by themselves are not gardens. We may appreciate the aesthetic character of an individual flower in a vase, and we could appreciate the same species' flower at a botanic garden. However, this does not make the single rose or the single plant a garden. (We will discuss *bonsai*, *penjing*, and the like, in the following chapter.)

Similarly, repositories of potted plants are not gardens. This is the case for many noncommercial nurseries, such as in botanical gardens, where plants are kept for conservation, study, or propagation. Most of these places are nonpublic and are not designed for any purpose other than efficient growth.

Virtual gardens—such as in the videogame *The SIMS*—are not gardens, even though they are designed and require ongoing care.

Although we saw a few planned communities that could be considered gardens, the majority of suburbs lack the characteristics we associate with gardens. A walk around the neighborhood brings us to multiple residential gardens, but it typically lacks the unification of design and purpose to be considered a single garden.

Tennis courts, baseball stadia, and the like have many of the features seen in our inventory, and many require a high degree of horticultural skill to keep them in pristine condition. Would it matter if the roof is retractable, if the turf is plastic, or if the court is clay? We believe most sports are played in something other than a garden, which makes intuitive sense and follows our inventory. These spaces are designed to be flat and clear, and their shapes are predetermined by established rules. The notable exception are golf courses, each of which involves an original design based on the existing character of the site. It is not uncommon for people to remark on the natural features of golf courses; very few people remark on the natural features of a football stadium.

Finally, aesthetic and art installations are not gardens. This includes any assemblage of artificial components, placed inside or out-of-doors. It would also include "sculpture gardens" that are simply collections of sculptures where the word "garden" is used metaphorically for the assemblage.

In the next chapter, we explore what characteristics may be common to each garden listed above. If such commonalities exist, it will be from those that a quality definition of "The Garden" may come.

2

Defining "The Garden"

In this chapter we move from thinking about how a definition of "The Garden" might be constructed—and the many gardens throughout time and across the globe to which such a definition, if it is worthy, should apply—and begin the work of actually constructing a definition.

Defining Features of Gardens

The ten features we present below, we believe, are representative of all bona fide gardens. That is, every bona fide garden exhibits to some degree each of the features described below. The temptation at this point is to retreat to modesty and claim only that these are necessary but not necessary-and-sufficient conditions for an object to be a garden, that they do not together constitute the definition of a garden. But we think that would be a mistake, and so we will advance the stronger claim that **taken as a whole, any object that exhibits each and every one of these features is indeed a garden.** The challenge will be for those who disagree to present counterexamples of either sort: an example of a garden (commonly accepted as a garden) that fails to exhibit one or more of these features, or an example of an object that exhibits all these features but fails to be a garden.

Most of the items listed below form the basis of more focused exploration as we proceed with the book. For the sake of articulating the list as a list, we delay this more in-depth conversation until later. Readers who may take issue with one or more of the items are asked to withhold final judgment until the book is read.

The first set of features connects to the idea of **enclosure**. There are counterexamples to the requirement that a garden must be enclosed by walls or fences. But the idea of enclosure that is so prominent in both the etymology and the ancient and medieval history of The Garden is not lost; it has evolved into "**boundedness**." This coheres with the inclusion of "demarcated" in Brook's

Maureen K. Chilton Azalea Garden, New York Botanical Garden, Bronx, NY
Multiple gardenists, 2011
Photo by Ethan Fenner, Spring 2017

The Art and Philosophy of the Garden. David Fenner and Ethan Fenner, Oxford University Press.
© Oxford University Press 2024. DOI: 10.1093/oso/9780197753590.003.0002

straightforward definition: "The definition of a garden I will be using is an enclosed or demarcated outside space with living plants."[1]

1. Spatially Three Dimensional

A garden is a place, one that exists geographically and three dimensionally. A garden is not an artistic (or otherwise mimetic) representation of a garden. A painting or photograph of, or a poem or story about, a garden is not a garden. Conceiving of gardens as three-dimensional means that we can and do think of them as capable of being entered and capable of surrounding us. Miller writes:

> Gardens . . . are inherently spatial. In terms of spatial organization, there are two classes of gardens, those that are meant to be walked through in a literal sense and those meant to be viewed from the outside without being entered physically. Yet the difference between the two is not as significant as one might suppose, for even those which are meant to be viewed from the outside . . . must be capable of being entered, in order to be maintained, and second, are meant to be entered imaginatively, to be recognized as three-dimensional spaces.[2]

We do not view gardens as framed and static, the way we do paintings, photographs, and the like, and this is even the case when the garden in question is not meant to be physically entered by visitors.

Gardens are *real* places. They are not metaphors, virtual spaces, or imagined spaces. While such spaces may be garden-like, they are not, in the strict sense of the term, gardens. This will become clearer as we proceed through the rest of the defining criteria. As real places, gardens have all the properties of other real places. They are subject to the laws of physics. They are highly complex, incorporating a range of properties, any subset of which may be stimuli for a particular perspective, the whole of which is too complex to be comprehensively attended to at any one time or by any one person. They do not depend, for their existence as places, on being the object of attention or perspective. They may present dangers to those who visit them or work in them.

---

[1] Isis Brook, "The Virtues of Gardening," in Dan O'Brien (ed.), *Gardening—Philosophy for Everyone: Cultivating Wisdom* (Chichester, UK: Wiley-Blackwell, 2010), pp. 13.
[2] Mara Miller, *The Garden as an Art* (New York, NY: State University of New York Press, 1993), p. 37.

2. Geographically Bound

Gardens are sited and fixed in space. While some gardens *apparently* have been moved from one location to another—such as the Royal Botanic Garden, Edinburgh, and the University of California Botanical Garden—gardens are location-bound. The Edinburgh and Berkeley examples are cases where the "idea" of the garden was moved; the reality is that the originally sited garden was eliminated and a new garden in a new location was begun. The name was kept because it designated a function or purpose, a design, and/or a history, but the original garden ceased to be and a new garden was created, and this is the case even if the vast majority or even if all the plants and hardscapes (and so forth) were moved.

Sites are unique in terms of their topographies, their climates, their hydrologies, how much sun they get, and so forth.[3] One cannot be swapped with another. And so the gardens that are on the sites they are on are wedded to those sites.[4] This implies that the garden is not its name, its design, or its history. A garden is not merely sited on a place; a garden is a place. We will say more on this point at this end of this chapter.

In complement, it is important to remember, too, that a garden is not merely its spatial location. Ross makes the point that a single location can have a plurality of different gardens populating it over time.[5] But while a garden is in its location, that garden is fixed to that location, as frequently illustrated by the fact that so many gardens bear the names of their locations.

Karesansui—Japanese or Japanese-style dry gardens—may seem an obvious counterexample. While some *karesansui*, such as we find at Saihō-ji, obviously connect to their sites in integral ways, it is less clear that the dry gardens at Ryōan-ji and Daisen-in do. Yet it would be a mistake to come to that conclusion for three reasons. First, these gardens are part of Buddhist temple complexes, and their identities are tied to those temples. As temples, these temples are bound to their sites; this is perhaps overdetermined, but it is easily seen by realizing that these temple complexes have other gardens with living elements that would be subject to the conditions of location mentioned above. Second, each instance of these *karesansui* gardens makes use, either subtle or overt, of the "borrowed scenery" of its surroundings (and we might include "borrowed sounds and smells" as well). Each is a walled garden, and in each case the contextual world is visible

---

[3] Hunt writes, "Landscape architecture is by its very nature a local art. Even botanical gardens, like Padua or Kew, make their collections or imagery derived from far-flung places relevant to local conditions (natural and cultural)." John Dixon Hunt, *Greater Perfections: The Practice of Garden Theory* (London, UK: Thames and Hudson, 2000), p. 248.

[4] Miller takes spatial location to be the defining feature of the continuity of the identity of a garden, given that all gardens are in a constant state of change. Miller, *The Garden as an Art*, p. 76.

[5] Stephanie Ross, *What Gardens Mean* (Chicago, IL: University of Chicago Press, 1998), p. 9.

beyond the wall. Third, dry gardens, like any other gardens, are necessarily open to the ecological processes of the site. While our images of these sites are of perfectly clean, raked gravel, there are leaves that blow on top, long-dormant seeds underneath, and mosses and lichen that make these rocks their home. When it rains on a dry garden, the percolation of water is a character unique to the site's soil type. A typhoon may be enough of a disturbance that some elements need to be reworked.

These reasons work for the *karesansui* that are part of temple complexes and open to ecological processes. These reasons do not work for dry landscape gardens that do not have such connections. The tiny gravel and rock "gardens"—complete with tiny rakes—that were popular a few decades ago, and the "*karesansui*" within certain shopping malls, do not constitute gardens proper (even if they provide the benefits of garden practices). If a dry landscape "garden" were not contextualized within or as part of a garden that is geographically bound in the ways we described, it would not constitute a garden either. Instead, at best it would be an aesthetic or art installation. A clear example was a mock *karesansui* at Epcot in Walt Disney World—not only was it placed indoors and designed to be a temporary exhibit, but the gravel was epoxied into place to prevent any change in appearance. This is decidedly not a garden, and Disney clearly meant it to be merely illustrative of *karesansui* in a context of the presence of many park visitors, some of whom may have wished to touch or walk upon the gravel.[6]

When it comes to immobility, we must address what might be called "liminal" cases of objects that may strike some as gardens but which are clearly mobile. We have in mind cases of *bonsai, penjing,* terraria, single potted plants including potted topiary or small specimen trees, or pots containing an assortment of plants but which are still effectively movable from one place to another, as we might find with some smaller alpine trough "gardens" and recently popular containers of succulents.[7] We claimed earlier that single-potted plants—as we would find lined up in a commercial nursery for sale or in a holding nursery waiting to be placed in a garden—do not constitute gardens, either singly or together. They do not incorporate design, and so they fail one of the most basic definitional tests of what constitutes a garden. But *bonsai, penjing,* and the like

---

[6] Kendall Brown writes that there are now more Japanese-style gardens outside Japan than inside (Kendall Brown and David Cobb, *Visionary Landscapes: Japanese Garden Design in North American* [Rutland, VT: Tuttle, 2017], p. 6). If, outside a temple complex, Japanese-style gardens contain a *karesansui* garden, then the conditions of being contextualized as described apply. If the *karesansui* "garden" stands alone, without such contextualization, we would think of it more as an illustration of a *karesansui* garden rather than an instance of one.

[7] Some readers may be tempted to include on this list the case where the bed of a pickup truck is loaded with dirt and plants are added in a fashion that includes deliberation and design. A decidedly gray area.

include not only design but many of the other characteristics that are on the list we are in the midst of describing now.

In these cases, it seems the correct answer to the question "are these gardens?" is "it depends." *Bonsai* and *penjing* rely on the same environmental conditions—sunlight, water, and so forth—that all gardens do, and so to move one of these "gardens" to another location is indeed to change the context that contributes to the character of the plant (or plants) in question. *Bonsai* are indeed moved from one location to another—historically, they have been given as gifts, put on rotating display, or hidden from destruction at the hands of the Cultural Revolution. Traditionally, however, these miniature compositions would by necessity be sited within gardens, and when moved, moved to another garden. In this way, we may think of them more as a prized plant that was dug out and transplanted to a new garden. In addition, as is the case with *karesansui*, *bonsai* and the rest rely heavily on the incorporation of borrowed scenery as their identity is established contextually. For instance, while an instance of *bonsai* set on a plinth in the midst of a surrounding garden may strike a viewer as a small garden room all on its own, an instance of *bonsai* that is lined up next to lots of other instances, perhaps in cramped quarters, each awaiting removal to a new and ideally permanent home, might strike no one as a garden unto itself. Change the scenery that forms the context of these potted compositions, and we may change their very identity.

It may be tempting to invoke the idea that a garden must be capable of being entered, and one cannot enter a single pot—or the idea that a garden must be capable of surrounding a visitor so that the immersion may be complete—but we should not succumb to such temptation as it might be just as possible to imaginatively enter the world of a single pot in the same way that we may imaginatively enter a garden where typical observation is only permitted through a window or from a viewing platform.

On the other hand, when we think of the typical garden room, we think of a garden that is a complete garden unto itself, just one that is nested within a larger complex of gardens. This may not be the typical way we think of *bonsai* and the rest. Instead, it may be more common to think about *bonsai* as a garden "accessory" or a garden component, as an object that adds depth to the features of a garden but is not such that if removed will be a garden unto itself. When a *bonsai* or *penjing* is brought inside, it functions more like a representation of a garden (or of a single tree, in the case of traditional *bonsai*), or an imitation of a garden in the same way that a small desktop rock garden constitutes not a *karesansui* garden unto itself but rather a representation of a *karesansui* garden. We do not typically think of a single potted indoor houseplant as a garden, and so to the extent that a *bonsai* is like that and placed similarly, we likely would not think of it as constituting a garden either.

A plant or plants that are contained within a structure that is mobile may be a garden, but each instance would have to be considered separately. In other words, *bonsai, penjing*, terraria, small alpine trough gardens, small potted topiaries, and the rest are not automatically gardens by virtue of their membership in these kinds. Given this reality, there is no reason to eliminate the criterion of "geographical boundedness" from this list, especially if such a criterion, as a matter of fact in a given instance, is employed as a lens through which to consider whether a particular object is a garden or not.

3. Human Scaled and Perspectivally Bound

While not all gardens are enclosed in a literal way, all gardens are constituted by a place that is bound in a human scale. That is, a person must be able to understand the garden as a "place"—and that implies that it must have spatial limits. Where gardens are massive, they are normally organized into "subset" gardens— "garden rooms" as we mentioned—that are more felicitous of human scale.

An interesting example of this is the great landscapes designed by Capability Brown. These landscapes might have stretched for hundreds of acres, but they are still identifiable as places: Stowe, Blenheim, Highclere. Some believe that because Brown included "ha-ha" walls into many of his designs, those ha-has constitute the boundaries of the garden. While it is possible that in some cases, the boundary established by a ha-ha wall could constitute an enclosure of a garden, it is common to think, in the case of Brown's designs, of the full landscape as the garden. Ha-ha walls were used by Brown as simply functional items to keep animals away from the main house. The fact that they were hidden from view by anyone looking out from the house is evidence they were not meant to mark the boundary of the garden—the sweeping vistas and extensive lines of vision characterize the extent of a Brown landscape.

In complement, we might consider the very small as well as the very grand. Alpine gardens, rock gardens, trough gardens, and "fairy gardens," for example, utilize miniature plants to create gardens on the smallest scale. They invite the viewer to imagine what it would be like to wander through these microcosms at the size of an insect. However, it is precisely this that makes these gardens human-scaled. We would not come to the same imaginative flights if the relationship of size between human and garden were any different. There is a limit to how small a garden can be—viewed under a microscope, a thumbnail of moss looks like a bosquet or bosco. But this is beyond the limits of our senses, and so we do not refer to the microscopic as gardens.

Gardens are bound to human perspective in another way, as well. Despite that we use the term "object" when referring to the ontology of the garden, gardens may be thought of as "objectless" since there is never anything like a fixed object which is the garden. There is never a moment when one can say "*this* is the garden." Appreciation of a garden, then, is always appreciation of a moment of the garden, a "time-slice" of the garden. One can say they know the *Mona Lisa* because they have seen it. But one can never say that they "know" a certain garden in a correlative sense. They may know that garden (or some portion of it) at a certain time—when it, say, had such-and-such plantings, when such-and-such was in bloom, in such-and-such a season or at such-and-such a time of day. To say otherwise is to take a liberty that may be justified for the sake of communication ("Have you been to Kew?" "Yes; it's incredible.") but which, on inspection, will reveal the limits of that acquaintance ("I thought the fields of bluebells were awe-inspiring." "Bluebells? You're sure you were at Kew? That area you're describing is just a meadow.")

> For some of us, creating those moments of perfection is the point of the garden: enjoying the illusion of timelessness, the illusion of paradise. But gardens are in a constant process of change.[8]

While some gardens have a fixed point from which they are meant to be observed, this is rare; even the range of Japanese gardens that are not meant to be entered physically by visitors still typically permit a plurality of visual vantage points. There are fifteen stones in the *karesansui* garden at Ryōan-ji, but no matter where one stands or sits on the observation platform adjacent to the garden, one can only see fourteen of them.

On the other hand, a garden may never be fully knowable simply through sensory acquaintance. Many gardens—though not all—have components that may not be experienced through the senses without taking heroic efforts. Apart from aeroponic gardens, gardens have networks of roots that are not typically seen; they may have soil that is hidden beneath mulch; and we typically cannot see the hydrology at work. Plants have beautiful microscopic features, flower colors in the ultraviolet range, and chemical processes which constantly release volatiles into the air—but we do not relate to these things during a garden visit. The Royal Botanic Garden at Kew has a treetop walkway so that one may experience the arboretum's canopy—Singapore's Gardens by the Bay has something similar, and other destination gardens do as well—but most times the canopy of a wood or forest or even a single tree is invisible to us.

---

[8] Mara Miller, "Time and Temporality in the Garden," in Dan O'Brien (ed.), *Gardening—Philosophy for Everyone: Cultivating Wisdom* (Chichester, UK: Wiley-Blackwell, 2010), p. 178. She makes this point in *Gardens as an Art* as well: p. 76.

The second set of features has to do with the **relationship of The Garden to nature** (without qualification about whether humans are or are not part of "nature") and the inescapability that this relationship is centrally characterized by **change**.

4. Dynamic and Time Dependent

Miller writes, "Nothing is more obvious in a garden than change."[9] Every garden changes, and this is essential to its character as a garden. Living components grow, expand, and die back through time—and they do so with different life-expectancies. Nonliving components erode, develop patinas, and darken or lighten through time. Above we talked about the three-dimensionality of gardens, but Miller makes the point that gardens are unlike all other kinds of three-dimensional artforms (like dance) because of the ever-present influence of time on them.[10]

> The better we know a garden, the more fully this tension between the present and the future versions asserts itself, to the point where the gardener himself is likely to lose all satisfaction with present beauties in his awareness of the possibilities to unfold in the future.[11]

It is likely better to say of gardens that they are four-dimensional places. Dynamism is the case even with the strictest instance of *karesansui* where (prima facie) no living component is present in the *karesansui* space. Even in that case, the stones themselves, as natural objects, are subject to the influence of the elements and of time, as are the elements of the borrowed scenery.[12]

Gardeners who work with plants must account for change in their work. A one-gallon tree from the nursery cannot be planted one foot from the house if it is expected to grow into its prime. The most talented gardeners visualize not only growth itself but rates of growth, lifespans, competition among plants, seasonal change in flower color, and the yearly succession of plantings.

Each garden is inexorably and integrally bound to time in terms of its being and its identity. Dynamism is not an accidental or contingent feature of gardens. Time is a defining characteristic not merely of The Garden as an abstraction but

---

[9] Miller, "Time and Temporality in the Garden," p. 178.
[10] Miller, *The Garden as an Art*, p. 38.
[11] Miller, p. 116.
[12] For more on nature and dynamism, see: David Fenner, "Environmental Aesthetics and the Dynamic Object," *Ethics and the Environment* 11:1 (2006), pp. 1–19.

of each particular garden.[13] Time thought of as the successive ordering of events is essentially about change, and even the most apparently static gardens participate in change. If they did not, they would not be gardens but rather, as we mentioned, installations.

5. Nature Interactive and Ecologically Bound

Every garden that appears in the inventory above includes living components—at least plants—immediately or contextually. While there are instances of gardens that include nonnatural components, these are only part of a garden that includes natural—and living—components as well. A "garden" made exclusively of nonnatural materials we would refer to as an installation; even if the nonnatural materials were shaped to look like plants and flowers, we would call it an installation that mimicked a garden. This echoes points made by Miller, Hunt, and Brook.

The Garden is connected to nature in integral and essential ways, not merely because nature forms its context but because The Garden interacts with nature and this interaction is constant through the entire life of a garden. Remove the sun and water, and living components die. Remove the wind and rain, and the natural erosion and evolution of the hardscape stops.[14]

When we write about living components or *life* in the garden, we use the term as it is used in modern Western science: living things propagate themselves and grow, like plants, fungi, and animals. But what is considered "living" is not standard among all cultures or periods of time. The stones in the gardens of China and Japan, for example, are typically seen by their creators and by visitors as possessing a life force (*qi*) or spirit (*kami*), which place these garden features in the same category as plants, animals, humans, and larger natural features. Those outside the culture, or those who do not appreciate this view, who come to the conclusion that *karesansui* is not a garden form because it is "devoid of life" misunderstand the importance of the stones that populate this garden.

The *karesansui* garden at Ryōan-ji was established in the mid-1400s. Over time, moss has taken hold and grown around the base of each of its fifteen stones, and the elements have softened the stones so they look even more natural and settled. Could *karesansui* be placed indoors? Some examples apparently exist

---

[13] Miller explores this relationship in her paper "Time and Temporality in the Garden." There she explores how the relationship can be articulated as she characterizes time itself in six different ways, each showcasing not only a different but together a cumulatively deeper relationship between a garden and time.

[14] "Climate is the result of complex interactions between time (the seasons) and the place (local topography and the larger geology). It may be the single most important determinant of garden style." Miller, p. 181.

inside Japanese shopping malls. But this may be a case where what populates a garden does not, on its own, make a garden. Removed from the context of the temple complex and the changes brought on by nature, we understand these *karesansui* to be closer to the dry gardens atop desks than to the one at Ryōan-ji.

As nature changes—either daily, seasonally, or in more temporally expansive ways—the gardens that participate in their particular natural contexts must necessarily change, too. One of the writers of this book has a garden in northern Florida, and with the advent of climate change, this garden has not experienced a hard winter in several years. Plants that used to die back no longer do; plants that used to benefit from a freeze have had to adapt. Additionally, as trees matured, died, and were removed, plants that thrived in the shade have either had to adapt or be moved. And plants that were modest in their expression have become with the added sun something quite different. These changes are all common and must be accepted as commonplace to the character of a garden that is ecologically bound. Gardens are built upon reactions to nature.

6. Unbounded in Time and Lifespan

While most gardens exist over a timespan easy to fathom in human dimensions, some gardens have persisted for hundreds of years. The ancient Mahamevnāwa Gardens in Anuradhapura, Sri Lanka, was founded sometime before 307 BCE and likely is the world's oldest extant garden. A garden was first created on the site of the Humble Administrator's Garden in Suzhou, China, during the Shaoxing period (1131–1162) of the Southern Song Dynasty. The gardens created by monks at Westminster Abbey are thought to be around that old. In 1545, *Orto Botanico di Padova* was created in Padua; it is thought to be the oldest surviving botanical garden in the world (*Orto Botanico di Pisa* was founded in 1544, but it was moved; Padua's claim is based on the fact that it never left its original site). *Orto Botanico di Firenze*, in Florence, was founded by Cosimo de Medici late in 1545. Iran's oldest extant garden, the Fin Garden, was created in 1590. France's oldest botanical garden, *Jardin des Plantes de Montpellier*, was created in 1593; Copenhagen's in 1600. In the United Kingdom, the Oxford Botanic Garden was founded in 1621. The Royal Botanic Garden, Kew—arguably the world's most important horticultural site—was founded by King George III's mother, Princess Augusta, as a botanical garden in 1759. The Missouri Botanical Garden, one of the United States' most respected gardens, was founded in 1859, and the New York Botanical Garden in 1891.

Even gardens that have significantly shorter lifespans still commonly include components that are investments in futures only imagined and only realized years after they are incorporated. Slow-growing trees are perhaps the best example. Many gardeners plant trees that they know will only reach maturity—will only really look like the trees they were originally envisioned to be—long after the gardener's death.

While gardens are not bound in time—they can theoretically exist for as long as there are gardeners to ensure their survival—we need to be careful in articulating this feature. The Garden is beholden to and controlled by time, a point to which we alluded just above. Saying that gardens are not bound by time is not to say they are not tethered to time. It is merely to state that they do not have fixed lifespans; they do not even have life expectancies, so long as there are humans committed to continuing to look after them, and replacement of short-lived plants such as annuals is factored in.

This is not to say that some gardens are not designed to be temporary, as is the case with London's Chelsea Flower Show or with demonstration gardens whose lifespans are designed for particular educational or display purposes. But the temporal limitation of the Chelsea Flower Show is not an essential characteristic of it. The Chelsea Flower Show could continue on indefinitely as a garden complex; it just does not happen to.

The Chelsea Flower Show is quite different from displays of cut flowers, such as we find in many flower shows and competitions, or displays of (mobile) potted plants one might see in succulent or carnivorous plant society competitions. The same is true of nursery trade shows and botanical garden demonstration tables. These are each meant to last only a short period of time and no longer; they are not designed to perdure as gardens, nor would they. Perhaps closer to the Chelsea Flower Show are the large-scale installations of plants that are found at the Philadelphia Flower Show or the immersive Orchid Show put on by the New York Botanical Garden every winter. While these certainly have aspects that are garden-like, they are not gardens. This is not because of their limited temporal duration, but because they fail to meet other garden-defining criteria, such as involving many of the gardening practices typical to maintaining gardens or involving the natural complexity or interactivity typical of gardens. It is commonly the case that these large shows use potted plants as their medium—attaching them to structures or dangling them from the ceiling—and if during the show one of these potted plants were to fade, it would be replaced by a new potted plant waiting in a greenhouse behind the scenes. These types of plant displays lack any long-term connection with the wider natural world, and they are closer to functioning as art installations rather than gardens. While the Chelsea Flower Show lasts only about a week, the entries in the show are tied to plots of land, and the "garden rooms" of which the show is composed involve preparation, gardening practices, and a clear connection with the ecological and climatic conditions of London. The Chelsea Flower Show's various garden rooms are experienced as small, short-lived gardens, with the whole show commonly thought of as a garden complex.

The third and final set of features has to do with the relationship of The Garden to humans. (It is likely apparent that this threefold partition is not particularly strict; the "human-scaled and perspective-bound" character of gardens could easily fit here rather than above.)

7. Designed

To be a garden, a garden must be seen to be a garden.[15] If not, it is simply nature. To put it in Hunt's terms, a garden must be seen to be Third Nature; if not, it is an instance of First or Second Nature automatically. In this vein, Hunt writes, "By insisting on naturalistic design, landscape architects run the risk of effacing themselves and their art . . . the necessary medium of nature must not be allowed wholly to subdue the evidence that it is itself being imitated."[16] This implies not only that the artifactuality be seen to be reality but that the artifactual nature of the garden is indeed reality. A garden must be the result of deliberate design and the implementation of that design, no matter how grand or how modest the intentional effort invested in either of those. (This all, of course, applies to "redesign" as well.) This is why, in part, "gardens" made by leaf-cutter ants, bowerbirds, and octopuses are not actual gardens.

Japanese tea gardens, some of the most naturalistic in the world, require some of the most dedicated care and certainly involve extensive and precise planning. The redesign of the New York Botanical Garden's azalea garden turned a garden that was becoming woodland into a garden that resembled woodland—this feat took multiple years and over five million dollars.[17]

An undesigned area may hold aesthetic fascination for a visitor, and an area created with straight rows of corn may exhibit clearly intentional design, but neither of these areas exhibits what Miller refers to as "an excess of form" or what Hunt describes as "more sophisticated, more deliberate, and more complex in their mixture of culture and nature." It is the amalgamation of nature and human design, where, as Miller says, the form goes beyond mere utility, where we find a garden. The character of this design, then, must be more than simply intentional or deliberate; it must be aesthetically engaging, and it must be formally coherent,

---

[15] This is not to say that a garden complex cannot contain a meadow or forest maintained to look "natural."

[16] Hunt, *Greater Perfections*, pp. 103 and 115.

[17] See: https://www.nytimes.com/2011/05/06/arts/design/azalea-garden-at-new-york-botanical-garden.html

where all the various components of the place fit together and reward the aesthetic attention focused on it.

8. Human Interactive and Requiring Maintenance

George Dickie included in his definition of art the requirement that every work of art be, in principle if not in fact, a candidate for appreciation.[18] The same is claimed here. Even if a garden is designed to be a "secret garden," the structure of that design incorporates at least one means by which the garden may be observed, appreciated, and maintained, if only by the gardenists themselves. This necessary structure provides in principle the means for others to observe and appreciate the garden as well.

Some gardens—most gardens—are interactive in the sense that one observes them by moving through them; they envelope a person; they engage all or many of the senses. But some gardens are designed not to be entered but rather observed from a location outside the garden; this is common with Japanese dry gardens, where to enter would mean disruption of the carefully raked gravel or sand. And with lawns marked with "keep off the grass." As Miller observes, all gardens must be entered by those who maintain them, even gardens meant to be observed through windows, portals, or from a viewing platform. Gardens embrace us as we move through them, literally or imaginatively, and as they do, we become part of those gardens for the time period of our visit.

Some gardens are places specifically focused on human activity—on relaxing, eating, playing, swimming, and so forth. This is the case whether the garden in question is a "function-forward" or "setting or framing" garden specifically designed for the purpose of this activity or if the garden is another sort that simply includes spaces appropriate for these sorts of activities. While some gardens have purposes "baked" into their design, all gardens allow visitors to bring to them their own purposes, some of which are regularized and some of which are incidental to a particular visit. One can ruminate or contemplate in any garden; one can celebrate or mourn in any garden; one can marvel or rest in comfortable familiarity in any garden.

In addition to interaction of the sort common to visits, all gardens require gardening.[19] They require it to maintain the original design, to rework the design, to stay alive and healthy, to stay free of unwanted elements, and so forth. Gardening

[18] George Dickie, *Art and the Aesthetic: An Institutional Analysis* (Ithaca, NY: Cornell University Press, 1974).

[19] It might be noted that even in Frances Hodgson Burnett's 1911 book *The Secret Garden* (Children's Classics, 1998) the first order of business, once the long unattended garden is discovered, is intensive gardening.

is itself one of the major ways in which humans interact with and participate in gardens. What practices are necessary to the maintenance of any particular garden differ with the style of garden. Miller writes that "gardens take more un-remitting effort than almost any other human activity."[20] Gardening ensures that the "wanted elements" of the garden—those items prescribed in the design or in an evolution of the design—thrive and that the "unwanted elements"—weeds, seedlings, fallen leaves, animal excrement, litter of any sort—are systematically removed. This happens on a continual basis; it is yet another reason to realize there is no such thing as an instance of a garden fixed in time. Any garden that reaches a perfected state, if only in the eyes of a single person, does not maintain this state for very long.

One might claim that gardens that are ignored or neglected—like an unat-tended American yard—do not seem to require gardening. But the very observa-tion by those who see them that they are in need of attention is evidence for this requirement. A garden left unattended for long enough ceases to be a garden, when "bringing it back" really entails starting over from scratch and creating a new garden.

Part of the effort that goes into maintaining a garden is accompanied by sweat and aches, but in general gardening need not be overly demanding through the course of the cycle of a year so long as one keeps up with both the efforts of fostering and culling, and especially when one utilizes plants appropriate to the climate. Part of what sustains gardeners in their unremitting efforts is the knowl-edge that there are peaks and valleys in terms of the attention a garden demands, and hard work is likely to be rewarded not only by results in which the gardener can take satisfaction but also in periods requiring only light attention.

One who claims that we may have begged the question by including "requiring gardening" as part of this criterion is easily understood, but, first, gardening practice is defined earlier and done so in a way that does not involve "slipping in" a definition of a garden. This is evident because we explicitly discuss "gardening practices" as a way to develop a definition of a garden. Second, the point we make here is about gardens requiring maintenance and attention; referring to these practices as "gardening" is a way to circumscribe what sort of maintenance and attention is necessary—but only that and no more.

9. Rule Governed

Every garden has rules that govern how humans should interact with it. These rules are not typically written—although sometimes they are ("keep to the path,"

---

[20] Miller, *The Garden as an Art*, p. 34.

"no dogs allowed").[21] The rules prescribe comportment, they specify a kind of ethic particular to interaction with a given garden. Reading a garden style correctly is usually sufficient to know the rules. One should not be boisterous while observing a Zen garden; one may be boisterous while in a *Biergarten*.

The rules are important insofar as following them demonstrates proper respect for the garden as that respect may be noticed by others visiting the garden or those who created and maintain the garden—regardless of whether they are present or not; that is, regardless of whether their notice is actual or merely theoretic. (A correlate is not stealing a piece of candy from a shop regardless of whether anyone is present with you.) Gardens cannot be offended, but the humans who are associated with them can, and so we follow the rules for the sake of not offending others or damaging the garden. Not following garden rules risks the longevity of the garden and how effectively it carries out its purpose. But the rules are tethered to a particular garden or garden type; it is because we are present in this type of garden or that type that we can know the rules and practice keeping the rules. While some garden ethos-rules can be grand or extensive, some may be extremely modest. When it comes to "setting and framing" gardens, the rules may only be that one is not allowed to molest the plantings or one should take care not to harm them accidentally.

The rules are not only injunctions against misbehavior. On the contrary, the majority of "rules" concern how to get the best possible experience out of our interaction with a garden, and how we can contribute to the ethos or context for others to get the best possible experience out of their interactions or visits. Cutting flower stalks, stepping on plants, and stealing rarities is considered bad form. For some gardens, a stroll rather than a race facilitates attention to detail, and quiet rather than noise fosters attention on what populates the garden. For other gardens, yelling to be heard over machinery integral to the garden is not only acceptable but a practical necessity. For others, running and shouting in the service of play is not only appropriate but expected. Many public places—museums, churches, playhouses, Chuck E. Cheese restaurants—have such implicit rules.

10. Aesthetically Engaging

Like every aesthetic object, gardens attract us, engage us, and hold our attention. We regularly return to visit them, to see what has changed but usually just to be

---

[21] "*No Ball Games*, or *Please Keep Off The Grass* being minimal modern versions of the more complicated 'garden law,' or *Lex hortorum*, which is was the Roman and Renaissance custom to post at entrances." Hunt, *Greater Perfections*, p. 123.

back in the garden. The reward one gains through an aesthetic experience, and that makes us return to them—and the reason many of us return to the same art gallery or museum time and time again—may be articulated in a variety of ways. What draws us back—the value we experience in a renewed visit—may be substantively plural. We may feel joy or some other emotion we want to revisit (like fear on Halloween). We may feel nostalgia when seeing a plant from our homeland or experience meaningful personal associations. We may take satisfaction in understanding and working out cognitive puzzles that present themselves or that we conjure up, puzzles like how the design of a garden works to present its plantings colorfully year-round, how a minimal hydrology is perfectly sufficient, or how a set of stones set in gravel make us aware.[22] We may experience escape or a sense of adventure as we duck under low-hanging branches in pursuit of where the path leads.

T. S. Eliot famously said, "humans cannot bear too much reality." Gardens present to us a new place, a place separate from our ordinary lives. They give us an opportunity to be "somewhere else," and to leave outside of that place anything—cares, worries—we do not wish to bring with us. They give us a chance to clear out what clutters our minds or dominates our emotions. They give us a chance to reset, rethink, and gain perspective. They provide a physical context that does not require us to focus cognitively in a certain way, and so they provide a mental freedom that allows us to think freshly.

What values gardens serve in our lives may be multifarious, but that they serve such values is unquestionable. It is the only explanation that makes sense in understanding why we revisit gardens and why we spend so much time in and on our own gardens. They reward us in ways that are not merely about impressing the neighbors, being pretty, or having a living space that transcends the walls of our home. Hunt writes:

> Paintings of landscaped sites lend themselves to registering [gardens'] magical, preternatural, even sacred meaning. . . . Gardens are special territories of children's explorations, hopes and fears, mysteries and rites of passage, sometimes forever forbidden, sometimes penetrated with trepidation.[23]

That gardens engage us aesthetically is not the same as simply appreciating a garden's design or simply appreciating that a garden is designed. Certainly, appreciating design may be part of aesthetic engagement, but design is part of the human side of the equation. As gardens are amalgams of both design and

---

[22] "The garden makes the humanized cosmos visible and intelligible." Damon Young, *Philosophy in the Garden* (Minneapolis, MN: Scribe Publications, 2020), p. 6.

[23] Hunt, *Greater Perfections*, pp. 167 and 176.

nature, so our aesthetic appreciation is of the complete amalgam. We can appreciate the design; we can appreciate nature itself (as we do when looking at a single flower); and we can appreciate how the two combine together to produce something the sum of which is greater than its two parts—for instance, how good design permits a garden to remain colorful across the span of a whole year, or how good horticulture brings out the characters of different species of plants. The sort of aesthetic appreciation we find with gardens is both multifaceted and is unique to what may be a uniquely amalgamated aesthetic form.

Any object, using that term broadly, that fulfills all these ten features is a garden. To use Hunt's expression, The Garden is a special sort of place-making. These ten features describe and circumscribe the character of that special sort of place-making. They do not describe wilderness, agriculture, or architecture, yet they do describe a place. Being "nature interactive and ecologically bound" and "dynamic and time dependent" separates the places that are *landscape architecture* from *architecture* per se, but this line is thin and appropriately so, as calling gardens "landscape architecture" is clearly meant to highlight.

To recap, gardens are:

1. Spatially Three Dimensional
2. Geographically Bound
3. Human Scaled and Perspectivally Bound
4. Dynamic and Time Dependent
5. Nature Interactive and Ecologically Bound
6. Unbounded in Time and Lifespan
7. Designed
8. Human Interactive and Requiring Maintenance
9. Rule Governed
10. Aesthetically Engaging

If this list represents necessary-and-sufficient conditions for something being a garden—if it represents a definition of "garden"—then the Wittgensteinian antiessentialism is not applicable in the case of defining The Garden.

Identity Conditions and Replication

We now turn to the final section of this chapter, only possible now that we have developed both our garden inventory and, from that, our set of criteria for what constitutes a garden. We begin with one of our criteria, the one focused on gardens being geographically (or locationally or place) bound. From this criterion, we recognize the implication that no garden can be replicated perfectly or copied fully. Miller writes:

> There are no forgeries of gardens. There are no full-scale replicas of gardens. . . . This is for three reasons. First, nearly all gardens include as major components large number of living plants. As living things . . . these plants are genetic individuals, and they are very much affected by environmental factors and by events in their life history. . . . The second reason for the uniqueness of gardens . . . is that both the quality of light and the particular contributions of the seasons of the year have a major impact on the appearance at a given viewing and on the way it will be designed. These are ephemeral, and although there are patterns of recurrence in the daily cycle and the yearly cycle, they do not repeat exactly. Finally, virtually all of the gardens that we consider works of art—and many that we do not—are designed for a specific site.[24]

Before proceeding further, it may be important for us to differentiate between two different senses of mimesis. While a garden cannot be replicated perfectly, a garden can be *represented*. The garden at the Canada Pavilion at Epcot is a representation of Butchart Gardens in British Columbia. It is meant to incorporate the salient features of this garden—which may reduce to fields of color created by a patchwork of different flowers and foliages in miniature. The garden at the Canada Pavilion evokes Butchart; it brings it to mind and so stands as a representation of it. The Epcot garden is not a copy of Butchart.

While the place-bound nature of gardens eliminates the possibility of the creation of an exact copy, a more modest notion of a garden-copy still seems possible. We have in mind an implication of the fact that some gardens (nominally) have been moved from one location to another. We have seen several examples in the course of this chapter. But in each case, it is not the garden that has been moved. The garden per se that was in the original spot ceased to exist as the garden placed in the new location began. This is the case, as we said, even if all the components of the original garden were carefully replaced to the new location, as Miller articulates in the quote directly above. Instead, what has been retained from the original garden to the relocated one are *aspects* of the garden, features such as name, purpose, design, history, and the like. These endure from the first location to the second. Hardscapes and other such structural elements may also endure, but plant character does not.

Commonly when a garden is "moved," the original garden no longer exists because, first, components of it that may be moved have been, and second, because it makes little sense to move a garden only to end up with the new garden

---

[24] Miller, *The Garden as an Art*, pp. 74–75.

and a shadow of the former garden, both presumably bearing the same name and participating in the same history. However, it is possible that "moving a garden" does not entail destroying the original garden. Instead of moving components, new components that are similar to the original ones could be purchased anew and installed. The "old" garden could remain untouched. This would be strange, but it is possible.

In such a case, we end up with two gardens that are very alike. They share many things in common, but not all. Like a clone where the original is not destroyed, we have a *twin* (albeit not similar in age). Twins are different in important respects; they have different experiences and occupy different places in space and time. Their "context-influenced" epigenetic profiles may be different, and that might mean they have observable phenotypic differences.

Similarly, a garden can have the same purpose, design, and to a significant degree the same history as another garden. They can contain similar plantings, have similar hardscapes, the same gardenists and gardeners, the same maintenance schedules, and the same administrative structures. In this way, we have garden twins.

If this sort of "copy"—which we can call a twin rather than a copy, for the sake of clarity—is possible, then it is possible to perform the operation of twin-creation where the intention is to pass off the twin as the original. This would be a **fake**. For instance, one could create a twin of Monet's famous Giverny garden in another part of France, a location where the climate, terrain, etc., is similar to the conditions at Giverny. The creator could then unscrupulously call the new garden "Monet's Giverny Garden" and charge admission to it; unsuspecting tourists might not be in the position to know they are being duped, and the fake will have achieved its unethical ends.

Capability Brown was the most popular landscape designer in England in the 1700s. After he was gone—and frankly perhaps a bit before that—there were other garden designers who copied his style. None were as successful with the style as Brown. Suppose one were to tell a potential client that either he represented Brown—and was a purveyor of original Brown designs—or even that he was Brown himself. Easy to imagine.

Something similar happened in the case of Frederick Law Olmsted and his son. Olmsted junior continued his father's design business, and some of the son's designs were accidentally credited to the father. No immoral intention drove this—the mistakes were honest (or at least no deception was the intent of the son; we cannot speak for later developments). A gardenist can certainly legitimately work "in the style of . . ." or "in the school of . . . ." It is only where the gardenist in question attempts to pass off their work as the work of another that we have a case that is ethically problematic.

Han van Meegeren, in the early part of the twentieth century, was a Dutch artist and an accomplished painter. He painted in the style of older Dutch masters, and critics found his work "derivative"—despite its technical virtuosity. Van Meegeren decided to address this problem by claiming that his paintings were actually painted by old Dutch masters; to prove his skills, he forged works of art in the styles of an assortment of artists, Johannes Vermeer the most famous. Van Meegeren's work was so good that his forgeries went undiscovered for many years. This same could happen with a style of garden associated with a particular gardenist or landscape designer. This would be a case of garden **forgery**.

Miller claims that a garden cannot be "forged," but we would say that the matter she is really addressing is a situation of a garden "fake." This is essentially a terminological matter; she uses one word where we would use a different one. More substantively, we would claim that while we fully agree that no garden can be replicated perfectly—which is what is essentially at the heart of Miller's claim—we can imagine cases where (1) garden twins exist, (2) imperfect garden copies are created—perhaps with unscrupulous intentions to pass them off as the originals, and (3) garden forgeries exist—as gardenists working "in the style of" another gardenist attempt to pass off their designs as those of another. None of this is to address the situation of representation or evocation, where a garden is created that is meant to bring to mind another garden. Representation of one garden by another is important, as we will see later, when it comes to considering (1) aesthetic features of gardens, and (2) aspects of gardens that encourage us to think of some gardens as works of art and The Garden as an artform. The next chapter takes up this very question: is The Garden an artform?

3

Gardens and Art

According to the definition we provided in the previous chapter, all gardens are designed in ways that prioritize their aesthetic features. In our inventory we saw gardens of all types, populated by all kinds of natural and artifactual features that contribute to the aesthetic experience of a garden. Even botanical gardens, whose primary goals may be conservation, collection, and research, are "aesthetic-forward," as are kitchen, vegetable, and other "function-forward" gardens that exhibit design that goes beyond the mere utilitarian. In this chapter, we approach the question of whether some of these may be considered works of art.

Whether The Garden is an artform or some gardens are works of art does not affect whether gardens may be and are experienced aesthetically. However, if some gardens are works of art, this can be illuminating in terms of understanding how gardens may be approached aesthetically. It may be obvious to some that "of course" some gardens are works of art, but it may be just as obvious to others that The Garden is not an artform proper and so no gardens are works of art. In her book *What Gardens Mean*, Stephanie Ross focuses attention on this question. Through a concerted exploration of eighteenth-century English gardens, she examines Walpole's claim that The Garden at that time was considered a "sister art" alongside poetry and painting. She makes a convincing case that Walpole was correct, but toward the end of her book she takes up discussion of the claim that those days are gone.[1]

How do we benefit if The Garden is an artform and if some gardens are works of art? In three ways.

- First, in such an exploration we are bound to discover aspects of the aesthetic appreciation of The Garden that strike us as revealing and useful.

---

[1] Stephanie Ross, *What Gardens Mean* (Chicago, IL: University of Chicago Press, 1998), p. 189.

Acacia Passage, Conservatory, Longwood Gardens, Kennett Square, PA
Gardenists under direction from Pierre S. duPont, 1921
Photo by Adam Dooling, Fall 2016

The Art and Philosophy of the Garden. David Fenner and Ethan Fenner, Oxford University Press.
© Oxford University Press 2024. DOI: 10.1093/oso/9780197753590.003.0003

- Second, works of art possess a kind of value that cannot be reduced to instrumentality or to money, a value that seems to preclude their owners from disposing of them in the same way they might anything else they own, a value that requires us to handle them in special ways. If it can be shown that some gardens possess this value or a value like the value that art possesses, this may alter the way we think about our interactions with gardens. The game, so to speak, may change radically if we value a particular garden as we value, say, a Picasso.

- Third, if a garden is a work of art, that opens possibilities in how we engage with that garden. For instance, a work of art is typically subject to interpretation and to criticism. Applying these activities to a garden enlivens at least our cognitive engagement with that garden, but if the garden is not a work of art, such activities may prove frustrating or a waste of time. If gardens do not reward such engagement with cognitive stimulation—that presumably enhances the depth of our appreciation of a garden or the reward we experience when attending to a garden—then time and effort should not be invested in these ways. This is a circuit, of course: if we show that some garden is a work of art, then it should, as is the case with all other works of art, engage us as we interpret and evaluate it. If it does not so engage us, then perhaps it was never a work of art in the first place. And, of course, there is one more option: it is a work of art but just not the kind that rewards interpretive or evaluative engagement. But building a case for that third option seems a steep hill to climb.

When a Garden May, or May Not, Be a Work of Art

We say, if any are, then "some" gardens may be works of art. But it seems a reasonable default position to claim that not all are. As mentioned in the last chapter, one of the two of us writing this book has a half-acre garden surrounding his house. This garden has extensive plantings. It is largely based on a collection of mature trees, mostly live oaks and laurel oaks, a set of sixty mature camellias, and large assortments of roses, azaleas, and a host of other plants. The beds are laid out as circles or curvilinearly, with St. Augustine grass filling in the lawn and walking areas. It is an attractive garden; once every few years it wins the neighborhood's "yard of the month" award. But despite enjoying this success, neither of us writing this book thinks of this garden as a work of art. It is aesthetically engaging, but by any argument or any definition of art that might be applied, it does not rise to the level of being a work of art.

A garden that merely represents a beautiful set of sights without "saying" anything about its materials, their powers and interrelations—that is, a garden that
is "mute"—remains a gorgeous garden but is very likely not a work of art.[2]

Whether a garden must say something to be a work of art—in the sense, perhaps,
of being profitably interpretable—may be debatable (something we address at
the end of this chapter), but at least the more general point Ross makes in this
quote seems correct. Some gardens, as aesthetically engaging as they may be, still
are not works of art. And this may have to do in part with whether they are interpretable. Roger Paden writes:

> I have used "landscape architecture" to mean both the practice of designing gar-
> dens to be works of art and the products of this practice. This allows me to con-
> trast "landscape architecture" with mere "gardening," just as Harries contrasted
> "architecture" with mere "building." Thus while utilitarian gardening—like util-
> itarian building—may not be entirely innocent of aesthetic intent, aesthetics is
> not central to its practice.[3]

We disagree with some of what Paden writes, but we are in sympathy with the
general trajectory. As some buildings do not rise to the level of being "architecture," so some gardens do not rise to the level of being works of art (although we
would not use "landscape architecture" to designate this class). Paden casts light
on the relationship between aesthetics and art to which Miller alludes.

> [T]o be great, art must meet two conditions: it must be formally excellent . . .
> and it must have important content. . . . By "great art," then, we mean art which
> meets four criteria: excellent form, significant human content, enduringness,
> and adequacy of form to content.[4]

What we are interested in is whether a garden can be a work of art at all, let alone
a great work. But Miller's definition is still helpful because in setting a high bar,
it allows us to reflect on the nature of the criteria of what counts as art in the
first place. The focus on (1) form that is sufficient to reward evaluative attention, (2) content sufficient to reward cognitive engagement, (3) a "test of time"
approach to understanding value in terms of consistent repeated experiential
reward, and (4) the internal integrity within the form of the garden to sustain

---

[2] Ross, pp. 15–16.
[3] Roger Paden, "The Ethical Function of Landscape Architecture," *Environmental Philosophy* 15:2
(2018), p. 146.
[4] Mara Miller, *The Garden as an Art* (New York, NY: State University of New York Press, 1993),
pp. 135 and 141.

attention placed on the meaning and meaningfulness of the garden—to use our version of Miller's criteria—gives us a sense of what we are looking for in that class of gardens that are candidates for being true works of art. And, by extension, this allows us to define the class of those gardens that are not candidates as well. The author's residential garden described above does not meet the second or fourth criterion in this criterial sketch; it may have a form substantive enough to permit evaluation, and it might reward sustained and repeated attention, but it does not have an interpretable content or a form sufficient as a base for interpretive attention.

The adequacy or correctness of Miller's criteria, or our version of it, is not at issue at present. We include it here to demonstrate that a mechanism can be created to separate gardens that may be art from gardens that may not, and by so doing, to meet our intuition that if there are some gardens that are works of art, there are some that are not. We will return to this point at the end of the chapter. The question about whether any gardens are works of art—about whether The Garden is an artform (which we take to be settled if some gardens are indeed works of art)—we take up now.

Application of Art Theory

There are many classes of objects and events we commonly take to be artforms: painting, printmaking, photography, film, sculpture, ceramics, literature, music, theater, dance, architecture, and so forth. There are some classes of objects (again, to stand for both "objects and events") we commonly do not take to be artforms even though they are "aesthetic-forward": crafts, hairstyling, sandwich-making, and so forth. And there is a wide assortment of objects that straddle the fence: culinary creation, perfume creation, interior design, cars, boats, topiary, tattoo, graffiti, and so forth. Our question is where The Garden fits.

We begin with the axiom that if The Garden does not fit into the first category (art), then it at least fits in the last. Surely there are many gardens that fit in the middle category and are akin to craft—but this is the same sort of claim one can make about any particular object that fits into the classes we recognize as artforms. Not every painting is a work of art, not every photograph, not every film, not every building. That there are objects that may correctly be classed as members of the sets of recognized artforms does not mean that they are automatically works of art because of that classification. So the existence of a garden that is not a work of art has no bearing on whether The Garden is an artform. A different conclusion might be reached if one were able to show that no garden ever has risen to be a work of art. In that case, The Garden's claim to be an artform would at best be theoretical, but more likely simply an empty one. The easiest way to proceed with this question is to look at the applicability of theories of art to The Garden. This we will do briefly and as a survey, as it has been done elsewhere, notably by Ross.[5]

Mimesis

From the time of Plato and Aristotle and for perhaps the next two millennia, it was accepted that art was essentially about the imitation of nature. Few continue to espouse this view, which is a shame for gardens since they have such a clear claim to be representative of nature. The artifice that is applied to gardens is about shaping what nature offers. Now, one might argue that it is because of this that gardens are not representative of nature. That is, one might claim that as gardens *are* nature, and apparently *cannot not* be nature, they are merely the thing itself and not an imitation of the thing. This claim does not survive much scrutiny, however. There is no clear threshold between the object-of-the-world and its representation that must be achieved in the creation of a work of art. One can easily imagine the presentation of an object-of-the-world as the presentation of a work of art—Andy Warhol and Marcel Duchamp offered as much[6]—the Duchamp readymades in particular erase the line entirely. And one can easily imagine that since a single object, as a member of a set, can represent all members of that set, so a patch of nature can represent all nature or at least all patches of nature sufficiently like that first patch. When we speak of "The Garden," we are speaking of an abstraction, and that is why we are consistent in capitalizing the first letters of both words—to demonstrate that we are naming a kind. There is no tangible thing which is "The Garden." There are only gardens, and so the only way to demonstrate concretely of what "The Garden" consists is to point to examples of actual gardens. This we did in the last chapter; our definition of The Garden is based on an inventory of actual gardens, and conditions in that definition are evidenced through citing actual gardens, or garden types that are then based on actual gardens. This is a representative function. So the claim that a garden cannot represent nature because it *is* nature is not a claim that will hold up. On the face of things, gardens are mimetic in the most obvious of ways.

Do gardens represent more deeply than merely standing for either The Garden or nature? While not all do—the residential garden described at the top of this chapter does not—some certainly seem to. The oft-quoted example is Stourhead, an English garden where a walk around the central defining lake is a retracing

---

[5] Ross, *What Gardens Mean*, pp. 10–24.

[6] In Andy Warhol's 1964 *Brillo Box*, which is formally identical to its real-world counterpart, and Marcel Duchamp's 1917 *Fountain* and 1915 *In Advance of a Broken Arm*, which were real-world objects that Duchamp "elevated" to be works of art.

of—a representation of—the journey of the Trojan hero Aeneas, as envisioned by Virgil. Components of many of the Kyoto *karesansui* gardens are representative of other things—oceans, rivers, mountains, boats, mythological figures, and so forth. Daisen-in is a great example of this, and Saihō-ji has, among other such things, a set of placed rocks that represent a cascading waterfall. Components of these gardens either represent these objects through their visual similarity or through evoking these objects by calling to mind not only a similarity in form but in the essential sense of the object represented. For instance, a stretch of gravel that extends widely enough to evoke a field rather than a channel may represent an ocean rather than a river; the rake marks encircling rocks, while static, still evoke the undulating ripples created as water laps against the shore. The evocations may be reducible to formal visual connections, but they need not, to still be thought of as robustly representative.

Expressivism

In the late 1800s and early 1900s, as mimetic definitions of art were coming up against examples of widely accepted works of art that challenged those theories or simply did not fit them, mimetically focused theories were replaced in popularity by theories that focused on artistic expression.[7] There is little question that art, if all art is artifactual (a claim that thus far has had no serious challenge[8]), is an expression of an artist. The question is: an expression of what? Those theories that arose at the turn of the nineteenth century into the twentieth focus on different conceptions of expressed emotions but also on expressed "intuitions" or ideas. To say that a work of art is an artistic expression is to say that the work is artifactual, but to describe the character of that expression, as that character is common to all such expressions, is to say something more interesting and provocative. Leaving aside choosing among theories of the content of such expressions, we still may ask: do gardens express? If we mean anything more than the question "Are gardens artifactual? Are all gardens designed?"—a question we meant to answer in the preceding chapter—then it seems clear that not all gardens express some content that is more than their own artifactual or designed character. The garden described at the top of this chapter does not. But are there gardens that do express in this more robust way?

---

[7] Such as the theories of Leo Tolstoy, Benedetto Croce, and R. G. Collingwood.

[8] John Cage's *4'33"* (1952) may be an instance of an art object that has no object and so the object cannot therefore possess artifactuality; yet this does not defeat the point that Cage's event is still a product of deliberate creation, and this seems enough to say that it fits the condition of artifactuality in spirit if not in letter. Salvatore Garau's *Io Sono* (2021) is an immaterial or intangible sculpture, but like Cage's work, it is still the product of deliberation and creation.

Perhaps every Capability Brown landscape—and there are many—expresses not only Brown's particular vision of what it meant for a piece of land to be quintessentially English countryside but also Brown's vision of what it is for humans to live in a natural setting that is idealized yet looks as if it were merely the product of natural processes and with only the most modest incursion of human intervention. The formal French gardens designed by André Le Nôtre commonly are said to express "man's relationship to nature" as they demonstrate the imposition of control and order. Perhaps still better examples are the *karesansui* gardens that express ideas that are worthy of consideration by Buddhist monks as they contemplate deeper meanings about ontology and relationship through focus on these gardens. If, for an object to qualify to be a work of art, an expressivist theory of art only requires the level of expression we have discussed, then some gardens do in fact meet this criterion.

Continuity Theories

Since the late 1970s, popular theories of art have begun to focus on concerns that seem social-scientific. Theories that focus on the institution of art as a collection of all those today we might call "stake-holders,"[9] theories that focus on the unfolding traditions of art,[10] on the histories of art, on the relationships of artworks created today with works created yesterday[11]—these sorts of theories attempt to track relationships as integral to the understanding of what makes an object a work of art, whether these relationships are among people (artists, audiences, critics, etc.) or among the objects themselves (including "tradition-relationships" as relationships among objects). These sorts of theories, perhaps the most popular today, present a challenge to thinking of The Garden as an artform because many contemporary theorists are comfortable with the view that while The Garden was in the eighteenth century a bona fide artform, it is not regarded that way today. If a "kind" is an artform today because it continues a tradition—that it was accepted as an artform and continues to be accepted as an artform, and that those gardens today are works of art because they possess a relationship to other gardens regarded this way—then it seems clear that the tradition of gardens-as-art was at some point after the eighteenth century broken.

---

[9] George Dickie, *Art and the Aesthetic: An Institutional Analysis* (Ithaca, NY: Cornell University Press, 1974).
[10] Arthur Danto, *Transfiguration of the Commonplace* (Cambridge, MA: Harvard University Press, 1981).
[11] Kendall Walton, "Categories of Art," *Philosophical Review* 79 (1970), pp. 334–367; Jerrold Levinson, "Defining Art Historically," *British Journal of Aesthetics* 19 (1979), pp. 232–250; Noël Carroll, "Historical Narratives and the Philosophy of Art," *Journal of Aesthetics and Art Criticism* 51:3 (1993), pp. 313–326.

On the other hand, if traditions (and relationships) can wax and wane, then perhaps the formal art designation The Garden enjoyed once might be restored. Think of it this way: at some point, there were kinds that were not accepted as artforms but eventually came to be. Film and architecture are two such examples. Film largely came about with the start of the twentieth century, and from its inception it has been accompanied by theory and criticism. There are certainly some who would say that film was more an artform in the earlier parts of the twentieth century than it is today.[12] But at some point, likely early on, film had to undergo being accepted as a member of the club. Its advocates had to argue that as a kind it deserved to be recognized and regarded as an artform. Architecture is a slightly different case. Buildings per se existed well before any particular buildings were candidates for elevation to the kind "architecture" and "architecture" to "art." The same case can be made for dance, photography, and ceramics. There had to be a point of acceptance of these forms as members of the "art club." Whatever the mechanism of their elevation, The Garden at least at one time had achieved membership. So, first, the reasons why it lost its membership—if indeed it was not merely deemed inactive, to push the metaphor—are important for us to explore. Second, what characteristics The Garden shares with so-called canonical artforms, characteristics that seem salient to their identities as artforms, need to be explored. This we do below.

It might be wise, at this point, to mention that there are several theories of art compatible with an aesthetic approach.[13] That is, several of the theories above are compatible with the notion that there is something endemic to the nature of art that includes reference to works incorporating aesthetic properties, and reference to art experiences being at least in part aesthetic experiences. This sort of approach one might have imagined would have been abandoned in the face of some iconic works of modern art where focus on their formal aesthetic properties will render a less valuable art experience—and/or render the object a less valuable artwork—than foci that are cognitive. A good example is Marcel Duchamp's famous 1917 readymade, *Fountain*, which is (or was, depending on your theory) a white porcelain urinal resting on its side. To appreciate it by focusing on its formal aesthetic properties likely is to miss the point of the work and so also to miss its value as a work of art and a contribution to the history of art. Yet defenders of aesthetic approaches to understanding the nature of art and to defining art (which are occasionally different ventures) are undaunted. Some do not admit that works like Duchamp's are art (the minority), others explain

---

[12] Such as Sergei Eisenstein, Andre Bazin, and Rudolph Arnheim (forgive the anachronism); see Arnheim, *Film as Art* (Berkeley, CA: University of California Press, 2006).

[13] The most famous of which likely is that of Monroe Beardsley as seen in *Aesthetics: Problems in the Philosophy of Criticism* (Indianapolis, IN: Hackett, 1981) and *The Aesthetic Point of View* (New York, NY: Cornell University Press, 1982).

how the salient artistic properties of the work are appreciable aesthetically.[14] Whether this second approach must expand what it means to appreciate a work of art aesthetically beyond simply appreciating its perceptually based aesthetic properties depends on the theory. In any case, the reason this is worth a mention is that gardens as a class of objects are already seen as aesthetic objects and rewarding of aesthetic attention, and so to the extent to which the marriage between art and aesthetics is successful, so much the better for the case that The Garden is an artform.

How The Garden Is Similar to Other Artforms in General Terms

In this section and the next, we explore how The Garden is like various artforms. We begin with points common to all works of art, and we continue by looking at specific artforms individually.

Design

There is no human expression that is purer than a work of art. Whether the human being who creates that work of art expresses a message, an emotion, a mood, an experience, an identification, an idea—whatever the expressed content may be—the focused deliberation that goes into works of art are expressions of humans for no greater purpose than for that of expressing themselves. This is not to say that a work of art cannot have additional purposes, that the expression must be received by an audience in the terms in which the artist created it, that an expression must be limited to a single artist, that other mechanisms besides the creation of a work of art cannot be used to express, that one cannot decide to retain or incorporate some happy accident that happened in the creation of a work of art, that the artist is in full command and control of the expression that forms the basis for the work of art, and so forth. But in general terms, as a class of objects (and events), there is no purer form of human expression than the creation of a work of art.

Every garden is designed. If it were not, it would be First (or Second) Nature rather than Third. Even if one rejects Hunt's designation schema of three different natures, as we saw in the first chapter there must be something that differentiates

---

[14] James Shelley, "The Problem of Non-Perceptual Art," *British Journal of Aesthetics* 43:4 (2003), pp. 363–378; Noël Carroll, "Non-Perceptual Aesthetic Properties: Comments for James Shelley," *British Journal of Aesthetics* 44:4 (2004), pp. 413–423.

a garden from, on the one hand, "wild" nature, and on the other, agriculture. We do not think of gardens as either of these and capturing not only that common intuition but also the common linguistic reference typically has to do with both the level and the character of human involvement—in other words, with not only the fact that gardens are designed but also the fact that these designs go beyond being merely utilitarian, as would be the case with agriculture. (This is why some vegetable gardens are gardens and some are simply miniature agriculture—it all depends on the character of their design.)

Every garden is the product of focused deliberation in terms of both its form and its content. There may be no greater affinity between the form of The Garden and the various forms of the canonical arts than the fact that all are human expressions whose design is the product of decision. As works of art are expressions, so are gardens. They may express their creator's values and commitments, their senses of relationship, their willingness to endure patience and ambiguity, and a host of other things, but no matter what, they serve as their gardenists' and gardeners' creative expressions.

Aesthetics

As we saw, Miller's definition of a garden begins: "A garden is any purposeful arrangement of natural objects . . . in which the form is not fully accounted for by purely practical considerations such as convenience."[15] The "purposeful arrangement" connects to the point made above, but we quote her here to place focus on the phrase "in which the form is not fully accounted for by purely practical considerations." Any aesthetic object can have any content—there are no boundaries—but all aesthetic objects have aesthetic properties (by definition) that work together to create the aesthetic form of that object. This form is purposed not merely in the modest sense that it is designed—although it certainly is that—but in the more robust sense that for the object in question to be one that rewards aesthetic attention, where the aesthetic experience has value, the arrangement of the object's (positive) aesthetic properties must cohere together in contribution to the object's overall aesthetic character. In other words, all objects, at least all objects capable of being sensed, have aesthetic properties— as evidenced by the fact that one can take an aesthetic point of view toward any (sensible) object—but those objects where the properties "fit" together are maximally worthy of aesthetic attention and valuable as aesthetic objects and the

---

[15] Miller, *The Garden as an Art*, p. 15.

foci of aesthetic experiences. This is the oft-mentioned concept of "significant form."[16]

While not every garden is a great aesthetic object, every garden has an arrangement of properties that make it a candidate for rewarding aesthetic attention. Every garden, even a botanical garden, is "aesthetic-forward" in the sense that it fronts its aesthetic features—as a result of its design, the fact that nature in itself is "aesthetic-forward," or any combination thereof. The degree to which its "aesthetic-forward" character rewards aesthetic attention differs, as is the case with all aesthetic objects, in terms of the success of the arrangement and fit of its aesthetic features. This is what is meant by talking about "garden form": the coherence of the garden's aesthetic properties to create a unified whole. (We take up the matter of "garden form" in detail in a later chapter.)

While the formalist may say this is all there is to understanding garden aesthetics—simply its form as composed of its (formal) aesthetic properties—a more contextually minded aesthetician is bound to point out that gardens also participate in those same broader aesthetic aspects that seemingly all artworks do. Beyond gardens having expressive properties, they sometimes possess representational properties. They engender and encourage imaginative and associative activity. They can hold one's cognitive attention and frequently one's emotional engagement as well. The person who sits in a garden, alone in contemplation, and as a result is refreshed and enlivened is evidence of this; as this situation obtains almost ubiquitously for those who undertake such garden-sited and garden-focused contemplation, evidence mounts as to the "aesthetic-forward" nature of The Garden as subject to the full breadth of aesthetic contextuality that is common to canonical artforms.

The purposed "aesthetic-forward" character of all gardens puts The Garden alongside other artforms with regard to their possession of that aesthetic character that some theorists believe is a necessary component of a work of art. If to be a work of art is to be "aesthetic-forward," The Garden as a class of objects certainly fulfills this condition.

Evaluation and Criticism

The Garden lends itself to the critical engagement common to all artforms. Even if there is not a professionalized mechanism by which garden criticism proceeds—even if garden criticism is not a regular part of each edition of the

---

[16] Miller discusses "significant form" in connection with Susanne Langer's work in *The Garden as an Art*, pp. 50–51, 124.

New Yorker (for instance[17])—there are among garden enthusiasts constant conversations focused on the strength of a particular garden's ability to reward a visit. The world is loaded with great gardens, and most of us can only visit some, and so we who are making plans for such visits—in the short term or long—visit websites, read garden books and periodicals, and talk to our friends, especially our well-traveled friends, to prioritize our plans for visits. While perhaps not as formalized as the criticism of some canonical artforms, the form is the same: we compare and prioritize based on the value—occasionally the historic or scientific value but usually the aesthetic value—of the particular gardens on our list.

The possession of value comes in varying degrees, and so we make comparisons and create lists of "good, better, and best." We do this with everything that has value, but when the value in question is aesthetic, the form of that comparison is the form of art criticism. Gardens are amenable to this kind of comparison, and this is one more reason to see them on a continuum with canonical artforms. (We discuss garden evaluation and garden criticism in detail in a later chapter.)

Art, Artist, and Audience

As works of art are created by artists and appreciated by audiences, so gardens have a similar relationship structure. They are designed, instantiated, and appreciated, just as all works of art are. Even the proverbial "secret garden" is appreciated by its gardenist and its gardeners. It has the form to be appreciated by audiences far and wide; its "secrecy" is not endemic to its character as a garden. The kind of appreciation that audiences have for gardens may vary in focus, but the most common is aesthetic, and again this puts The Garden on a continuum with canonical artforms by those who advance theories that possession of an aesthetic character is necessary to a work of art. It is the particular kind of appreciation at issue—and the fact that a garden is designed to engage such appreciative attention and reward it—that separates the "creator-created-used" relationship from being a mere utilitarian one to one consonant with "artist-art-audience." Cooper writes:

> Gardenworld is replete with figures of the same kind that populate Artworld— creative designers, craftsmen, critics, connoisseurs, and so on—and with similar organizational structures—competitions, shows, dissemination through photographs, and so on.[18]

---

[17] It should be noted that articles about gardens appear in the *New Yorker* from time to time, and the focus of such articles is largely about their aesthetic properties and contexts.

[18] David E. Cooper, *A Philosophy of Gardens* (Oxford, UK: Oxford University Press, 2006), p. 25.

The creative path followed by any artist moves along familiar terrain. There must first be an openness to creativity. One must be prepared to be creative by having the right frame of mind, the time and inclination to pursue where creativity moves, and the openness to be inspired. Then comes the spark, the idea, the impetus that will lead to the construction of a vision and ultimately an expression. After the inspiration, the planning starts, and as planning proceeds, the initial design for the instantiation begins. Once the critical process of developing the idea and developing the design begins, one undergoes an iterative process between new inspirations and insights and moves toward a settled design. Then comes execution of the design in some form that can be sensed. One acquires a medium or media and starts to shape and mold those media to fit the design. At some point, the execution ends, and the work is completed.

This process, which is common to the creation of any work of art, is the same process that a gardenist (or team of gardenists) pursues in the creation of a garden. There is no difference apart from the fact that the garden is never in a "completed" state and the fact that the gardenist must be prepared to adjust their design (or have their design adjusted) as the pragmatics of the site requires and as the processes of nature take hold and begin to change the living components virtually the moment they are installed. Gardenists may work with media that is much grander in several ways than the media employed by other artists, and this again speaks to the requirement that garden creation is very resource intensive. It may also help soften the reality that once a garden (of any magnitude) is installed, it is fixed to its place; this is common to much art that is physically large but is a necessary aspect of all gardens. If we were to pursue the path of defining a garden by understanding it to be the result of gardening practices—which we chose not to do in the last chapter (in part because the application of gardening practices can be employed in ways that do not result in gardens, in accord with our definition)—we might be tempted to say that a work of art is the result of "artistic practices" (which likely would include the creative process sketched above and more—since creativity alone seems insufficient to define the whole of "artistic practices"). This project may not succeed, but what one can know is that if the process of creating a work of art is similar in all salient respects to creating a garden, this stands as yet more reason to think of The Garden as an artform.

What makes a great gardenist? Miller defines what goes into making a great work of art, as we saw; her definition focuses on properties of the object, on the uptake of those properties by those attending to the garden aesthetically, or on contextual properties relevant to approaching the garden aesthetically. Miller's definition might be expanded to include reference to a great work of art being produced by a great artist, and so a great garden being produced by a great gardenist. Such a definition has intuitive appeal because it connects with common practice. There are few great gardens where we do not know the responsible gardenist(s)—where cases such as these exist, it is usually because historical records break down, or because those responsible for the garden's creation were from laborer classes and, because they were not regarded as professionals, their names were never recorded in the first place. Knowing the identity of these gardenists is not only of historical or art-historical interest; and it is not even so that we can map out the style of a particular gardenist or focus attention on their full oeuvre. We seek to know who the gardenists were for the same reason we want to know who great artists were (or are, in both cases): evidenced in the designs and resulting works is some degree of artistic genius, and that aspect is salient to evidencing the case for the artistic quality of the resulting work under consideration. This is an argument against aesthetic formalism and for contextualism, but it is common practice in the assessment of a work of art—and in the construction of a fully rounded aesthetic experience—to want to know genetic facts of the creation of a work of art and then fold those in as data points in our judgments and in the constitution of our experiences. While formalists will complain that we thereby commit an intentional fallacy in allowing such externalities to color our judgments and our experiences, the burden of proof lies with those who argue that common practice errs.

A great gardenist (1) fully engages with the creative process, (2) is sufficiently inspired to develop a compelling vision, (3) has and employs the critical skills to develop that vision, (4) has and employs the skills to create a plan to instantiate that vision into a sensible expression, (5) carries out the instantiation of the vision in an expression, and (6) is receptive to the suggestions of nature as cocreator. While the hardest part may be the final instantiation, it is the driving vision that is the most important part, and it is usually the vision of the gardenist that is most cited in a case not only for the genius of that gardenist but as evidence in a case for the quality of the resulting garden. Great gardenists exhibit a pattern of genius of vision throughout their gardens and garden designs.

Style

We have used the words "garden style" to mean the sort, kind, or type of garden that a particular garden is, and we have talked briefly about the introduction of "garden genre" as a synonym.[19] We may be taking liberties here with the aesthetic terminology. An artistic style might better be thought of as possession of a form that has distinctive aesthetic markers that identify it as coming from the domain

---

[19] Ross sees "garden style" and "garden genre" as different things—for her, a "garden style" denotes a kind of garden that accords with a particular "author, period, place, or school" (following Nelson Goodman) and a "garden genre" as a more generalized kind that captures the defining essence of an object. Ross, *What Gardens Mean*, pp. 81–84.

of a particular vision for that form. If this definition more adequately captures what style is about, then we should ask two questions: should we use the term this way in talking about gardens? And if this is the correct usage, do gardens possess style?

The first question is answered by a consideration of functionality: it is important for us to have a term that denotes a sort, kind, or type of garden that is stable in its denotation as we worked out the taxonomy that informed our inventory; so we will stick to our use of "style" in that regard. While "genre" seems an appropriate term for the largest taxa in our inventory—the six primary divisions—and it may even work for the next level down, it seems less appropriate when discussing the more detailed taxa because those more detailed levels do not admit of the same sort of "grouping" structure as the larger ones.

The second question—concerning whether gardens possess style—is more apropos now, and the answer to the question is "yes." This can be seen readily in a survey of André Le Nôtre seventeenth-century French gardens, Capability Brown's eighteenth-century English landscapes, or Frederick Law Olmsted's nineteenth-century American metropolitan parks. Each of these designers had a distinctive artistic style that not only can be identified with ease but can be replicated (with of course varying degrees of success) by following certain formal markers or characteristics. Just as Han Van Meegeran was able to forge Johannes Vermeer paintings convincingly by painting in Vermeer's style, so the same is possible with Le Nôtre, Brown, and Olmsted gardens.

This point is reinforced as modest residential garden creators talk about their style of gardening. The residential garden described at the top of this chapter has some French elements and some English cottage-garden ones—and these ascriptions can be evidenced through pointing out features of the garden. Such style ascriptions can be advanced and defended through citing garden features. While we may be back to conflating the two different senses of "style" used in these paragraphs, the point remains that even on a narrow definition of artistic style, gardens possess these traits (although, as apparently with everything, in varying degrees).

How The Garden Is Similar to Other Specific Artforms

We want at this point to switch gears slightly and survey particular (canonical) artforms with which The Garden has been compared or is comparable. One caveat first: Cooper warns against "assimilationism"—against the idea that we can treat appreciation of the garden as we would appreciation of any other aesthetic form.[20] He argues that such assimilation is to miss or ignore aspects of The

[20] Cooper, *A Philosophy of Gardens*, pp. 24–34.

Garden that are either unique to The Garden or, still more to the point, exist only when we appreciate The Garden as an amalgam[21] of nature and design, inextricably combined, with important aspects only truly understood by thinking about The Garden as such an amalgam. Yet, while avoiding this trap, it is illuminating to appreciate the close connection between appreciation of gardens and appreciation of other artforms. In the survey below, we attempt to distill some of the more salient points of connection. We begin with the most discussed: the connection between gardens and painting.

Painting

There has been a long exchange between gardening and painting.[22] The mostly monochrome Song Dynasty landscape paintings had a significant impact on the development of Chinese and Japanese gardens we know today.[23] Claude Monet wrote about his famous garden at Giverny "I cultivated [waterlilies] with no thought of painting them. . . . And then suddenly I had a revelation of the magic of my pond. . . . From this moment I have had almost no other model."

Of course, there are differences between painting and gardening, some of which are obvious. But it is unclear they make significant aesthetic differences. For instance, gardens are four dimensional and paintings are two dimensional. Paintings are framed, and while gardens are spatially bounded and humanly scaled, the perspective one takes to a garden, even one meant to be viewed from a platform or window, generally shifts. If, however, one adopts the idea of the "mobile frame" as Lee discusses it in connection with the garden theory of Karl Heydenreich,[24] one can account to some degree with the minimization of aesthetic difference between the two-dimensional framed painting and the (albeit phenomenally bound) visual uptake common to viewing a garden while strolling through it (we examine this more thoroughly in the following chapter). Humans typically see the spatial world three-dimensionally, and the depth of objects, depth of field, and depth of focus may be different from what can be captured in a

---

[21] "Amalgam" is our word for this phenomenon; we think it captures the spirit of what Cooper describes.

[22] We will skip "photography" being a separate discussion as many of the insights about the relations between gardens and painting apply to photography, too. We end this section with considering how gardens and film are related, and insights that apply to photography are there, as well.

[23] Francois Berthier, *Reading Zen in the Rocks: The Japanese Dry Landscape Garden*, ed. and trans. Graham Parkes (Chicago, IL: University of Chicago Press, 2000), pp. 100–102. See also Elizabeth Barlow Rogers, *Landscape Design: A Cultural and Architectural History* (New York, NY: Abrams, 2001), pp. 284–287.

[24] Michael G. Lee, *The German "Mittelweg": Garden Theory and Philosophy in the Time of Kant* (New York, NY: Routledge, 2013), p. 110. For a further exploration of Heydenreich's commitment to viewing gardens as akin to painting, *pace* Kant, see pp. 62–64 and 70.

painting (or photograph for that matter), so there are perceptual differences that may lead to aesthetic differences. To the degree that these aesthetic differences are significant is the degree to which they make an experiential difference, and that may vary from subject to subject.

From Kant to Walpole, there is a tradition of the aesthetic consideration of gardens being the same or importantly similar to the aesthetic consideration of paintings. Later we will talk about the Picturesque approach to garden design; that approach was/is based essentially in understanding the landscape as a three-dimensional scene, an extended painting, and the virtues of gardens created in the Picturesque style were painterly virtues—the aesthetic compositional virtues that painters up to that time had sought to include in their works.[25] As we saw earlier, Kant's categorization of "painting proper" alongside "landscape gardening" is because both present the "beautiful" representation of nature, either through painterly portrayal or through "arrangement of its products."[26]

> For although gardens delight all our senses, their primary appeal is visual, and gardeners are necessarily aware of such "painterly" concerns as color, texture, balance, form, perspective, and light and shade.[27]

When the design of a garden involves central consideration of its aesthetic features, it is easy to follow the path that Ross describes above and think in terms of visual components.

- How will the composition of garden components be seen from this angle or that?
- How can I ensure that every perspective includes some color (defined by so many gardenists and gardeners as "other than green")?
- How do I place "color" so that viewers will find it harmonious?
- How do I balance large leaves and large blooms with smaller ones of both sorts?

---

[25] Rogers, *Landscape Design*, pp. 251–254; Isis Brook, "Wildness in the English Garden Tradition: A Reassessment of the Picturesque from Environmental Philosophy," *Ethics and the Environment* 13:1 (2008), pp. 105–119; Roger Paden, "A Defense of the Picturesque," *Environmental Philosophy* 10:2 (2013), pp. 1–21, and "Picturesque Landscape Painting and Environmental Aesthetics," *Journal of Aesthetic Education* 49:2 (2015), pp. 39–61.

[26] Immanuel Kant, *Critique of Judgment* (Indianapolis, IL: Hackett, 1987), pp. 322–324. For more on Kant and painting, see Ismay Barwell and John Powell, "Gardens, Music, and Time," in Dan O'Brien (ed.), *Gardening—Philosophy for Everyone: Cultivating Wisdom* (Chichester, UK: Wiley-Blackwell, 2010), pp. 135.

[27] Ross, *What Gardens Mean*, p. 130.

- How do I lighten the shady areas of the garden with color, and how do I manage to use my "full sun" areas to best advantage?
- How do I account for vistas that are stable or at least aesthetically stable as perspectives of those strolling through my garden change?
- How do I draw emphasis toward certain features and obscure others?

Lynden Miller, designer of the Central Park Conservancy Garden, began her career as a painter and considers such "painterly" questions in the design of her urban landscapes. One of her signature components is the use of deep red- or purple-leaved shrubs to make the other plantings brighter by contrast. Lynden Miller's approach is echoed in the approach of many gardenists who, even if they are not explicitly working with a Picturesque approach, still take into account in a focused way the same formal aesthetic qualities painters do.

Sculpture

While a less popular artform to discuss in connection with The Garden than either painting or architecture (the next item on our list), we include sculpture here because it has similarities to The Garden that transcend painterly ones. First, sculpture is typically spatially three-dimensional as gardens are spatially three-dimensional, and second, many sculptures are as site-bound as gardens seem to be. Being immobile, of course, is not an essential property of sculptures, even particularly large ones. Michelangelo's *David* was famously moved from its original location to an indoor location to keep it protected from the elements. Perhaps one of the best examples of sculptural immobility is Richard Serra's 1981 sculpture *Tilted Arc*, a very large piece of curved metal—raw steel, 120 feet long by 12 feet high—installed in the forecourt of the Federal Plaza in New York City. In 1989, after complaints and a public hearing, it was removed; Serra claimed that the piece was created to be site-specific, so to move the piece was to destroy it. We claim to move a garden is not actually to move it but rather to create a new one in a new place and (usually) eliminate the original one; in this, gardens and *Tilted Arc* are alike.

Sculpture and gardening share a unique trait: the process of their creation sometimes involves removal as well as addition. A clear example is marble sculpture, where rock is chiseled away to reveal the artist's vision. Gardenists, in the same way, remove weeds, dead branches, misplaced plants, and other unwanted objects. They may reroute waterways or flatten hills. Gardenists typically do not begin with a "blank canvas" since there are almost always natural features, including soil type, that must be incorporated or worked around. This is similar to a sculpturist working with the character of the rock, wood, or other material.

Architecture

The terms "landscape architecture" and "landscape architect" are used with fre-
quency by those who not only wish to demonstrate the breadth of scale with
which gardenists and gardeners sometimes must contend but also to demonstrate a certain professionalism[28] and, most pertinently for our purposes
here, to show a connection between building a building and building a garden.
Building anything follows a path that includes inception, design, planning,
preparing, installing, and then maintaining, and in these things architecture
and The Garden are similar. In both The Garden and in architecture, challenges
are addressed in similar ways, with "form following function" and attention to
pragmatics.[29] Moreover, in both cases, the end result is a place that either can
be occupied literally or can be occupied conceptually as in the case of spatially
three-dimensional gardens that are meant to be viewed as if they were generally
two-dimensional ("generally" because as we mentioned even "framed" gardens
allow for different perspectives and entering for the sake of maintenance).

As all works of architecture are functional—because to be a work of architecture that is not functional is to be a work of sculpture[30]—virtually all gardens
are functional in the sense that each typically serves a range of purposes. The
functions of some gardens are fore-fronted, as we saw with many of the items on
our inventory. Yet even those gardens that do not seem to be "function-forward"
serve purposes as those who visit them use them as places to meditate, have a
cup of tea or a sandwich, make a phone call, check email and texts, play a game
of chess, walk the dog, throw a frisbee, or just rest and rejuvenate. A garden devoid of purpose—either a designed purpose or an incidental purpose occasioned
by a visitor—is the exception rather than the rule. A "purposeless" garden—
in the sense of a garden that is exclusively the focus of disinterested aesthetic
attention—is certainly possible, but it is very rare. Once a visitor brings an incidental purpose to that garden, it immediately ceases to be a "purposeless" garden.

Installations

An art installation, like a garden and like a work of architecture, is site-specific
and site-bound. While they may, like a work of architecture, be theoretically or
even practically mobile, this is counterintuitive to their nature. That is, we can

---

[28] Hunt, John Dixon, *Greater Perfections: The Practice of Garden Theory* (London, UK: Thames and Hudson, 2000), p. 1.

[29] This is a point Lee attributes to Friedrich Schiller. Lee, *The German "Mittelweg"*, pp. 138–139.

[30] David Fenner, "Pure Architecture," in Michael H. Mitias (ed.), *Architecture and Civilization* (Amsterdam, NL: Rodopi Press, 1999), pp. 43–57.

in principle and in practice move a building, but each time we do this we tend to marvel at the act. The same seems true of an installation (though perhaps with a bit less marveling). While we think of objects that might count as gardens except that they contain no living components or are not parts of larger wholes that contain living components as installations—like a set of plastic flowers and foliage in an assemblage inside a building; or like Cadillac Ranch in Amarillo, Texas; or perhaps even like Stonehenge in Wiltshire if we think of Stonehenge in aesthetic terms—the principle connection between The Garden and installations does not seem to cast any more revealing light on the question of whether and how The Garden is an artform than do the previous discussions about sculpture and architecture. So we will leave it there for now.

Poetry

Walpole, in naming the "sister arts," places gardening among painting and poetry. This is a principal focus of Ross—as is understandable given that her task is to understand The Garden as an object of meaning.

> [A] garden can often convey the same content as a poem; furthermore, it can do so in part by exploiting the fact that gardens, like poems, are experienced over time. . . . To "read" [Alexander] Pope's garden ensemble requires the very same skills as reading a poem. . . . How such meaning is conveyed by ensembles of the sort just described is a problem for art in general, not one newly raised by gardens and by the claim that gardens must be read.[31]

If gardens represent or have elements that represent—or perhaps more precisely, possess representational content—then to understand a garden is to understand the relationship between some aspect of the garden (or the garden as a whole) and the thing it represents. This kind of interpretive activity is most apparently at play in reading poetry, where interpretation of the associations of the poet's expressions to other things—ideas, emotions, places, times, and so forth—is not only key but the dominant activity while reading a poem. If the capacity of an object to lend itself to interpretive activity—to the sort of cognitive engagement that is endemic to working out the mysteries of meaning of the object—is not only a hallmark but a necessary condition of an object's type being an artform, then thinking of The Garden in relation to poetry makes eminent sense.

Poetry and The Garden share another trait, just as interesting as that focused on meaning. Poetry incorporates nonrepresentational aspects—rhyme,

---

[31] Ross, *What Gardens Mean*, pp. 54 and 58.

meter, rhythm, structure—that contribute to the mood of the piece and are determined by word choice and sentence structure. In the same way, garden form may employ nonrepresentational elements that contribute to the mood of the place. A straight path may be "read" as if punctuation were absent; our "reading" of the landscape slows down when our path is punctuated with winding, uneven steppingstones that require our attention. In a similar way, plants or other elements may be utilized to provide structure throughout, eschew structure, coalesce, contrast, and otherwise contribute to the atmosphere—to borrow a term from Cooper—of the garden.

Gardens are "aesthetic-forward," and the properties they present as their obvious aesthetic properties are the ones—as Ross says in a quote above—that are primarily visual. While poetry may evoke all manner of images in the mind of the reader—with diction, meter, rhyme, and the like, building a sense of pacing and mood—a garden's aesthetic properties require only the most minimal application of taste and aesthetic sensibility to the "base objective" properties of color, line, shape, and the rest. A robust appreciation of gardens can be achieved with no knowledge of the language, so to speak. Yet many gardens are better appreciated with an understanding of the cultural or religious context in which the garden exists—for instance, by knowing the allusions, allegories, and symbols of the statuary and other incorporated structures. Mara Miller, in talking about European estate gardens, suggests that these landscapes require a reading knowledge of Latin and an education in mythology to be fully appreciated.[32] Her observation is true about symbolism and allusion in all art, including painting. While understanding these references may offer clues as to the artist's intentions, in both painting and gardens the meaning is likely secondary to how these aspects visually work together in the composition of the scene. Secondly, we must remember that the gardens Miller suggests require education to be read are those that have had the wealth and prestige to secure their survival. These were created by the educated upper class, and their primary audience would have been the educated upper class. More humble gardens, made by and for the less-educated, would have had an even greater emphasis on the visual aesthetic properties over the meaning of individual components.

Performance/Dance

In "The Garden as a Performance,"[33] Mateusz Salwa outlines an interesting approach relevant to our question about The Garden as an artform. The notion of

---

[32] Miller, *The Garden as an Art*, p. 36.
[33] Mateusz Salwa, "The Garden as a Performance," *Estetika* 51:1 (2014), pp. 42–61.

connecting The Garden to performance—specifically to dance—is mentioned by
Hunt as well:

> But prime among those paradoxes of garden art . . . is its constitution by two
> prime constituents: what we loosely call "nature," but which are really the un-
> mediated ingredients and processes of the physical world, organic and inor-
> ganic, and what we call art or culture, by which those "natural" elements are
> mediated. Gardens are arguably unique among the arts in this combination—
> only the dance and body painting come to mind as arts that actively involve a
> living, organic, and changing component.[34]

While there are other performance-based artforms beyond dance, and while
we might add tattoo alongside body painting, Hunt's point should not be
eclipsed: gardens are living—or, as we have argued, have living components associated with them—and so insights derived from thinking about the garden
as informed by all the implications of the fact that it is alive may be valuable in
finding (canonical) artforms with which to compare The Garden. Salwa offers a
first sketch of what this might look like:

> [W]hen we say that gardens are like performances . . . we have a creator of the
> garden, performers of the garden, and audience members participating in the
> garden . . . the landscape architect who is the creator but not so much of the
> garden itself but of its project or plan . . . a plan of a garden is, then, similar to a
> musical score . . . anonymous garden workers (whose role is often neglected and
> sentenced to oblivion) who correspond to musicians playing in an orchestra . .
> [and] visitors who are like an audience.[35]

This model does not seem adequate though, because it "reduces animate and
inanimate nature to the role of a passive instrument which is manipulated by
people and therefore is only a vehicle for their ideas."[36] So he emends the model:

> Were we to maintain this analogy, we should treat the natural elements of the
> garden as actors and the garden workers as technical staff, while such factors
> as weather, temperature, and light would play an analogous role to that of, say,
> stage lighting. . . . We can push the metaphor further and compare the garden
> even to dance: the garden architect would be the choreographer (garden

---

[34] Hunt, *Greater Perfections*, p. 9.
[35] Salwa, "The Garden as a Performance," p. 52.
[36] Salwa, p. 52.

workers, then, would be "subchoreographers"), whereas the plants would be like dancers acting in the way intended by the gardeners.[37]

[W]e cannot grasp the essence of the garden unless we assume this anti-anthropocentric perspective and treat natural elements of the garden as performers. . . . [W]hen we visit a garden we participate in an event gathering ourselves and nature within a frame offered by the garden's creator. . . . [I]t means seeing gardens as constant "processes" or "actions" performed by nature. . . . Describing gardens as processes, one treats any stage in their history as equally important as other stages, giving priority to none of them. . . . Lastly, it implies that there is no fixed meaning of a garden, in the sense that a garden is not a medium conveying a meaning established by the architect or gardener—its meaning is at best created individually on the basis of his or her suggestions. . . . [T]reating gardens as performances allows one to see that they are what they are thanks to a variety of creative moments born out of owner-architect, architect-garden workers, garden workers-flora, flora-visitors relations, to name only a few.[38]

Salwa's model takes into account not only the interplay between a living artwork and a living audience but also the dynamism that each, garden and audience, brings to their interaction. It fits each player into an appropriate role, and while there is a risk of anthropomorphizing nature in this model, the motivation of the model is to do the opposite: to recognize the role that nature plays in both the creation and instantiation of the artwork.

There is another insight to be had about the relationship of The Garden to dance. Artworks like literature, poetry, operas, symphonies, and plays all derive stability from mechanisms like scores and scripts—so that one edition or performance is much like another, so that the author/artist's original vision or expression is preserved, and so that identity conditions among various instantiations can be checked for maintenance of authenticity. But this is not commonly the case with dance. The way that a particular dance is learned and then taught by one dancer to another is through memory. While notation systems exist, they are cumbersome and not "industry standard." Memory is the mechanism of transmission, and this allows changes to creep in as memory may not be perfect and as a particular memory of a dance will incorporate contingencies like stage-size, company-size, or the strengths of dancers. This is further complicated by the fact that how one dancer is trained may be different from how another dancer—trained elsewhere by a different teacher—was taught to execute certain movements. Highly formalized dance—as we find in the ballets of George

[37] Salwa, p. 53.
[38] Salwa, pp. 53, 55, and 56.

Balanchine—are best performed by members of a company that were together throughout their training; it is the best way to ensure consistency of movement execution. In addition, a certain dance—say, a ballet like *Swan Lake*—may be instantiated in radically different ways and still be recognized as an instance of *Swan Lake*. A case in point is Matthew Bourne's version from the mid-1990s that utilized male dancers in the corp. Audience members who may have been inclined to claim that Bourne's version was not *Swan Lake* would have been hard pressed to establish their claim, since dance admits of a wider variation— variation that is not only tolerated but celebrated by audiences—than other artforms, even clearly allographic ones. The wide variety of gardens—how they are conceived, designed, instantiated, maintained, re-created—how they are appreciated by a particular visitor, at a particular time of year, at a particular time of day, with a set of particularized climactic conditions—makes dance an even more interesting and apt way to think of The Garden as an artform.

Finally, a fascinating connection between gardens and performance— specifically "performance art"—is that these are the only two artforms (accepting The Garden to be an artform ex hypothesi at the moment) that are both dynamic and autographic. All other artforms that are dynamic are allographic—or at least arguably so—but an instance of performance art exists as a unique event.

Music

In "Gardens, Music, and Time,"[39] Ismay Barwell and John Powell consider how The Garden is like music. Their focus is not on music as a performative but rather on the essence of what music is.

> [W]e will develop a suggestion that gardens present the passing of time visually in a way that is analogous to the way in which music presents the passing of time audibly. . . . Aesthetic appreciation of gardens that is confirmed to the way they appear only at particular moments and that does not take account of the relation between these appearances and those that precede and succeed them misses out a whole dimension of aesthetic experience.[40]

Like music, gardens have natural rhythms not only guided by the seasons but also employed as central to their designs. No serious gardenist thinks of the garden as fixed in time, and it is evidence of limitations of a particular gardenist that plants are planted too close to one another. The serious gardenist has a sense

[39] Barwell and Powell, "Gardens, Music, and Time," pp. 135–147.
[40] Barwell and Powell, p. 136.

of the mature form of the plant in question, but more important than that, the serious gardenist knows—at least in general terms—how that plant will manifest itself through the course of a year's seasons and through the course of its life-time. Such knowledge is not absolute. We had a friend who was convinced that camellias exhibit free will—it is the only one way he could explain the breadth of unexpected variety in the flowers each winter. But in general terms we know that camellias in northern Florida will bloom in late fall to early spring, that the "deb-utante" variety may bloom first, and that if the spent blooms are not collected, they will contribute to the problem of petal blight in the future. And so we plan both the placement of plants and their maintenance accordingly.

Experiencing The Garden as an object over time comes in two categories: the dynamism over years, season, and hours—as discussed—and the changing experiences one has moving through a garden during a single visit. As with the punctuation and sentence structure in poetry, the tempo of music can be compared with the movements one makes through a garden. In addi-tion, there are other nonrepresentational indicators of mood: the use of major or minor chords, the fullness or sparseness of the composition, consonant or dissonant harmonies, the use of distortion pedals, and so forth. As music has rhythms, repetition which brings order and predictability to the enjoyment of music, so gardens incorporate elements that do the same in a visual field. Garden aesthetics, as Barwell and Powell make clear, requires a sensibility that is not "time-sliced" and bound to a particular moment. The relationship of The Garden to music makes this clear, and so from this relationship we may glean the insight that as The Garden is a form that is temporally dynamic in ways that are ordered, so too is music, and music—to make the point—is without question a canonical artform. Dynamism need not be a problem for understanding The Garden to be an artform.

Music that is improvisational mirrors the cocreative processes in a garden of the imposition of human design—and maintenance that allows that design to endure—and of the natural dynamic processes that shape that garden from the ground up, so to speak. As one force reacts to the other, the unique garden amalgam emerges. Improvisational music results as musicians hear what their colleagues play and react in ways that are complementary and augmentative. While nature's reactions cannot be said to be deliberate without that description verging on the inappropriately anthropomorphic, we can understand the char-acter of the gardener's actions and reactions, as they tend a garden, to be much like those of the improvisational musician. And we can understand nature's reactions, in turn, to be shaped and guided by the decisions and actions of the gardener.

Film

Hunt writes:

> Apart from such residual spaces as roof terraces, balcony gardens, window boxes, or Wardian cases, gardens have always been designed to be walked through while looking—"an intertwining of vision and movement." Movies would well serve as a prime visual model for such landscape experience and design . . . but they are rarely invoked.[41]

One of the most illustrative contributions consideration of film can make toward understanding The Garden as an artform is through appreciation of the fact that theorists about film have been deliberately focal about the categorization of those components of film that go into the creation of its aesthetic form. While theorists may differ on the detail, "film form" commonly has been understood as having five elements: its visual contents (the film's "mise en scène"), its photographic aspects (the film's cinematography), its narrative or textuality (the film's plot or, more broadly, story), its sequencing of shots and scenes (the film's editing style), and finally its sound (both diegetic and nondiegetic). Understanding "film form" provides a platform not only for principled evaluation of films but also for developing insights into ways to interpret films and understand their meanings. In addition, through the analysis of a film's form, we can come to appreciate its appropriate categorization in terms of genre and style (whether that style is focused on the film's director or whether we mean by "style" the utilization of editing patterns).

If such a treatment of The Garden's form could be established, this would go a long way not only to understanding methods of approaching The Garden aesthetically but to understanding how The Garden relates, formally, to other artforms. Many of the topics covered above, focused on the relations of The Garden to canonical artforms, include insights about this very thing, and in the next chapters we will try to glean from those insights further revelations about The Garden's aesthetic characteristics. The brass ring, so to speak, is to develop for The Garden what we have for film; we attempt this in the chapters devoted to garden interpretation.

Film today is an amalgam of sight and sound.[42] Past the advent of the "talkie"—in 1927—sound has been thought of as integral to film. Film is not unique in being an aesthetic amalgamation—dance typically includes both music and

---

[41] Hunt, *Greater Perfections*, p. 136.

[42] Film theorist Erwin Panofsky coined the expression "coexpressibility" to refer to the integration of sight and sound in film as simultaneous conveyors of the film's aesthetic properties.

movement; opera typically includes both music and acting—but its amalgamated character allows us to appreciate better The Garden as both designed and natural, an ongoing theme of garden theory and of this book. As "factorizing" appreciation of dance or opera into its separate components is bound to fail as an approach to understanding their amalgamated character, so "factorizing"—to use Cooper's expression—of garden appreciation into appreciation of its artful (designed) character and its natural character will fail as well. This is a prominent theme in Cooper's book.[43]

In the next chapter we begin to make a case for The Garden to be considered a bona fide artform, moving from the set of common characteristics discussed above, through arguments for why The Garden should not be considered an artform and how The Garden is different from other artforms—both generally and specifically. We then finally settle in to offer our case.

---

[43] Cooper, *A Philosophy of Gardens*, especially pp. 51, 54, and 59.

4

The Garden as an Artform

In the previous chapter we looked at a range of ways in which The Garden is like other artforms. In this chapter, we want to do the reverse: to look at how it is different. If the situation is that these differences are too great, then perhaps The Garden is not an artform. We believe, however, that the differences are not great enough to eliminate The Garden from contention, and we argue that there are good reasons for considering it as an artform. We end the chapter with consideration of the conditions to be fulfilled for a particular garden to be a work of art.

Arguments against The Garden Being an Artform

Let us begin by considering reasons against the claim that The Garden is or can be an artform.

Kant and Hegel

We begin by noting that Immanuel Kant did not deny that The Garden is an artform. Unfortunately, however, it was in part because of his writings that The Garden was pushed to the edges of the conversation. In the *Critique of Judgment*, Kant writes:

> Painting, as the second kind of formative art, which presents the sensuous semblance in artful combination with ideas, I would divide into that of the beautiful Portrayal of nature, and that of the beautiful arrangement of its products. The first is painting proper, the second landscape gardening. For the first gives only the semblance of bodily extension; whereas the second, giving this, no doubt, according to its truth, gives only the semblance of utility and employment for ends other than the play of the imagination in the contemplation of its

Joju-in Garden, Kiyomizu-dera Temple, Kyoto, Japan
Unknown gardenist, starting 17th century
Photo by Marta McDowell, Fall 2009

The Art and Philosophy of the Garden. David Fenner and Ethan Fenner, Oxford University Press.

forms. (It seems strange that landscape gardening may be regarded as a kind of painting, notwithstanding that it presents its forms corporeally. But, as it takes its forms bodily from nature—the trees, shrubs, grasses, and flowers taken, originally at least, from wood and field—it is to that extent not an art such as, let us say, plastic art. Further, the arrangement which it makes is not conditioned by any concept of the object or of its end—as is the case in sculpture— but by the mere free play of the imagination in the act of contemplation. Hence it bears a degree of resemblance to simple aesthetic painting that has no definite theme—but by means of light and shade makes a pleasing composition of atmosphere, land, and water.) The latter consists in no more than decking out the ground with the same manifold variety (grasses, flowers, shrubs, and trees, and even water, hills, and dales) as that with which nature presents it to our view, only arranged differently and in obedience to certain ideas. The beautiful arrangement of corporeal things, however, is also a thing for the eye only, just like painting—the sense of touch can form no intuitable representation of such a form. In addition, I would place under the head of painting, in the wide sense, the decoration of rooms by means of hangings, ornamental accessories, and all beautiful furniture the sole function of which is to be looked at; and in the same way the art of tasteful dressing (with rings, snuffboxes, etc.). For a parterre of various flowers, a room with a variety of ornaments (including even the ladies' attire), go to make at a festal gathering a sort of picture which, like pictures in the true sense of the word (those which are not intended to teach history or natural science), has no business beyond appealing to the eye[1]

Through claiming that gardens are simply the arrangement of things already present in nature, that they are not "plastic" in the sense of being thoroughly moldable according to human designs, that they are like common household decorations, and that they may only be appreciated visually, Kant effectively moves gardens to the very edge of the world of art. Yes, they can be appreciated aesthetically, but not in ways that are significant or like the rest of the arts. When comparing gardens to painting, Kant is not elevating gardens to the level of aesthetic appreciation of paintings-as-art. In separating "landscape gardens" from "painting proper," he is instead describing gardens as simply a collection of pretty things and nothing more.

Kant's treatment of gardens as aesthetic objects is shaped in large measure by two of his central views. First, to appreciate an object aesthetically—or perhaps more precisely, to enable oneself to develop a true judgment of taste—one must be disinterested in the object under attention. We will discuss the ideas of disinterestedness and engagement later in the book. Kant's experience of gardens did

---

[1] Immanuel Kant, *Critique of Judgment* (Indianapolis, IN: Hackett, 1987), pp. 322–324.

not permit a distinction between a garden as a pure aesthetic object, capable of being experienced disinterestedly, and a garden considered only in terms of its utility. For Kant, gardens have functions, and their instrumental dimensions as places of provision for food and medicine are inescapable. Even those that do not participate in these functions are too closely related to those that do, so that contemplation of them necessarily is corrupted by consideration of their instrumental natures.

Second, true aesthetic consideration of an object results both from and in what Kant describes as a free play between the imagination and the intellect/understanding. Gardens do not present occasion for this, thought Kant. As we contemplate a garden, we can delight in its visual aspects, and our imagination can be engaged, but gardens do not engage the faculties of understanding. We cannot understand in the garden object its purposive nature, how its aspects taken all together present us with an opportunity for seeing the object as having an internal order that strikes us as designed as if having a purpose (in the absence, of course, of having an actual purpose, such as food or medicine production—this goes back to the disinterestedness). If gardens cannot be appreciated according to the dictates that govern the possibility of development of a true judgment of taste, of subjectively being able to discern whether an aesthetic object is truly worthy, then gardens must be removed from the center of such discussions.

If Kant is an antagonist in the story of The Garden—and some would contest this, perhaps Michael Lee chief among them—it is because, first, he set up theoretical structures that limit the capacity of The Garden to be appreciated and evaluated aesthetically, and, second and more pertinent right now, he not only accords landscape gardening a low ranking as an art, but he places it alongside an assortment of aesthetic forms that few today would consider appropriate to be considered art—decoration of rooms, ornamental accessories, rings, snuffboxes, ladies' attire (though many would include furniture under architecture today, in part thanks to Frank Lloyd Wright). He does The Garden no favors, but in the end he does not explicitly deny that The Garden is an artform.

Neither does Hegel, although if there is a primary antagonist in the story of The Garden, it is Georg Hegel. In his *Aesthetics: Lectures on Fine Art*, he writes:

Now the question arises of what interest or end man sets before himself when he produces such subject-matter in the form of works of art. . . . If in this matter we cast a glance at what is commonly thought, one of the most prevalent ideas which may occur to us is (a) the principle of the imitation of nature. According to this view, imitation, as facility in copying natural forms just as they are, in a way that corresponds to them completely, is supposed to constitute the essential end and aim of art, and the success of this portrayal in correspondence with nature is supposed to afford complete satisfaction. This definition contains, prima facie, only the purely formal aim that whatever exists already in the external world, and the manner in which it exists there, is now to be made over again as a copy, as well as a man can do with the means at his disposal. But this repetition can be seen at once to be a superfluous labor, since what pictures, theatrical productions, etc., display imitatively—animals, natural scenes, human affairs— we already possess otherwise in our gardens or in our own houses or in matters within our narrower or wider circle of acquaintance. And, looked at more closely, this superfluous labor may even be regarded as a presumptuous game.[2]

For the architectural work of art is not just an end in itself; it is something external for something else to which it serves as an adornment, dwelling-place, etc. A building awaits the sculptural figure of a god or else the group of people who take up their home there. Consequently, such a work of art should not essentially draw attention to itself. In this connection regularity and symmetry are pre-eminently appropriate as the decisive law for the external shape, since the intellect takes in a thoroughly regular shape at a glance and is not required to preoccupy itself with it for long. . . . The same thing is valid too for that strict kind of gardening which can count as a modified application of architectural forms to actual nature. In gardens, as in buildings, man is the chief thing. Now of course there is another kind of gardening which makes variety and its lack of regularity into a rule; but regularity is to be preferred. For if we look at the variously complex mazes and shrubberies continually diversified in their twistings and windings, the bridges over stagnant water, the surprise of gothic chapels, temples, Chinese pagodas, hermitages, urns, pyres, mounds, statues—despite all their claims to independence we have soon had more than enough; and if we look a second time, we at once feel disgust. It is quite different with natural regions and their beauty; they are not there for the purpose of use and gratification, and may come before us on their own account as an object of consideration and enjoyment. On the other hand, regularity in gardens ought not to surprise us but to enable man, as is to be demanded, to appear as the chief person in the external environment of nature.[3]

The comment about looking a second time at the objects that populate some gardens and feeling "disgust" is a hard hit. The reason for including the first quoted paragraph above is to demonstrate what seems to be Hegel's motivation for his disdain for gardens: they add little but a sense of wasted effort in the imitation of nature. Better simply to view nature itself instead of a copy of it. Whatever virtue gardens have lies in the demonstration of human mastery in harnessing

---

[2] G. W. F. Hegel, *Aesthetics: Lectures on Fine Art* (Oxford, UK: Oxford University Press, 1975), pp. 41–42.

[3] Hegel, *Aesthetics*, p. 248.

and shaping nature, and that virtue is shown in (and perhaps only in) the regularity and symmetry that a person can impose upon nature.

Hegel's rejection of The Garden as fit for aesthetic attention, first, is wedded to his acceptance of the notion that art is principally about mimesis or imitation. As we mentioned above, this view has a history that begins with Plato and Aristotle and extends for hundreds of years, but it is not a view that constrains us today. So Hegel's argument focused this way should not concern us much.

A second reason, embedded in the second paragraph quoted above, for Hegel's rejection of The Garden is connected to the notion of disinterestedness. A work of art should be an end in itself. If it is instead primarily considered in terms of its utility, attention is directed toward that functionality and away from the thing itself. This is the case, says Hegel, for architecture and for gardens.

It may be of interest to note that Hegel does not take The Garden to be a subset or relation of painting, as Kant does, but rather of architecture. And so it may be to Hegel that credit is owed for our categorization of garden creation as landscape architecture, which started in the middle 1800s. (Kant's third *Critique* was originally published in 1790 and Hegel's lectures on aesthetics were given in the 1820s and compiled and published in the 1830s.) Kant and Hegel have the effect of throwing a bucket of cold water on the efforts of the British taste theorists as those earlier theorists navigated the aesthetic value that was shown (largely) by people of means and presumably of taste as they invested a great deal of time, focus, and resources on the creation (and many times re-creation) of their sometimes vast and complex gardens. Interestingly, as we shortly will see below, it is German philosophers—largely following in the influence of Kant—who take up and further discussion of garden theory. Contemporary literature on garden philosophy, or abstract garden theory, can be seen in some measure as motivated through an attempt to answer critics like Kant and Hegel, not only in terms of offering a kind of "apologetic" for garden philosophy but through offering theory whose very success casts a shadow on the claims of those who think The Garden not worthy of such attention.

While Hegel does not deny that The Garden is an artform—for indeed this was widely accepted as settled when both he and Kant were writing—he denigrates gardening so deeply, alongside architecture, that it is difficult to see what good his lack of denial amounts to. Cooper writes, "For Hegel, gardening is an 'imperfect art,' and gardens, while welcome if they provide 'cheerful surroundings,' are 'worth nothing in themselves.'"[4]

---

[4] David E. Cooper, *A Philosophy of Gardens*, (Oxford, UK: Oxford University Press, 2006), p. 9.

The Museum Conception of Art

As is the case with other recent pioneers in the philosophy of gardens, Cooper offers insightful comments on why attention to gardens has been sparse.

> *The Philosophy of Gardens* would be a fraudulent title: for here there is no discipline as yet to be introduced to.[5] ... While I have urged that a philosophy of gardens should not be restricted to the aesthetics of gardens, it is nevertheless within the domain of aesthetics that one might above all have expected discussion of the garden to have been more prominent than it has been. There are, I suspect, four reasons why this has not been. To begin with, gardens are typically and strikingly different in a number of respects from those artworks which have tended to be regarded as paradigmatic by philosophers of art. ... Second ... gardens have suffered as a result of those complex contingencies and historical developments, economic ones included, that by the nineteenth century had helped to favour what John Dewey called the "museum conception of art." ... A third reason for aesthetics' neglect of gardens is surely the deprecatory attitude toward the garden by some of the founding fathers of the discipline, above all Hegel. . . . A final factor contributing to aesthetics' continuing neglect of the garden reflects the degree to which the agenda of contemporary aesthetics has been set by "late modern" developments in the arts—developments, however, to which the garden has largely been immune. A common perception is that, in contrast with painting, people no longer "make statements" in the medium of gardening.[6]

Hegel's regard for aesthetic attention directed at The Garden we visited above, and how The Garden differs from other art and aesthetic kinds forms the backbone of this book, so we should not presume to try to answer that in a potentially reductionist way here. So let's move to his other two reasons.

What does Cooper have in mind when he talks about "the museum conception of art"? John Dewey believed that art appreciation is based on aesthetic appreciation, and aesthetic appreciation is based on seeing within ordinary everyday experiences the internal coherence that is indicative of the aesthetic.[7] In other words, for Dewey, aesthetic experiences—a species of which he famously termed "*an* experience"[8]—exhibits an organic wholeness that every experience has in

---

[5] Cooper, p. 1.
[6] Cooper, pp. 8–10.
[7] John Dewey, *Art as Experience* (New York, NY: Perigee, 1934).
[8] "*An* experience" for Dewey may have been a broader class of experiences than merely aesthetic ones, but aesthetic experiences are the core of what Dewey seems to have in mind when he describes "*an* experience."

some degree. Those experiences that have this quality in strong measure—those whose narrative arcs have a clear beginning, middle and end; those we recount in our memories as having this boundedness—include the ones we call "aesthetic," but they are not different in kind from run-of-the-mill experiences. The "aesthetic" is simply at one end of a spectrum of all experiences. The appreciation of what we find valuable in art is based, for Dewey, on the appreciation of the experienced quality of the aesthetic, and so, transitively, the appreciation of art is based first and foremost on a particular sensitivity to the ordinary. To attempt to sever the aesthetic from the ordinary is to misunderstand it, and so to attempt to sever the appreciation of art from the ordinary is similarly problematic.

A museum is a mechanism by which we remove works of art from their originating contexts and place them in a metaphorical blank slate for appreciation.[9] Many works of art are built for this destiny: they are mobile; they are not so large that they cannot be experienced during the course of a single (say) visual encounter; they are framed to stand as single items; they are complete and not subject to further modification. Of course, not all works of art—not even all paintings—share these traits. But they are indicative of the sense of how we commonly, perhaps as a matter of cliché, understand works (including photographs, prints, sculpture, and so forth) that ultimately populate galleries and museums. This "separating-off" is common to museums; it is what Cooper seems focused on when he says that gardens suffer from not sharing with the kind of works that populate museums this character, and as they do not share it, so it is easy to think of them as not participating in the possession of the character necessary for being works of art. Cooper clearly has more in mind, as he talks about "complex contingencies and historical developments, economic ones included," as to why his reference to Dewey is apt, but the "separating-off" phenomenon at the heart of Dewey's lament seems core to Cooper's explanation for why gardens are held at an aesthetic arm's-length from artworks readily embraced as art.

Cognitive Engagement

Cooper's final reason for why gardens have not enjoyed the aesthetic attention they might or once did tracks the art world and the cognitive turn it has taken over the last number of decades. Surely there are plenty of opportunities for aesthetic engagement—understood here as having a central focus on sensory connection between subject and artwork—in art venues today, despite the reality that representational aspects of works of art have gone decidedly toward the abstract. But when reflecting on what history likely will remember about the art world of

[9] David Fenner, *Art in Context* (Athens, OH: Ohio University Press, 2008).

the twentieth and twenty-first centuries, it is those objects and movements that focus on cognitive engagement that first come to mind. While some gardens do make "statements" and engage their audiences in ways that are primarily cognitively focused, the majority of gardens in our survey do not. And even those that do make statements have pronounced "aesthetic" characters in addition. This is out of sync with the spectrum of art central to the art world today, and so it is easy to see gardens as similarly out of sync. And because of this, it is easy simply to leave gardens to one side, approaching them in terms of their artistic qualities and characters either as novelties or as ancillary or even metaphorical to "true" art. Cooper identifies this problem with what both Kant and Hegel express about The Garden: "For both writers, the garden cannot legitimately engage the 'thought' component in serious appreciation: once its pretensions go beyond affording pleasure and cheer—to the attempt to convey truths about nature—it necessarily fails."[10]

Cooper responds, "the criticisms levelled against gardens would equally apply, *pace* the authors' intentions, to landscape paintings."[11] Cooper goes on to point out that in addition, not only may elements of a garden stand in a representational relationship to other things, contemplation of which rewards cognitive engagement, but that whole gardens may do this as well, and he points to the *karesansui* gardens around Kyoto as examples. Cooper does not take this to be a full solution to the problem, however, because representation—by a garden element or a garden proper—is a contingent aspect of The Garden. Some gardens may represent or have elements that represent, but this is not a feature necessarily present in all gardens. But he does not abandon the idea altogether and instead talks about The Garden's proclivity for "evoking" and "intimating" representationally relational aspects concerning such things as how, when appreciating a particular garden, one can enter into the mood, sensibilities, and vision of the gardenist, or how one can appreciate the intimacy of the relationship of humans and nature as they are so manifestly evidenced in every part of every garden.[12] The degree to which Cooper is successful in his rebuttal to Kant and Hegel in their dismissal of The Garden's ability to engage in ways that transcend mere appreciation of their "pleasure and cheer" may depend on his readers' response to his descriptions of how gardens "evoke" and "intimate," but whether readers react with full-throated acceptance of his proposal or not, he makes the point that at least for some—at least for him and for those who share his views—The Garden does reward contemplative engagement, and this is enough to show that Kant

---

[10] David E. Cooper, "In Praise of Gardens," *British Journal of Aesthetics* 43:2 (2003), pp. 103.
[11] Cooper, pp. 103.
[12] Cooper, pp. 103 ff.

and Hegel's dismissal is ultimately unsuccessful if their claims are meant to be taken as universal.

Cooper's concern is shared by Miller, who points out that gardens are plainly "material-forward." All are physical places composed of physical objects undergoing physical change. The depth of their tangibility as four-dimensional objects can interfere with the ease of our cognitive or imaginative engagement with them, and this sort of "thought" engagement—to use Cooper's word—has become increasingly important to understanding the world of modern art. Miller believes the demotion of The Garden from being an artform is occasioned in part by the cleaving of art from aesthetics—where art is not principally identified in terms of the aesthetic features of works, examples of which are common in the twentieth and twenty-first centuries. No longer are aesthetic virtues automatically artistic ones, and as gardens are "aesthetic-forward," their aesthetic virtues do not necessarily contribute to their identities as works of art. As we have seen among comments by every garden scholar considered thus far, while appreciation of The Garden may not reduce to aesthetic appreciation (even thought of as broader than simply formal aesthetic appreciation), garden appreciation seems to begin with aesthetics.

Some point to the presentation of Pablo Picasso's 1907 *Les Demoiselles d'Avignon* as the start of modern art, and they cite as the painting's most salient feature in this regard its abandonment of the pursuit of beauty as its primary goal. Whether this claim pans out as one regards the history of art, it is still illuminating: that much modern art does not take as its aim, or at least its primary aim, the production of beauty. While there are theories of art, as we saw above, that continue to hold that an object's possession of aesthetic properties is necessary to its identity as a work of art, theories such as this are met with a plethora of examples, as again we saw earlier, that seem to evidence the view that taking a perceptually based aesthetic approach to some artworks is to miss their true value. Yet despite this, while all gardens are appreciable aesthetically, some experimental gardens—some of which are included in our inventory—are consonant with modern art in that they prioritize cognitive engagement over their aesthetic virtues. If cognitive engagement of this sort is necessary to the identity of a modern work of art, some gardens do in fact clear this hurdle.

While a perhaps disproportionate amount of attention (as compared to the overall literature on garden philosophy) has been given to whether a garden has the capacity to engage cognitively one who attends to it,[13] it seems that in at least

---

[13] Robert Stewart and Roderick Nicholls, "Virtual Worlds, Travel, and the Picturesque," *Philosophy and Geography* 5:1 (2002), pp. 83–99; Jane Gillette, "Can Gardens Mean?" *Landscape Journal* 24:1 (2005), pp. 85–97; Susan Herrington, "Gardens Can Mean," *Landscape Journal* 26:2 (2007), pp. 302–317; G. R. F. Ferrari, "The Meaninglessness of Gardens," *Journal of Aesthetics and Art Criticism* 68:1 (2010), pp. 33–45; and, of course, Stephanie Ross.

some cases, the capacity of particular gardens to do just that and to do it in full-blown ways is beyond question. Stourhead Gardens in Wiltshire is the default example.[14] Those who claim that a garden's ability to engage us aesthetically—by which they tend to mean in terms of that garden's perceptually based aesthetic properties (e.g., being beautiful)—depends on the garden making no intellectual demands on us either fail to understand the character of aesthetics and art from at least the last century or fail to see that counterexamples like Stourhead—which is remarkably beautiful—exist. While we claim that some gardens are particularly good at being places where one is not forced to engage cognitively, and that this allows the mind freedom to move as it may, this is not only incidental to particular gardens but is more a description about the state of the subject than of the garden.

We devote two chapters to discussion of garden meaning and interpretation—because we believe it to be an important component not only of approaching The Garden aesthetically but also as an artform with comprehensive claim to being such—and in those chapters we focus on this question of cognitive engagement. There we talk about how cognitive engagement with gardens is typified by working out the connections of the various aspects of the form of a garden in understanding how those parts work together to create an integrated, coherent whole. As we work through these connections, we come to understand the garden's meaning—or meanings, if we take a pluralist approach—and perhaps also come to understand a garden as the unique artistic expression of those who designed and created it. We claim that such an approach can be employed by anyone about any garden, although some gardens obviously reward that sort of consideration more strongly or deeply than others.

Expense and Resource Commitment

Ross argues that one of the factors that has led to a diminished focus on The Garden as an artform has to do with the expense gardens require.[15] She claims that as gardens on the scale necessary to sustain the kind and depth of attention that works of art typically engender become rarer—and as a consequence focus on gardens becomes rarer—it is increasingly easy to think of gardens either as purely functional (as we might think of "setting and framing" gardens, consumable gardens, or even botanical gardens) or as unusual diversions (as we

[14] Stephanie Ross, *What Gardens Mean* (Chicago, IL: University of Chicago Press, 1998), p. 175: "In working out the iconographic program of a garden like Stowe or Stourhead, we are clearly solving an intellectual puzzle."

[15] Stephanie Ross, "*Ut Hortis Poesis*—Gardening and Her Sister Arts in Eighteenth-Century England," *British Journal of Aesthetics* 25:1 (1985), pp. 17–32.

might think of destination gardens that require airplanes and multiday holidays to visit).

The gardens that are Ross's primary focus are massive English landscapes. These indeed require enormous wealth to maintain, and as resources may become less available, or the owner's inclination to commit resources to their gardens lessens, corners that were cut mean erosion of the original gardenists' designs—likely with a reduction in value because both designs and resources eroded. While the point may be criticized because the evidence set for her claim may be limited, the truth is, to varying degrees depending on the garden in question, Ross is certainly correct that gardens require resources, almost always significant, given their scale.

Agriculture is thought to be purely functional in terms of the provision of food and acquisition of profits, and so agricultural ventures should pay off; the resources that go into them are offset by returns that are greater. But for the vast majority of gardens—large Brownian landscapes to small residential gardens—the resources that go into their creation and maintenance deliver rewards that are primarily intangible. This is the case even if we include the produce of fruits and vegetables; the chances are high that the resources—time, talent, and treasure—that go into the "creation" of a garden tomato are far greater than those required to purchase one at the local grocery store. When the estate is stressed for finances, the garden is likely the first to go.

Consider a parallel. The existence of the National Endowment for the Arts may be justified on the grounds that, first, the cultural identity of a society (arguably the only truly lasting element of a nation's identity) is tied in important ways to the production of its art (among other cultural artifacts), and, second, the production of some works of art—or, better, the production of works in some artforms—is similar to the construction of roads and bridges and the provision of national defense in the sense that it cannot be achieved except corporately, as citizens pool resources to make it happen (albeit perhaps a coerced pooling, depending on how one regards taxation). If a society wants to have opera (for example), it likely will have to be subsidized; ticket sales will not provide enough support—or so the argument goes. Those who argue against such subsidy, and believe the National Endowment for the Arts (NEA) should be disbanded, argue that if there is not enough popular interest in an artform, that artform either needs to retool to attract sufficient interest or, as in the case of a business that no longer produces items people want to purchase, it should cease to exist. One could make the same argument concerning gardens. Perhaps the existence of some gardens—such as the United States Botanic Garden—should be subsidized as an important part of the cultural identity of the nation, but if there are other gardens that cannot survive without some level or kind of subsidy, they should either put in some roller coasters or shut down.

The argument goes: if one wishes to invest one's own resources in the instal-lation and maintenance of a garden—as is the case with so many of us—then that is our choice, and we should bear the cost of that ourselves. The practical results of this sort of argument entail that only the wealthy will have the sort of gardens that Ross considers as she surveys seventeenth- and eighteenth-century Europe or that, for most of us, gardens will be modest. Neither of these avenues is likely to produce gardens that rise to the level of being works of art. Some of course may, but they will be so few that The Garden as a form will hardly be large enough to attract the sort of attention required for The Garden to be regarded in any significant way as an artform.

Preservation of Original Gardenist Designs

Ross's point about expense is amplified in the realization that as maintenance resources diminish in the case of large gardens that do not continue to enjoy the resources they once did, it becomes difficult to preserve the original designs of those gardens. Museum-style art tends to be static—paintings, sculptures, photographs. Performance art (with the exception of dance) typically works from scripts and scores that provide stable continuity over its various allographic instantiations. Other allographic arts—like film and literature—are even more stable. But The Garden is not. Gardens evolve—daily—and so original designs lose their integrity quickly without constant vigilance. If the status of The Garden being art is tied to gardenists being artists—or if that is only one factor but still a necessary one—then as original designs are lost, so the claim of a particular garden being a work of art will be lost.

Erosion of the Status of Gardeners

Ross writes:

> But the profession [of gardening] continued [in the time of Le Nôtre] to operate like a manual trade or mechanical art throughout the seventeenth century, and the gardeners themselves didn't band together to seek the higher status that might come through guilds, societies, academies, and the like.[16]

Happily, we may find some relief for the unhappy implications of the lack of pro-fessionalization experienced or pursued by those working with gardens in the

---

[16] Ross, *What Gardens Mean*, p. 47.

seventeenth and eighteenth (and following) centuries. To achieve respect is to earn it; one cannot simply command respect and expect it to be forthcoming. Nonetheless, without expressing a claim, recognition is hard to achieve. Ross's point, we take it, is that as respect as an artform was not sought by those who more frequently populated the world of The Garden, those on the outside stopped regarding it as worthy. This is tied, of course, to the resource-hungry expansive gardens Ross studied; the labor involved was tremendous, and laborers are not commonly thought of as artists. Concomitantly, gardenists themselves were becoming a smaller class of individuals as resource-hungry gardens expanded and as fashion dictates resulted in only some gardenists being sought out. Happily, those working on gardens today have experienced a resurgence in professionalization. Credentialing has become much more important to both gardenists and gardeners—not to mention their employers—and credentialing itself has become deeper and richer over the last number of decades.

Differences between Traditional Artforms and The Garden

We turn now to Miller.

> Stephanie Ross has suggested that the decline is attributable to incidental factors such as the expense of gardening on the grand scale and the difficulty of preserving gardens in their original condition. I believe the problem lies deeper than that, in the fundamental conflict between the nature of gardens on the one hand and certain features of what we choose to call art on the other.[17]
>
> [G]ardens enjoy an ambiguous status in a number of different respects—between poles of "art" and/or the "artificial" on the one hand and "nature" on the other, between art and craft, and between fine and applied art. . . . [G] ardens violate a number of implicit preferences upon which most theory of art is premised—preferences for a single final form of a work of art (for uniqueness and perdurance), for artistic (or authorial) control by a (single) (human) agent, for immateriality, and for what is known as "disinterest" or "distance" or "autonomy."[18]

Miller's central argument is that The Garden has not withstood the many differences with traditional or canonical artforms, that a particularly strong difference or a set of differences is enough to thwart the claim that The Garden is an artform. In the next section, we explore these differences, keeping an eye on the

---

[17] Miller, *The Garden as an Art* (New York, NY: State University of New York Press, 1993), p. 69.
[18] Miller, p. 72.

possibility that some such difference or differences may be enough to eliminate The Garden from consideration as an artform.

Function and Disinterest

An argument used to distinguish between bona fide artforms and nonartforms focuses on whether members of the set of whatever form we are considering can be, or typically are, appreciated "for their own sakes" and not for any instrumental relationships in which they might enter. The literature on food is a good example; since food has a nourishing function, it has been argued that food, as a kind, cannot be an artform.[19] A similar sort of argument could be made about The Garden; since gardens have purposes and functions, they do not typically lend themselves to being appreciated "for their own sakes," or, in other words, in terms of exclusive appreciation of their traditional or narrow aesthetic properties. This argument has its roots in the disinterestedness tradition crafted by the British taste theorists that came to full glory in the work of Kant. This is what we meant above in saying that Kant "set up theoretical structures that limit the capacity of The Garden to be appreciated and evaluated aesthetically." What is ironic is that the concept of disinterest came into being and popularity at precisely the same time and largely in those European places where The Garden was commonly accepted as an artform.

If we do not include recent aesthetic formalists here, the last strong proponent of disinterestedness was Jerome Stolnitz in the 1960s. Since that time, the tradition has waned to the point that few aestheticians see instrumentality, function, and "engagement" as unacceptable either for aesthetic evaluation or experience, and those who continue a focus on traditional or narrow aesthetic properties do not focus on disinterestedness per se but rather simply on formalism.

Gardens are typically aesthetically rewarding in ways that do not involve disinterest or its psychological cousin, psychical distance. We will save a full discussion of gardens and disinterest for a later chapter. There we look at disinterest and "engagement" (Arnold Berleant's expression that he derives from John Dewey's work), but we also look at the detail of the traditions in which disinterest was articulated and advanced, and we look at the detail of arguably its most sophisticated rendering, in the work of Kant, to see whether and/or how

---

[19] Marienne Quinet, "Food as Art: The Problem of Function," *British Journal of Aesthetics* 21 (1981), pp. 159–171. See also Elizabeth Tefler, "Food as Art," in Alex Neill and Aaron Ridley (eds.), *Arguing about Art* (London, UK: Routledge, 2002, second edition), pp. 9–27; and Carolyn Korsmeyer, "The Meaning of Taste and the Taste of Meaning," in Alex Neill and Aaron Ridley (eds.), *Arguing about Art* (London, UK: Routledge, 2002, second edition), pp. 28–49.

disinterestedness and The Garden as an art could be envisioned as coexisting harmoniously. It should be noted, though, that in the final analysis—following Dewey, Berleant, and Miller—disinterest ultimately must be jettisoned as the concept is a hinderance not only to aesthetically appreciating The Garden but to aesthetically appreciating a wide range of objects. This costs us nothing since an argument for rejecting The Garden as an artform that is based on disinterest and function likely is a nonstarter today.

If we accept

1. that some gardens are cognitively engaging and reward cognitive attention;
2. that concerns about resources, access to original designs, and gardener esteem are incidental factors (to use Miller's characterization)—that exploration of them goes a long way to explain why The Garden today no longer enjoys uncontested status as an artform but are in the end incidental;
3. that Miller's account of the differences between The Garden and uncontested artforms only functions as an argument for why The Garden cannot be an artform if one or more, or some combination of the set, constitutes that argument—in other words, the fact of differences among gardens is not a reason to eliminate The Garden's claim—as must be abundantly evident given that Miller's *The Garden as an Art* is an argument for the claim; and
4. that disinterestedness is not a true impediment, as we will argue in detail later;

then the arguments surveyed here do not make an adequate case for why The Garden cannot be considered a bona fide artform. They may explain why The Garden has not been seen, at least over the last century, as an artform, but they do not present sufficient reason for the conclusion that The Garden should not be seen as an artform. What we need to examine now are the differences between The Garden and other artforms.

How The Garden Is Different from Other Artforms

Some of the items addressed below we simply may have to accept as differences between The Garden and other artforms—and if those differences seem great enough to eliminate The Garden from being a bona fide artform, then so be it. Some will demonstrate opportunities for us to stretch our views in principled ways about what art is; that is, some of the differences cast light on issues in art that may have hitherto gone unnoticed but which, once seen, offer opportunities to understand the nature of art more deeply. And some differences may require us to expand our art theoretical views, to enlarge our concepts of what properly counts as art and perhaps to alter our definitions or theories accordingly.

A Final Single Form and Objecthood

Let's begin with the differences Miller identifies, and let us begin with the one that ultimately presents the greatest challenge. Miller points out that gardens do not admit of a single final form that is unique and perduring. The Garden, unlike works created in many artforms (though not necessarily all), is never in a completed state; it is never at a point of perfection that is stable enough to say that a particular instantiation of a particular garden is the artwork which is that garden. The Garden is "objectless" in the sense that we cannot point to a single tangible "time-sliced" version of a garden as "the garden."

Part of the problem lies in the influence of "Edenic Culture," the notion that we both do and should think of The Garden as continuous in concept with notions of the Garden of Eden and the idea of everlasting life. In the Hebrew text describing Eden, we get a strong sense of that garden being fixed in time: Adam and Eve were originally meant to live forever, partaking of the fruit of the Tree of Life, and Eden was meant to be a constant context for their lives. Eden was to be a perfect place, so essentially it could not change (even though there was movement and growth in the garden), for a change would mean that perfection had been breached. Those cultures that were inspired by the Eden story—and that certainly includes movements within all the Abrahamic monotheisms— were (or are) burdened with a notion of perfection as being permanent and unchanging, or being an objectifiable "time-slice." This notion has made its way into our aesthetic conceptualization—that once perfection is reached, it must be frozen to remain perfect. But the garden does not do that and is not that. The garden's perfection is about change, about decay as well as growth, and that must be embraced as we jettison the opposite notion, a notion especially plaguing the West as we read the metaphorical as the literal.[20]

Miller writes:

> [T]he environmental nature of the garden is in direct conflict with its status as a work of art for two reasons. First, art depends upon virtuality, the creation and manipulation of illusion, while as we have seen, the essence of an environment is actuality. In spite of the conflict between these two contradictory sets of demands, however, both illusion and actuality are essential, inseparable aspects

[20] David Fenner, "The Aesthetic Impact of the Garden of Eden," *Contemporary Aesthetics* 20 (2022), contempaesthetics.org/2022/03/24/the-aesthetic-impact-of-the-garden-of-eden/.

of gardens. . . . Second, the environmental nature of the garden violates one of the most cherished preferences for a discrete *object* of aesthetic experience.[21]

It is the second reason on which we most will focus, but let's begin with the first. It is the fact that The Garden is Third Nature and not First that demonstrates that The Garden is indeed an amalgam of actuality (nature unto itself) and virtuality (the imposition of organization following the vision that steers the artist's (s') expression). Absent design, a "garden" is "wild nature" and no garden at all. Even the most modest gardenist incorporates, eliminates, and organizes what components there are in their garden in ways that are purposeful. They may not have a design worthy of interpretation and criticism—and the modest garden may not rise to the level of being a work of art—but the organization applied creates something that was not there before: a garden. Amplify those same efforts, and the result can be truly impressive—amplify and rarify the purposefulness of the organization to the point where it constitutes a vision that drives an instantiation that we recognize as a true expression, and we have a work of art.[22] In other words, the purposeful organization of nature moves First Nature into Third, and the purposeful organization that may elevate a garden to be a work of art is on a continuum with all purposeful organization that, through application, brings a place from First to Third Nature. When Miller talks about "illusion and actuality" being "essential, inseparable elements of the garden," she contributes to the definition of The Garden; when that illusion is taken forward, we have the possibility of a work of art.

The second insight—that The Garden is objectless—presents the more formidable challenge to the claim that The Garden can and should be regarded as an artform. Throughout this book we have used the word "object" to refer to a garden, and we will continue to—the reason for this is that we need some ontological term to identify the discrete and bounded "thing" that the garden is. Even here, "thing" is a mere placeholder for "object." Yet "object" seems ontologically richer than we can afford if we understand by "object" a thing that is stable and endures change only beyond the typical human's ability to notice it. Everything changes—a point Heraclitus made—but to know a thing is to know it as stable and unchanging—a point Plato made. So in perhaps a trivial way, there are no objects that may be known through one's senses that are Platonically stable. This is especially the case for plants. If we are asked to picture a redwood, we call to mind ancient towering trees along the California coast, and not the redwood as a sapling or germinating seed, even though both images are equally valid. But

---

[21] Miller, *The Garden as an Art*, p. 57.

[22] Recall Miller's definition of "great art" as possessing "excellent form, significant human content, enduringness, and adequacy of form to content." Miller, p. 141.

we still use the word "object" to denote things that change more slowly than presents a risk to our referring to them or knowing of them or about them—with apologies to Plato. Yet even in this modest sense, to talk about a garden as an object is to take liberties.

Besides using the term "object" to refer to a garden throughout this book, we are also using "object" to stand for both "object and event" in reference to aesthetic and/or art objects. An event is a series of changes to three-dimensional objects as they move through the fourth dimension, the key word being "change." If a garden is not an object in the important sense of the word, is it an event? Likely this is a reference with problems of its own. For an event to be the focus of an aesthetic experience, it must include boundaries in time—in large measure so that the internal coherence of its passage through time can be assessed. Aesthetic experiences—following Dewey—have narrative arcs that have starts, middles, and ends. So a *visit* to a garden may count as an event and may—probably typically—count as an aesthetic experience. If we push this approach in thinking about a garden (as a place) then we may end up with that garden being an event whose start was occasioned with its original planting (supposing that kind of garden) and whose end occurs when, absent care, it reverts back to First Nature (or some semblance thereof). The chances that humans would even notice that end—much less be able to pinpoint it in time—are remote. Worse, very few gardens are ever experienced wholly through their lifespans. Perhaps some pavement gardens, some allotment gardens, telegardens—but not many more. So pressing the view that a garden is an event may present more problems than it may solve.

In an earlier chapter, we talked about how gardens are experienced as bound in human scale, both geographically and phenomenally or perspectivally—which is to say, for this second type of boundary, that one's experience of a garden is both bound in time and bound in terms of what sensory stimulations can be attended to. For instance, our field of vision is limited, and only so much may occupy it, and as we move through the garden, our visual perspective only lasts in time as long as we linger in place or until such time as we subjectively identify our visual field as different from what it was a moment ago. In addition, as we learned from Stolnitz,[23] attention is selective. What presents itself to our visual field is not what we see; what we see is what we attend to in our visual field. This is to think of the "object" that is the garden as the "object" or focus of an experience. Essentially this is a riff on thinking about a garden as an event, but it locates the "object" functionally (rather than ontologically) as "*of an experience*" rather than of "what-there-is."

---

[23] That is, "learned" in philosophically important ways. Jerome Stolnitz, *Aesthetics and Philosophy of Art Criticism* (New York, NY: Houghton Mifflin, 1960).

Does this help? Ontologically, it does not. Ontologically the best we might hope for is to classify a garden as a place, one in three dimensions, moving through the fourth, but bounded in all. This does some good in terms of understanding the nature of The Garden, especially as we amplify Hunt's insight about "place-making."

But our focus now is not on ontology but rather on the claim of whether The Garden can and should be thought of as an artform. Thinking of a garden in phenomenological terms may be useful to thinking about that garden in aesthetic ones. Working from the view that to be a work of art includes some necessary reference to an object possessing aesthetic properties, and the possession of aesthetic properties includes some necessary reference to the application by humans of interpreting objective features to discover (or create) aesthetic ones (at least perceptually based or narrow formal ones), capturing a garden as a focus of an experience may be useful.

If through realizing there is no fixed, stable, unchanging object which is a garden—because of the necessary dynamism of its character—we are still able to identify a garden in these functional or phenomenological ways, we may preserve our ability to talk about a garden as an object in just those ways necessary to understanding The Garden as possessing enough "objecthood" to allow for (at least some) of its tokens to be thought of, at least potentially, as works of art.

Put another way, if the claim is that a garden cannot be a work of art because it does not have stable objecthood—in Miller's words, a "cherished" preference "for a discrete *object* of aesthetic experience"—the reply can be that the object of an aesthetic experience is not a four-dimensional bit of discrete ontological existence but rather the perspectivally, attention-bounded sensory focus of an experience—an object "virtual" ontologically, but perfectly real experientially. And it is the experiential object that is important to the aesthetic point of view.

An implication of this is that we might be forced to accept the damage this approach may entail to our efforts to compare gardens one to another. If a garden is, in terms of its objecthood, simply a phenomenon whose existence is entire in an experience, it is unclear how we are to make comparisons among gardens for the purpose of evaluation or for adjudication of the adequacy of a particular interpretation of a garden. This approach may cost us normativity with regard to judgments about the aesthetic quality among tokens of the type The Garden, and that would be in itself a tremendous blow to the case for The Garden being a bona fide artform. Moreover, we do not see how to avoid this problem. That is, the presence of the problem is not occasioned by our proposed solution to the objecthood problem but rather is the result of Miller's original assertation—which we take to be eminently correct—that the garden is not an object in the stable ways needed for aesthetic analysis and disposition.

Our proposed solution to this problem is to embrace indexing judgments to identifiable experiential patterns. While no visit to a garden will be like any other visit, if we take the implications of the phenomenology seriously, (1) some visits will be more like other visits than third sorts of visits, and (2) subjective taste patterns can be identified sufficiently. Consider the following scenario:

"Have you been to Kew?"
"Yes; it's incredible."
"I thought the fields of bluebells were awe-inspiring."
"Bluebells? You're sure you were at Kew? That area you're describing is just a meadow."

The mystery is solved by indexing the conversation to a certain time of year. So the ensuing judgment that works for the sake of comparison is similarly indexed: "Kew is wonderful in the spring time" or "Kew is wonderful on a sunny day in the spring." Further, this sort of indexing likely will include reference to the taste profiles of the persons involved in the conversation: "I know you love botanical gardens. . . . Kew is wonderful on a sunny day in the spring."

What is lost is absolute normativity, but absolute normativity was lost the moment we recognized that there is no unchanging object to which we can refer when talking about any garden. This loss is not catastrophic. Aesthetic judgments are typically or at least frequently indexed to taste profiles. This is the case with even the most seasoned critics we may read; they do not all agree, and we may agree with some but not others. The best explanation is divergence in taste, and this is readily accounted for by indexing comparisons, and the judgments that result from them, to groups who share similar tastes. A botanical garden is different in important ways from a pleasure garden, and some people prefer one over the other. We can account for that in the way we frame our recommendations. Second, we commonly index our comparisons among aesthetic events in terms of features we find salient to judgments of quality. One event may be like another of the same kind in important ways—same performers, perhaps, same artistic interpretation, same venue—but as events, they lack the stability that objects (per se) possess. So in offering a recommendation, we may include caveats such as "I think you would love listening to Berlioz's *Requiem*, but make sure you get the recording from X, performed by Y."

Granted that the phenomenal nature of a particular experience of a visit to a garden involves more "moving parts," more factors that contribute to the character of that particular experience. But even with such complexity, patterns are still identifiable that can then be used to underwrite the strength of recommendations—the greater the experiential similarities, the stronger the pattern, and the stronger the recommendation that is based on the pattern. Given that as a matter of fact we compare gardens, and we apparently do it without hesitation, the normative problem only requires a theoretic rather than a practical solution. Indexing judgments may be that theoretic solution, and we continue this thread in the following chapters.

Before leaving discussion of the garden's objecthood as a consequence of the "nature of nature," we should add one more point. Nature is not monolithic. In his essay "The Nature of Urban Gardens: Toward a Political Ecology of Urban Agriculture,"[24] Michael Classens, in discussing the "hybridity of socio-nature," writes:

> Less intuitively "nature" is not wholly "natural," and the supposed separation between things "natural" and things "social" is a great deal murkier than it appears.... Water is certainly scarce in many places in the world ... which is undoubtedly a pressing issue—but how that scarcity is dealt with, who continues to receive water, who does not, and who gets to decide this are all equally important. The people, institutions, policies and regulations that emerge to sort out crises like water scarcity reveal how (socio)nature, is mobilized as a means of reproducing conditions conducive to the reproduction of systemic urban inequality.[25]

So while a garden may not be an ontologically stable object, its context—nature—may not be relevantly stable either. That is, beyond the change that we expect in nature, the very way we conceptualize nature may itself be subject to change depending on the human conditions that contextualize that bit of nature salient to our focus. There is little to be done about this apart from embracing the increased complexity and attempting to account for it.

In sum, the fact that The Garden is "objectless" ontologically should not be a hindrance to the claim of The Garden to be an artform if we can redefine what we mean by The-Garden-as-object in aesthetically relevant experiential terms. The cost of this, as we saw, is that we must adapt other aspects important to The Garden's identity as an artform (if our case is successful)—namely the ability to create normative judgments about the aesthetic quality of a garden or the interpretation of a garden—by indexing our judgments (in epistemically naturalist terms) to discoverable patterns. There are complexities to account for, but complexities of this sort—even if not to this degree—are present in other canonical artforms, and their capacity for supporting judgments is not curtailed. Furthermore, given the commonality of comparing gardens one to another, the

---

[24] Michael Classens, "The Nature of Urban Gardens: Toward a Political Ecology of Urban Agriculture," *Agriculture and Human Values* 32:2 (2015), pp. 229–239.
[25] Classens, "Nature of Urban Gardens," pp. 229–231.

problem is clearly not with common practice but rather the theory that might explain it.

The Place and Role of the Artist

Miller points out that gardens do not admit of control by a single artist with full autonomy over the results of their creation, who can be viewed as the possessor (and controller) of a unique vision or expression that was the exclusive impetus for the coming into being of a garden. In addition, she points out that gardens are amalgams of the artificial and the natural, unlike most artforms which typically are entirely "plastic."

While gardens have a similar structure to works of art insofar as they involve an object, a creator of that object, and an appreciator of that creation, this is only a part of the story. A more precise rendering would be something closer to a five-part relationship: the garden is still the focal object, and the audience is still in the role of appreciator, but the creator of a garden is more complex than we find with typical artworks. The creator role is shared by three: first, the human designer, the gardenist, who creates the plan and perhaps is instrumental in installing the garden originally. Second is nature itself; the gardenist exercises only limited control over the fashioning of the garden. This is evident as the gardenist plans but also as they execute. Nature's role in serving as cocreator (we will qualify this expression below) is further evidenced by the fact of the existence of a third creative party: the gardener, the person who maintains the garden, who with nature steers the evolution of the garden, and who may one day be responsible for re-creating the garden, in part or in whole, in accord with the original gardenist's design or on occasion entirely anew. The gardener is not only cocreator with nature; the gardener is partner with nature throughout the life of the garden as this person serves as gardener. It is not as though "control"—such as it can be in a garden—is shared between the gardenist and nature and only between them. Gardens require more than two creators; they generally require many as gardeners come and go, as single gardeners become plural gardeners, and as gardens are redesigned.

The happy news for the case for The Garden as artform is that this structure—where there are a plurality of creators, not all of whom can even be identified at a given point in time during the life of the garden—has precedent to some degree in the artform of film. While there is some generalized support for the concept of the film "auteur"—the film's author or artist, typically understood to be the film's director—and while we can point to instances where film directors, through their signature styles and technical capacity for realizing their visions in their films, have risen to the level of "auteur," the reality is that the film theoretic literature has not done enough with this concept to make it more than an attempt to capture our intuitions about great directors who create great films with recognizable styles. That is, thus far no argument has been put forward that the concept of the "auteur" is a reliable mechanism for understanding the "artistic (or authorial) control by a (single) (human) agent" that Miller mentions. In reality, no film is created without a plurality of creators; early and "golden age" (or "classic Hollywood") films used to have credits—ones that could be seen in a few shots—that named those who partnered together to create a film. Today the "credit crawl" at the end of a film, even ones not particularly "big-budget," can go on for many minutes. While one might object that such a list includes caterers, drivers, and the like—and these folks should not be considered creators of the film—the vast majority of names on the "crawl" are of people who were intimately connected with the film's creation. As the "auteur theory" is still an amorphous matter in film—and in some circles downright suspect in its attribution of singular artistic control to the film's director—any attempt to translate that sort of approach to The Garden is doomed. Though we credit great gardenists in recognition of the genius of their designs, the influence they had on future garden design, or how their designs became signature styles of particular times and/or places—and while they rightly deserve such reputations—they are but one piece of a puzzle and arguably do not exert as much control over the instantiation of their (already nature-informed) design plans as do film directors.

Does art require single artists who exert exclusive control over every aspect of their creations? Archibald Alison believed that the greater an artform, the more control over the medium is exercised by the artist.[26] But Alison's point was not definitional about art; it was hierarchical. We have precedents for pluralities of artists working together to create single works of art—Andy Warhol's "Factory" comes to mind, as do bands who write and perform music cooperatively. Paul McCartney wrote and sang "And I Love Her," but it was George Harrison's guitar additions that made the song memorable. It would be especially hard to attribute the music of an improvisational jam band to a single author. An enhanced version of this model is at play with The Garden.

Does a plurality of artists working together dilute the original vision and expression of the artist who conceived of the idea for the work in the first place? That is, is it right to call a thing a work of art when the artifactual nature of that object did not arise through deliberation and decision by a single mind? It should first be noted that the requirement that a work of art arise consciously and volitionally from and through a human mind largely rests on the idea that to be a work of art is to be a human expression. This is what is key. But this requirement

[26] Michael G. Lee, *The German "Mittelweg": Garden Theory and Philosophy in the Time of Kant* (New York, NY: Routledge, 2013). pp. 106–107.

does not entail that a human expression must be singular in its conception, its instantiation, or its evolution. It may well be inertia from Enlightenment thought that we believe that values connected to the production of art, to epistemic justification, to the possession of rights, to political voice, to ethical responsibility (and so on) must all rest exclusively with the individual human being. We in the West today may see focus on the individual as self-evidently justified in all these sorts of cases, but as different approaches have existed in the past or exist in other places, the "self-evidency" erodes and, if we choose to hold "human-particularism" as our preferred approach, we need to argue for it. Absent such argument in the case of what counts as art, we are free to say that whether the human expression that is the work of art is singular or plural (or must be one or the other) in its inception and instantiation is an open question.

Must control over the medium be absolute for the resulting product to be a work of art? One of the writers of this book has a modest art collection that includes

1. a photograph that as it was coming out of a printer was caught in the machine and badly smeared—the resulting image is arrestingly vibrant;
2. a work of Japanese calligraphy in a "grass style" that includes drops of ink scattered around the rice paper as the brush moved quickly in the creation of characters;
3. a watercolor where the dilution of the pigment was such that the color ran down the paper;
4. a monoprint where, as the print was being made, the glass to which the paint had been applied shifted slightly;
5. an assortment of prints on paper that was not entirely rectilinear (and so gave framers challenges);
6. a work of stained glass where a small gap exists where the lead connecting two panes was not enough to fill;
7. numerous recordings of acoustic guitar music where the slide of the player's fingers along the strings is clearly audible;
8. a sculpture where one piece was inadvertently not attached properly and so "floats" on the piece;
9. a collage where a tiny bit of glue is visible; and
10. a few charcoal or chalk drawings where stray charcoal or chalk on the artists' hands created smudges on the paper.

One can read this list and conclude that the collection includes a lot of mistakes or is not high-quality, but while we cannot ensure the former is untrue, we can guarantee the collection is indeed of decent quality. Accidents and imperfections are frequently allowed by the artist to exist in the final work—in some cases they are unavoidable and expected; in some cases they enhance the aesthetic quality of the work. One could argue that leaving "accidents and imperfections" behind is still a form of artistic control when the artist notices them and chooses to leave them. But in those cases where the accident does not enhance the aesthetic or artistic value of the work, the better explanation is that the artist believes that the value of what was not an accident outweighs the decrease in value of what was—which is to say, the artist would not have included the accident had they been able to exercise full control. To read this as an instance of an artist retrospectively saying "I meant to do that" is more far-fetched than the artist realizing that something went wrong but choosing to embrace the lack of control the accident evidences.

In some cases, giving up control to the process is an essential part of the work. Gerhardt Richter has several series of paintings in which objects are dragged over a painted canvas to create less predictable textures and color combinations. The Tate Modern houses such a series, inspired by the music of John Cage, who invited a range of accidents and unpredictability into his compositions (scores that instruct musicians to tune to various radio stations, for example). Amplifying this point, Francois Berthier writes:

> A traditional condition for successful landscape painting in China is *qi yun sheng tong*, which refers to the artist's ability to let his work be animated by the same qi that produces the natural phenomena he is painting. So rather than attempting to reproduce the visual appearance of the natural world, the artist lets the brushstrokes flow from the common source that produces both natural phenomena and his own activity. This condition was easily adapted to the art of garden making, where the very elements of the artist's craft are natural beings, which are then artfully selected and arranged in order to reproduce harmonies in the natural world.[27]

Gardenists and gardeners make predictions based on patterns, and usually their predictions come true, but not always. In ideal circumstances, nature can be predictable to high degrees, but gardens infrequently present ideal circumstances. Gardens, to use the vernacular, have too many moving parts; all sorts of various forces are at work, and even the most seasoned gardenist or gardener cannot account for every possible eventuality. One could say this is a case of the gardenist/gardener having to share control with nature, but that borders on the unpleasantly anthropomorphic. It may be better simply to say that in a garden, control is an illusion.

[27] Francois Berthier, *Reading Zen in the Rocks: The Japanese Dry Landscape Garden*, ed. and trans. Graham Parkes (Chicago, IL: University of Chicago Press, 2000), p. 102.

"Sharing control" suggests too a kind of creative dualism between two op-
posing forces, and while some garden theorists have written along these lines,
a dualist conception of the relationship between "the human" and "nature" is
rarely useful.[28] First, humans are part of nature, and this immediately muddies
the waters. Second, nature is largely if not perfectly predictable, and humans
work through their knowledge about natural patterns. Humans do not fight
nature in gardening; we cooperate or adjust—or else "nature" will ensure there
is no garden. Enlightenment-era conceptions of humans bringing order to nature, subduing nature and making it perform as we desire were as wrong-headed
then as they are now—not because they cannot be, through massive effort, efficacious in constructing gardens of great beauty (as we perhaps see in Le Nôtre's
gardens), but because they cannot bring about the same degree of beauty without
cooperation with nature. So we relinquish absolute control for cooperation, in
much the same way that the artists who created the works described at the top
of the preceding paragraph relinquished absolute control over their works. No
one doubts that the works described there are bona fide works of art—at least
no one with the opportunity to inspect them and know their histories—and so
this approach of "cooperation-over-control" that is endemic to successful gardening should pose no serious problem for us as we think about The Garden as
an artform. Cooper writes:

> Such remarks [concerning how nature is so prominent in the shaping of a
> garden] hint at the breadth and depth of the thought that there is something
> profoundly wrong with a familiar image of the artist—the Promethean image,
> engrained among romantics and modernists alike, of the artist as an autono-
> mous "genius" stamping an entirely novel, individual and creative vision upon
> the world.[29]

It may well explain why The Garden has not received the attention, as an artform,
that other canonical artforms have received in that

- a garden does not have a single human being that stands in the role of
  "artist" but almost always a plurality;
- while likely a mistake to refer to "nature" as an "artist" (because this would
  imply a view of nature that is strongly teleological—even perhaps theolog-
  ical if works of art are purposed expressions—that may not be the case) or

---

[28]   See Cooper, *A Philosophy of Gardens*, p. 47.
[29]   Cooper, "In Praise of Gardens," p. 110.

even as "cocreator" (since this might suggest that nature has identifiable points at which a thing, like a garden, comes into being—rather than nature simply being continuous and constant), still "nature" stands in a role that has important equivalencies to "artist" and "cocreator;"

- a garden's design is not the product of the exercise of absolute control by the artist or artists;
- the design is only a generalized template to the instantiation of a garden as (1) accommodation to nature occurs and (2) natural change occurs immediately and constantly; and
- "control" as a concept may not be useful when it comes to nature and so by extension to gardens.

Yet none of these points stand as reasons for why The Garden may not be thought of as an artform. Other canonical artforms variously incorporate each of these aspects, and when in a particular case they do, we do not say that the resulting object is not a work of art on those grounds. So while these truths about the "artistry" of The Garden are relevant both to the apologetics of the situation and as they form a context for insights about the nature of The Garden and the nature of art, they do not stand as impediments to the claim that The Garden is or can be an artform.

Craft and Labor

Amplifying the point made directly above, and in a way that resonates with Ross's concern that gardening suffered the loss of prestige, Miller highlights the fact that gardens are places where the work involves as much "craft" as "creativity," or where, to put it slightly differently, time allocated to gardening is consumed by more sweat-inducing labor than fine artistry. Ross's concern with the loss of prestige is being addressed with increasing efforts toward, and norming of, training and credentialing in horticulture and landscape architecture, as we mentioned. But nothing will eliminate the fact that gardens require, in both their installation and maintenance, much back-breaking work.

Yet while some artforms do not require this, and their execution might even be described as delicate, there are artforms that require sweat-inducing labor not dissimilar from that required in the garden. Anyone who has ever attended an "iron pour" in the service of sculpture, the building of a kiln, or the throwing of clay knows that sweat is not foreign to the creation of these artforms. Those who erect the buildings that come to be known as works of architecture experience the same. Setting up props and lights, holding boom mics, and dragging cables around a film set also takes a physical toll. And so the presence of physical labor is not sufficient to exclude The Garden as an artform.

The same sort of case could be made if instead of focusing on the hard work associated with gardening, we instead focus more narrowly on the character of the work as involving more craft than creativity. The answer is the same as in the case of hard work: many artforms involve technical manipulation that sometimes dwarfs the creative effort invested in the creation of a work of art. The same examples apply here: metal sculpture, ceramics, construction, and filmmaking.

Fiction

Miller writes, "For insofar as the garden represents an ideal world or an ideal life, it seems by its very existence to prove that world valid."[30] She also writes that a garden is always an actuality—one conjoined necessarily to an "illusion," but still one that by its very existence shows the world, shows nature, as it is. It would seem that no garden can be a fiction and still be a garden. Gardens can represent what is unreal (unicorn topiaries, mythical places, or *kiku*-style chrysanthemums trained to have 1,000 blooms, for instance); they can express ideas that are fictional (as Little Sparta and/or the Garden of Cosmic Speculation may); they can, like Stourhead, be emblematic of a fiction (Virgil's *Aeneid*). And they can certainly be the subject of other artworks—paintings, novels, poems, and so forth—that conjure up fictional gardens as their subjects or parts of their fictional contexts. But a garden itself cannot be a fiction; it is a four-dimensional reality. And if we were to talk about a garden being fictional—perhaps as we express some appreciation of its magical quality—we employ those terms in less than literal ways. The artifice, the illusion, the virtuality that a garden may display or convey may be very important to how we experience that garden and how we construct an aesthetic experience from a visit to a garden or time spent in a garden, but once that garden is instantiated as a work of nature and in nature, it is real—despite that it retains its ability to engage us fantastically, it is real enough that to call it a fiction per se is to say something beyond what is warranted. This is a case that seems an opportunity for art to rise to meet The Garden. Not all works of art can be fictions, and if this is only the case for The Garden among all other artforms, then that is simply a fact.

---

[30] Miller, *The Garden as an Art*, p. 60.

Negation

Miller writes:

> [T]here are no perverted gardens, as imagination would seem to require: The garden has no way to trap man in his body to his discomfort and humiliation, because the terms are set up and resolved on the physical level. "Perverted gardens," gardens which are set up to be horrid, to exemplify the opposite of everything we want and expect in a garden, might seem a good idea—but in fact malice is attributed only to fictional woods, not to gardens.[31]

On the face of things, it seems easy to imagine a scary wood (that resulted from design), a garden set up like a haunted house, a garden with dead and dying components (as was sometimes incorporated in the Picturesque style of eighteenth and nineteenth century), or a garden composed only of black and dark purple components. It seems easy to imagine a postmodern gardenist taking Miller's words as a challenge, too.

Despite this, nature does not come imbued with values, good or bad. Nature simply is. To quote Hamlet, "there is nothing either good or bad, but thinking makes it so." The unwanted elements that all gardens have are not unwanted by nature (so to speak anthropomorphically about nature); they are unwanted by humans. The Alnwick Garden in Northumberland contains a Poison Garden whose cast iron gates have "THESE PLANTS CAN KILL" written alongside skull and crossbones. But what is deadly to us is to the plant a means of survival. When a plant dies, nature does not mourn; only humans—who spent money on the plant and fertilized and watered it—do. When a disease comes in, nature does not take pains to stop it; humans do. When winter leaves the trees leafless and kills the perennials back to the roots, not one of the plants is sad. To think otherwise is either to think of nature as highly teleological or simply anthropomorphized. So if a claim is advanced that a particular garden is perverted or expresses some negation, that claim rests exclusively on a subjective contribution, on the subject's point of view and transference of an attitudinal property to the garden (as object). Unfortunately this commits us to the complement: that there are also no "good gardens," but rather there are gardens that we think to be good. (More— much more—on this topic later.)

Works of art commonly are thought to be expressive in themselves—we do not say only that we feel sad listening to a piece of music but rather that a piece of music is sad. We attribute the expressive property to the object in the same way we attribute representational properties to them. This may be, to the formalist, taking liberties, but it is common. This opens two paths to us: one is to say that

---

[31] Miller, p. 129.

all expressive (and likely representative) properties of works of art are subjective contributions to what is objective, or to say that gardens uniquely must rely on subjective contributions for all ascriptions of properties that do not derive exclusively from their objective properties. The former claim is too big to tackle here; the second may put gardens a bit out of sync with other works of art. Given Miller's claims and our own inclinations, we elect the second path. While other works of art may express negation (as a property of the work itself), gardens cannot.

Immobility and Physical Context Dependency

While Cooper says he sees no reason why a garden cannot be moved[32]—he uses the example of the Chelsea Flower Show and cites his intuition that once it has run its course as a show it might be moved to a different part of England—much of the rest of his thought about the interrelations of The Garden with nature seem strongly to imply that gardens really cannot move without being different gardens. The Chelsea Flower Show is an interesting example because its immobility is overdetermined by virtue of its name: the "Chelsea" Flower Show is in Chelsea (technically, the Royal Borough of Kensington and Chelsea, London). But even absent this fun but perhaps less-than-significant point, once the various gardens that make up the Show are moved, they will grow differently than they did in Chelsea. Different soil, different sunlight, different climactic conditions, different surroundings—these will render the two gardens different in many discernable ways, some of which may alter the character of particular "garden rooms" or subgardens. The claim that "this *was* the Chelsea Flower Show" seems legitimate, but the claim "this *is* the Chelsea Flower Show" seems not only theoretically implausible but intuitively not right; it hits the ear oddly and is likely to provoke simple questions like "then why is it not in Chelsea?" Some works of art cannot move. The massive art installations that Christo and Jean-Claude create cannot move; works of architecture rarely ever move, and some cannot; some works of sculpture—Richard Serra's *Tilted Arc*, as we mentioned earlier—cannot move. So while the immobility of a particular garden may strike one as exceptional in the world of art where artworks commonly are taken from one place to another, it should give us little cause for concern that gardens cannot move.

Modification

With most works of art, especially autographic ones—as The Garden is—modification once the artwork is completed is prohibited. Gardens are singular

---

[32] Cooper, *A Philosophy of Gardens*, pp. 15–16.

creations, and so it might be expected that one would argue that once a garden design is instantiated, there should be no modification of it. But, of course, the one overwhelming constant in The Garden is change. It is not only that the essence of nature is change, but that the original design of a garden, through human hands, will change as well. Weeds, seedlings, damage from the elements, plant diseases, shade appearing as trees mature, shade disappearing as trees die—all sorts of garden events occasion times when gardeners must make changes to a garden, even changes that alter in noticeable ways the garden's original design. Creation in the garden is not a "one-off" but rather a process that continues through a garden's full lifespan. Given this, prohibitions against modification must be reduced only to generalities about keeping intact a garden's motivating purpose and keeping intact as much of an original garden plan or design—especially an exceptional one—as is practically possible, with the key word being "practically." New York City's High Line, whose planting scheme was designed by Piet Oudolf, is set in/on a disused elevated rail track. Oudolf took as his inspiration the "novel ecosystem" of weeds that took root on the tracks and, with a team of landscape architects, modified the track into a one-and-a-half-mile linear park. The High Line is currently New York's most popular tourist destination, with eight million visitors in 2019. In fact, it has brought so much attention to the Chelsea neighborhood of Manhattan that a new community has developed around it, and now the developing luxury high-rises threaten to shade out the very reason they were built. The High Line plantings will inevitably change to accommodate the change in environment, or else the High Line will cease to exist as a garden.

Dance undergoes evolution and we embrace modifications in that artform when such modifications are necessary to maintaining the aesthetic quality of a performance given the size of the stage, the size of the company, the strengths of the dancers, and so forth. Interpretation of a score by musicians or conductors and interpretation of a script by actors or directors may qualify as modest modification, as well—although in those cases it might be better to think of artistic interpretation as a necessary complement to the score or script rather than a modification of the work of art.

What does it say about The Garden that prescriptions against modification must reduce to either conceptualities or be constrained by practicality? Here we see another case where "art" must rise to meet The Garden and not vice versa. To claim that since modification of artworks is not a fully tenable concept when applied to The Garden does not seem an adequate argument to discredit The Garden's claim to be an artform—modification prohibition is an epiphenomenon of art, not a motivating force, and it may be a relic of what Cooper, following Dewey, terms a kind of "museum-mentality" about art. Indeed, there are works of art produced over the last few decades that incorporate elements that involve change—such as Chris Ofili's 1996 *The Holy Virgin Mary* that incorporates elephant dung—and there may be more that involve not merely predictable change but rather audience-participant alteration; it would be odd indeed if such a trend were to change the way we classify those objects as works of art.

Completeness

No work of art presents an entire world sufficient unto itself. This is the case despite the strong intuitive appeal of theories that describe aesthetic value in terms of the capacity of an experience to engage us in so many ways—aesthetically, emotionally, cognitively, psychologically—that it constitutes for us another world to which we can be transported.[33] Worlds presented by artworks are incomplete. They must be because they must fit within the pages of a novel or poem; within the frame of a painting, photograph, or screen; within the span of time of the performance or film run; and so forth. The only way they can constitute for us new worlds, complete enough for us to be transported to, is for us—the audience members—to contribute enough to round out that world sufficiently so that it may seem a complete world to us. This happens frequently with characters portrayed in art. They are not complete human beings; instead, we fill in those parts of the personality that the art object does not present with aspects that seem reasonably consonant with what is presented. Sometimes this means that we fill in with aspects of our own lives and personalities or with those of people we know who seem to fit the mold of the character presented.[34] This sort of "personal identification" with characters—but also with various other aspects of the artwork—is common and it not only enlivens the experience for us by employing our own associations but it completes (to a sufficient level) the world presented so that we may find it occupiable.[35]

Can gardens do this? Do gardens present themselves as opportunities for us to be transported to new worlds? Easily and readily. And perhaps even more so than the case with other works of art. Miller writes:

> Gardens present their illusions with a strength of conviction that is unique among the arts. This peculiar conviction is derived most importantly from the fact that they are demonstrably successful in a special way: they are—like us—alive. This fact, along with several other features—their vulnerability to their environment, their inability to express negation, the commensurability of their temporal and spatial scales with those of normal everyday life—make

---

[33] Alan H. Goldman, *Aesthetic Value* (Boulder, CO: Westview Press, 1995).
[34] The film theorist Andre Bazin discusses this phenomenon.
[35] David Fenner, "Imagination and Identification in Film and Photography," in Keith Moser (ed.), *Imagination and Art* (Leiden, NL: Brill, 2020).

the illusions presented by gardens more compelling than those of many other arts.[36]

The argument might be made that since gardens are actual places, and since thereby they include, within their human-scaled "attention" boundaries, everything sufficient to make full and complete worlds, they do not require the subjective contributions common to other canonical artforms. While some may find such an argument compelling, we find more compelling that many—perhaps most—gardens present ready opportunities for "escape" from the "real world" in terms of laying aside everyday cares, in terms of inspiring imagination, and in terms of presenting opportunities to be cognitively, emotionally, aesthetically, and psychologically engaged, even arrested. We imagine ourselves to be explorers as we duck and weave through a densely planted trail; we imagine ourselves to be Aeneas as we walk around the lake at Stourhead; we imagine a ring of mushrooms growing in a lawn or an unplanted bed to signal the presence of fairies; we imagine what it is to be an aristocrat as we survey a Brownian landscape or the king of England as we look out the windows at Hampton Court Palace or the king of France at Versailles; we lie on the grass and imagine we are small dogs or cats or even bugs; we recall the emotions of playing a game of croquet with family; we think of the purple impatiens that used to ring the oak trees in the front yard many years ago and we are transported back to those days. Though they are instantiated as physical places, complete with a full array of physical objects and obeying the full slate of physical laws, gardens as Third Nature all present an illusion, a virtuality, that provides the platform for transportation. That transportation is only accomplished for the subject, as an experience, through the subject's contribution of imagination. Absent that contribution, Third Nature just looks like Second—or worse, First—Nature.

This is why, incidentally, the "doppelganger" argument can only go so far. This argument says that there could exist a piece of "untouched wilderness" physically identical to a garden, and without knowing the external fact that one was the result of design and the other was not, we would not know the difference. The argument may be used to show that a garden is not a place but rather an intention taken toward a place.[37] While a site that is Third Nature may be physically identical to one that is First Nature—an idealized situation that may itself never be practically instantiable, frankly—this only shows that a garden is not merely its physicality. On the other hand, one should be careful about imposing upon the object which is the garden properties that are better understood as subjective contributions, as we have been discussing. Gardens are amalgams of nature

---

[36] Miller, *The Garden as an Art*, p. 178.
[37] Cooper, *A Philosophy of Gardens*, p. 18.

and artifice, of physical reality with subjective uptake and contribution (as are all works of art), of a place regarded in a certain way. To claim that the garden itself is both real and virtual is more than we need to claim; to claim that a place is only ontologically what it is due to how it is regarded is also too much. An amalgamated, cooperative approach best solves the theoretical problems and more accurately captures in functional and pragmatically useful terms the creation, sustaining, and appreciating of gardens. Arguments for this claim abound in this book.

Is the Kind "Garden" an Artform?

As we saw, Miller's treatment of the question of whether The Garden is an artform focuses largely on the differences between The Garden and other artforms. If any of these differences, or combination of these differences, is insurmountable, this may be grounds to abandon the case that The Garden is a bona fide artform. Claims of differences we examined—many of which were discussed by Cooper, Ross, and Miller—included the following:

1. Gardens, or at least good gardens, engage us merely in (formal) aesthetic terms and do not engage us cognitively.
2. Gardens defy our preference for decontextualizing works of art by placing them in museums and the like.
3. Gardens require so many resources that great gardens—ones with perhaps the best chances to be regarded as works of art—become so rare as to be relegated easily to the fringes of the world of art.
4. Gardens are so susceptible to design change that it is difficult to ensure the continuity of the original designs of great gardenists, designs that might have been central to their cases to be classed as works of art.
5. Gardeners, as a class of individuals, have enjoyed reduced prestige as their work has been seen to involve more sweaty labor and craft than the delicacy and creativity that might be expected of artists involved with "fine art." This is exacerbated by the fact that gardens are "material-forward."
6. Gardens are typically "purpose-forward," most possessing a variety of purposes, both designed and incidental, that make it difficult to appreciate them in the absence of recognizing their instrumentalities. If art is meant to be appreciated "for its own sake alone," or from a disinterested perspective, this approach does not work in the cases of most gardens.
7. Gardens do not admit of final single forms. Gardens are necessarily dynamic, and this puts them out of sync with the common expectation that works of art, once finished, do not change in significant ways. This connects to the expectation that with typical works of art, modification,

once the work is complete, is not allowed—a concept at odds with garden dynamism.

8. Gardens do not admit of control by a single artist with full autonomy over the results of their creation, who can be viewed as the possessor (and controller) of a unique vision or expression that was the exclusive impetus for the coming into being of a garden. In addition, gardens are amalgams of the artificial and the natural, unlike most artforms which typically are entirely, or at least primarily, "plastic."

9. Gardens are real places in four dimensions. While a garden can be the subject of a fiction, a garden cannot itself be a fiction. This connects to the insight that if we believe that the value of works of art is either wholly or partly about their ability to suggest and perhaps transport us to new worlds, a garden may be seen as lacking this capacity since the garden is a complete real world unto itself.

10. Gardens cannot express negation.

11. Gardens cannot be moved; they are dependent on, and fixed to, their physical contexts.

We have attempted to address some of these points, and we have promised to address others more thoroughly as we move through this book. Some of the items, however, may strike some readers as irreconcilable to the "normal" character of canonical artforms. Yet none of these items seems ultimately problematic to the claim that The Garden is and should be regarded as a proper artform. They either are so peripheral to the essential character of what makes an object a work of art (if there is such an essential character) as expressed in various theories about the nature of art, or they are so minor in comparison with other aspects of such theories, that they do not seem to present much of a challenge.

On the other hand, this may be an opportunity for growth in terms of how we regard the nature of art itself. The appropriate response may be to say that works of art can indeed include, for example, objects and events that permit of modification—an easy thing to imagine—and do not permit of the expression of fictionality or negation. The latter two seem of greater importance than the first, given that they may exclusively apply in the case of gardens and given that acceptance of the matter requires us to limit the full range of what an artist may express through a garden. If works of art are human expressions par excellence, to limit their range seems unfortunate at the least.

Throughout this chapter we have repeatedly retreated to the notion that gardens as works of art must be thought of in terms of objective and subjective amalgamations. To think of them as either one or the other—in the absence of the other—is to mischaracterize them. Such mischaracterization results in paradoxes in some cases, but it results in a lack of appreciation for the reality of the situation in all. Given this approach, we need not limit artistic expression in The Garden; we need only think of it as part of the subjective contribution offered by, in some cases, the gardenist or gardener (if we accept intentionalist approaches to understanding art or its meaning) or by the audience member— the garden visitor—who makes the garden the focus of an experience, an experience that is "added to" by that experiencer. Gardens can be scary—because thinking makes them so—and gardens can through the addition of imagination become platforms for all sorts of fantastical engagements. While thinking in these amalgamated terms may be special, even unique, to the way we think about gardens, there is every possibility that the best way to think about all works of art is in amalgamated terms of objectivity coupled with subjectivity.

So far, we have encountered the following in support of the claim that The Garden is an artform.

1. There are instances of gardens that arguably fit mimetic theories of the nature of art.
2. There are instances of gardens that arguably fit expressivist theories of the nature of art.
3. The Garden was considered an artform from the high renaissance into the nineteenth century in modern Europe.
4. The Garden was and continues to be regarded as equal in artistic esteem to other artforms in the East, from the Mid-East to the Indian and Pacific Oceans.
5. 1 and 2 imply that The Garden can fit into contemporary "continuity theories" of the nature of art.
6. There are instances of gardens that engage us cognitively, that reward the sort of cognitive engagement that typifies interpretation and the discovery of meaning, and whose value may be partly connected to their capacity for, and in some cases design for, cognitive engagement. This more fully will be evidenced in the forthcoming chapters on interpretation.
7. Some gardens are such that they can be compared to other gardens or to some garden idealization in ways that are typical of art evaluation and in art criticism. While already introduced, this too will form the basis for a forthcoming chapter.
8. Being careful not to fall into the "assimilationist" trap that Cooper warns us about—reducing appreciation of The Garden to appreciation of some other artform—we found that The Garden shares in common with other artforms the following items:
   a. Gardens, like all other artforms, are designed. Their "artists" are gardenists and gardeners. The creative path garden artists travel is structurally the same all other artists traverse.

    b. Gardens, like all other artforms, have audiences.

    c. Gardens, like many other artforms—and perhaps all, depending on the theoretical approach adopted—are "aesthetic-forward" and all reward aesthetic attention to a significant degree.

    d. Gardens exhibit style, with "style" denoting possession of a form that has distinctive aesthetic markers that identify it as coming from the domain of a particular vision for that form.

Miller's treatment of the claim that The Garden is an artform includes the following:

1. Gardens exhibit the sort of contrastandard properties—to use Kendall Walton's characterization—that works of art, as compared with previous works of art, include: "the role that breaking with precedent plays in gardens is much closer to the role it plays in other arts that in other kinds of social institutions."[38]

2. Gardens do not exhibit evidence of the skill that necessarily went into their creation; this lack of evidencing Miller takes to be common to art.[39]

And most importantly,

3. Gardens exhibit what Miller occasionally refers to as an "excess of form," and she connects this to Susanne Langer's description of "organic form" that is at the heart of Significant Form, which Miller takes to be "the characteristic feature of art."[40]

She writes:

Yet some gardens are unquestionably fine art—created by persons recognized as artists . . . , highly creative and self-conscious and deliberate in design and execution, slave to no purpose, masterfully exploiting the artistic materials to full aesthetic effect. If even these gardens—Stowe, Ryōan-ji, Stourhead—are ignored in discussions of fine arts, there is something wrong with the way we are understanding—or at least defining—art.[41]

---

[38] Miller, *The Garden as an Art*, p. 16.
[39] Miller, p. 84.
[40] Miller, pp. 15, 50–51, 124.
[41] Miller, pp. 90–91.

In the final analysis, we believe there is greater reason to accept The Garden as an artform—and say that some gardens are in fact works of art—than to do the reverse. We began this chapter by pointing out the benefits of regarding The Garden as an artform, and since we see no overwhelming reason to reject The Garden as an artform, it seems value is served by considering it to be one.

When Is a Garden a Work of Art?

A garden is a work of art when

1. its aesthetic properties, broadly construed, cohere together in a form that optimally rewards aesthetic attention, with "reward" understood in the highly plural terms that typify the many various ways in which audience members find value in aesthetic attention (more on this in the chapter on evaluation and criticism);
2. it exemplifies the character of art as described in theories of the nature of art, for instance mimetic, expressivist, or "continuity" theories;
3. it rewards typical interpretive attention and in at least that sense is cognitively engaging (more on this in the chapters on interpretation);
4. it exemplifies "style" as defined above; and
5. it is profitably the subject of structured and principled art-critical attention.

Not every garden that is a work of art will participate in these five criteria to the same level or in the same ways, but every garden that is a work of art will connect to each of these five in some degree. Gardens that are great works of art will be exemplary in some combination of these criteria.

The first criterion is the most basic, and on the surface it may seem as though it likely carries little weight since we already claimed in our definition of a garden that all gardens are "aesthetic-forward," a claim we repeated at the top of this chapter. Nonetheless, this criterion is significant if (1) we focus on the integrity of the form wherein all the aesthetic features of a garden cohere; (2) we separate those theories that include a required focus on aesthetics in their characterization of the nature of art from those that do not; or (3) if we take the "aesthetic-forwardness" of gardens to mean that all approaches to gardens are like all approaches to any work of art (or potential work of art) insofar as we at least initially focus on those properties of the object or event that are immediately present to our senses (or so derived). The point of this first criterion is not to repeat the claim that all gardens are aesthetic-forward but rather to claim that some gardens exhibit formal coherence of their aesthetic features in ways that provides the sorts of rewards we expect—and typically receive through attention to—works of art (or at least those whose aesthetic character is important to their overall character as art). (As we use the terms here, "form" and "style" are closely related, with the later dependent on the former, but since they mean different things, we include the fourth criterion as a separate one.)

The second criterion may strike some as unnecessary, in the technical sense of the term, since theories of the nature of art seemingly are always being challenged by the introduction of work that is accepted as art but fails to meet the conditions of the theory under consideration. Yet most teachers of philosophical aesthetics default to some theory of this sort when answering a student's question about whether some object or event is a work of art. This move may be replicated when one asks about whether a particular garden is a work of art, and so, because it is so common and intuitive to make such an appeal, we include this criterion on our list. In an important sense, for a garden to have passed the test of the applicability of some theory of the nature of art is to have already proven the claim at issue.

The last criterion may be tautological of the first and third. That is, if the garden in question is both aesthetically engaging and interpretively/cognitively engaging, then it is automatically "art-critically" engaging.

We court controversy by attempting the creation of a list of gardens that qualify as works of art, yet some examples are perhaps less controversial than others. We would include on such a list the following (alphabetically by country or US state):[42]

1. Butchart Gardens, Canada
2. Zhuozheng Yuan, China
3. Sissinghurst, England
4. Stourhead, England
5. Claude Monet's garden at Giverny, France
6. Versailles, France
7. Columbus Park, Illinois
8. Villa Medici, Italy
9. Saihō-ji (also known as Kokedera), Japan
10. The Het Loo Palace gardens, Netherlands
11. The MET Cloisters, New York
12. The High Line, New York
13. The Garden of Cosmic Speculation, Scotland
14. The Alhambra, Spain
15. Robert Smithson's *Spiral Jetty*, Utah

---

[42] This list was constructed partly with the aid of Linda Chisholm's *The History of Landscape Design in 100 Gardens* (Portland, OR: Timber Press, 2018).

There is little question that each of these gardens is remarkably beautiful (even *Spiral Jetty*), but what sets them apart from being "merely" beautiful is, first, each is highly integrated, with ranges of formal features that cohere to a high degree, and, second, each rewards cognitive consideration. While being interpretable may not require more than an account of how a garden's formal features cohere to a high degree, in each of the cases on this list, interpretive accounts may be created that reward sustained cognitive consideration. Of the five criteria above, interpretability may be the most central in marking out a garden as a potential work of art (a point alluded to earlier in this chapter). While interpretability cannot be reduced to "saying something" or communicating a message, all interpretation involves an articulation of meaning—or, to use Cooper's expression—an explanation. Again, more on this matter in the chapters on interpretation.

In the next chapter, we dig into the principal focus of this book and consider what it means to appreciate a garden aesthetically. There we find that our greatest challenge is one we introduced in this chapter: how do we appreciate an "object" that is so dynamic that its very "objecthood" is perpetually unstable? This is a focus for our chapter on evaluating gardens, as well. In this second chapter we began to sketch a way to deal with the objecthood problem through focusing on the garden as an object of experience rather than as an object-in-the-world. This approach entails a variety of challenges, and our proposed solutions require a level of complexity that we do not encounter with nondynamic or "dynamically closed" aesthetic objects, but our view is that this complexity cannot be avoided if our common-sense intuitions about aesthetically appreciating gardens are to be accepted. Our goal for the next chapter is, where possible, to keep complexity to a minimum, and where the complexity cannot be avoided, to explain it and justify it thoroughly.

5

Formal Aesthetic Properties of Gardens

Appreciating a garden aesthetically—just as in the case of considering whether The Garden is an artform—has characteristics that are like appreciating anything aesthetically and characteristics that are special to gardens. In this chapter, as in the last two, we want to cover both.

The case we attempted to evidence in the preceding chapter ultimately was a positive case for The Garden being an artform. The differences between understanding The Garden as an artform and understanding any other canonical artform that way amounted to matters that were explainable by exploring the character of that difference, realizing that some differences do not rise to enough importance to disqualify The Garden from being an artform, recognizing that even though differences exist they are not salient features in applicable theories of art, or adopting an approach to The Garden that takes to heart that gardens must be understood in terms that combine both the objective and the subjective. This last point will be relevant to the examination of experiencing gardens aesthetically as well.

Just as each garden is a unique amalgam of nature and design, so each is also an amalgam of the objective—the physical, four-dimensional garden itself—and the subjective, how we who create and experience gardens regard them. As both these syntheses are important to defining The Garden, so they continue to be important to understanding how to approach the garden aesthetically. That is, if we have established that The Garden has a character that is stable across all versions—the task of Chapters 1 and 2—then that character will be relevant to all considerations. Evidence of this is that we had to invoke the objective-subjective amalgam to begin to address the problem of the "objecthood" of The Garden that arose in the last chapter, a matter we return to below.

We begin our examination of what it means to experience gardens aesthetically by first asking what it means to experience anything aesthetically. We begin this way because we need a framework through which to approach the chapter's motivating question. The story we tell in sketching this framework is meant

Pond Garden, Hampton Court, Molesey, East Molesey, UK
Multiple gardenists, developed c. 1535, redesigned c. 1890
Photo by David Fenner, Summer 2011

The Art and Philosophy of the Garden. David Fenner and Ethan Fenner, Oxford University Press.
© Oxford University Press 2024. DOI: 10.1093/oso/9780197753590.003.0005

to capture common experience and describe it comprehensively. It is only one possible version of the story, however, and on occasion we will note where the story might proceed on a path different from the one we take. The theoretical commitments inherent in the path we take likely will not have a large impact on the points we make about garden appreciation. Nonetheless, we must begin with a framework, and ours is described below.

What Does It Mean to Aesthetically Appreciate Anything?

"Appreciation" is a decidedly subjective thing to do—with "subjective" referring not to the presence of partiality, as it is sometimes commonly used, but rather, as Kant used the term, to indicate a location, in this case the "appreciator." We use the word "appreciate" in both the common senses aestheticians use the word: first, as a synonym for "register" as in "I appreciate the gravity of the situation." Second, we use the word to refer to positive assessment or an acknowledgment of value, as in, when an astronaut who has been on the international space station for six months and is newly back on earth says, "I appreciate the gravity of the situation." Or perhaps better, "I appreciate the kindness you have shown." The two senses are conjoined, with the latter reliant on the former as both are about registration. It is more common for aestheticians to use the word in the former sense, without prejudicing the case as to whether the registration in question is of positive value, negative value, or neutrality.

When it comes to The Garden, to notice a garden's aesthetic features is primarily to notice its positive features. This does not mean that a visitor cannot notice a parterre curve that does not arc elegantly, missed weeds that muddy the clarity of the presentation of the flowers, dead leaves and faded blossoms, and the like. But these features, while doing the opposite of contributing positively to the aesthetic character of a garden, exist only as blemishes against a backdrop of positive aesthetic features. In other words, to notice the negative is first to have noticed the positive. Gardens that do not first present themselves as the objects of positive aesthetic value—say ones that are severely neglected or make the visitor feel uneasy—are generally not experienced as gardens at all (despite what they may be called). We will explore this claim in a later chapter. So, if this claim is true, "appreciation" when it comes to The Garden can be used in both common ways.

To aesthetically appreciate a thing is first to be prepared to appreciate it aesthetically. This can happen in a plurality of ways, two of which are:

1. one decides—consciously and volitionally—to adopt an aesthetic point of view about some object or
2. an aesthetic experience breaks in upon a receptive mind.

No one late for a court appearance is likely to have an aesthetic experience; no one rushing to the hospital is. But, on the other hand, to demand of the aesthetic phenomenon that it is only the result of deliberate attention volitionally applied to some object in a certain way is to ignore the experiences that many of us have had where "beauty breaks in," as we notice, say on our drive into work, a brightly colored cloud formation as the sun rises.

There are a set of theories, beginning in the early 1700s with Anthony Shaftesbury, that hold that one must adopt an "aesthetic attitude" in order either to be appropriately prepared for judgment of an object's aesthetic worth or to be appropriately focused on the correct aspects of an object to experience it aesthetically. Shaftesbury believed the only way to access the property of beauty in an object is to adopt an attitude of disinterest in how the object might fulfill an instrumental role in service of some purpose, thereby ridding ourselves of personal interest in the object.[1] Francis Hutcheson, for slightly different reasons, believed the same,[2] as did Joseph Addison[3] and Archibald Alison.[4] Kant gives the tradition its most sophisticated rendering, embedding disinterest more deeply into his overall theory of aesthetics. As with those theories that came before, disinterest begins with an absence of consideration of any function or instrumental purpose to which the object may be put. Furthermore, to be disinterested, for Kant, means chiefly to take no interest in the actual existence of the object under consideration; we care only about the object as it is an object of our attention. In addition, Kant says that being disinterested includes avoidance of bringing the object "under any category"—we must consider the object merely as a collection of phenomenal properties arranged in a way that is pleasing; to consider the object "under a category" is to bring the object under external relations with other objects. Instead, we must consider only the internal relationships the properties of the object have with its other properties; this is the way we may come to see the formal order of the object, how its properties form a coherent whole that suggests the object is purposive in design without consideration of any purpose to which the object may be employed. Arthur Schopenhauer[5] continues the disinterestedness tradition after Kant. We find in the twentieth century a variety of aesthetic attitude theories, the two most important of which arguably are

---

[1] Anthony Shaftesbury, *Characteristics of Men, Manners, Opinions, Times* (New York, NY: Bobbs-Merrill, 1964).

[2] Francis Hutcheson, *An Inquiry into the Original of Our Ideas of Beauty and Virtue* (New York, NY: Garland Press, 1971).

[3] Joseph Addison, "On the Pleasures of the Imagination," Nos. 411–421, in Robert Allen (ed.), *Selections from The Tatler and The Spectator* (New York, NY: Holt, Rinehart and Winston, 1957), pp. 411–21.

[4] Archibald Alison, *Essays on the Nature and Principles of Taste* (Boston, MA: Cummings & Hilliard, 1812).

[5] Arthur Schopenhauer, *The World as Will and Idea* (London, UK: Routledge & Kegan Paul, 1896).

Edward Bullough's theory of "Psychical Distance,"[6] and Jerome Stolnitz's reformation of disinterestedness.[7] It is when, says Stolnitz, we are attending to an object in the absence of purpose—when we are attending simply to the phenomenal properties of the object for their own sake—that our experience will be aesthetic. The most important critic of the disinterestedness tradition is George Dickie, who argues that attention need not be divided into different kinds, aesthetic and not;[8] consideration of an object can incorporate at the same time attention to its aesthetic features and to features having to do with, say, the moral point of view being expressed by the object or perhaps by the artist. (This review of disinterestedness will be important later.)

Theories of the "aesthetic attitude" tradition are all of the "deliberate-volitional" sort. None explicitly treats the phenomenon of "beauty breaking in" and, so, no theory of this sort is likely to explain in successful terms the gamut of "aesthetic preparation." In addition, it is possible to object that the aesthetic attitude tradition is really not describing aesthetic preparation per se. One could argue that to adopt a perspective of disinterest, one must first have had to decide that the object to which this attitude might be taken is a good candidate for such attention. That is, our first encounter with an object is not one of disinterest; the first encounter is one colored by induction and the categorization of the object into a kind. Volitionally adopting a particular attitude toward the object may be at best a secondary act. This might entail that no volitional approach can qualify as aesthetic preparation, yet it certainly seems a common phenomenon that we can "turn on" and "turn off" aesthetic attention at will.

Arthur Danto holds that to see an object as a work of art is first to interpret that object as a work of art.[9] This interpretive activity takes place before the subject attends to the artistic character or features of the object; it is essentially work of categorization. Since this interpretative work is a matter of conscious and volitional effort, it is possible to understand it as constituting an appropriate preparation for aesthetic experience (if we are not conflating aesthetic and artistic experience too badly).

Being prepared for an aesthetic experience may mean being expectant that whatever value one finds in aesthetic experiences will manifest—whether that

---

[6] Edward Bullough, "'Psychical Distance' as a Factor in Art and as an Aesthetic Principle," *British Journal of Psychology* 5 (1912), pp. 87–98.

[7] Jerome Stolnitz, *Aesthetics and Philosophy of Art Criticism* (New York, NY: Houghton Mifflin, 1960); "On the Origins of 'Aesthetic Disinterestedness,'" *Journal of Aesthetics and Art Criticism* 20 (1961), pp. 131–143; and "'The Aesthetic Attitude' in the Rise of Modern Aesthetics," *Journal of Aesthetics and Art Criticism* 36 (1978), pp. 409–422.

[8] George Dickie, "The Myth of the Aesthetic Attitude," *American Philosophical Quarterly* 1:1 (1964), pp. 56–65.

[9] Arthur Danto, *Transfiguration of the Commonplace* (Cambridge, MA: Harvard University Press, 1981).

value is transport to new worlds, escape from this one, rewarding cognitive engagement, rewarding emotional engagement, delight in mimetic representation, an expansion of one's horizons by appreciating the expression of another human being, and so forth (values we will explore in the chapter focused on evaluation and criticism). We do not invest time and attention in ways that do not provide some sort of reward—wasting time for the pure sake of wasting time does not motivate us[10]—and so the expectation of achievement of that reward may be an important part of what it is to be properly prepared for an aesthetic experience.[11] This approach involves volition, but the initial move is not necessarily the adoption of a special attitude. This approach simply takes seriously that aesthetic attention must be motivated and then takes that motivation as what accounts for being prepared to experience aesthetically.

Typically, once appropriately prepared, aesthetic attention is engaged by determining the object's classification or kind:

- "What am I looking at?"
- "A performance?"
- "A dance?"
- "A ballet?"
- "A formalist ballet?"
- "A George Balanchine ballet?"
- "A Balanchine ballet performed by the New York City Ballet?"

Once we have classification detailed to a point that seems sufficient to us, we next typically turn to the representational aspects of the work of art. (We do this with nonart aesthetic objects—as in seeing elephants, turtles, and Mickey Mouse in cloud formations—but we do this less typically than we do with art objects.) Some works of art are abstracts, but humans are consummate masters at finding patterns and so in tending to classificatory matters first, we search out even the greatest of abstractions for representational qualities. On occasion, our minds add them in: "Look at Jackson Pollack's *Lavender Mist*—don't you see a face right here?" Determining the classificatory and representational properties of a work of art prepares us to be selective to what we attend in that work. Certain properties fade from our attention as we consider them less important or not at all important to the consideration of the work in terms of what kind of work it is. For instance, the person focused on the proscenium arch as the New York City

---

[10] John Lennon is thought to have said, "time you enjoy wasting was not wasted." This is our point—there must be reward that motivates.

[11] David Fenner, *The Aesthetic Attitude* (Atlantic Highlands, NJ: Humanities Press, 1996), the last chapter.

Ballet dances or on the frame surrounding *Lavender Mist* is the exception rather than the rule.

After deciding on what to focus our attention, we begin the task of establishing our experience of the aesthetic properties of the object. "Formal" aesthetic properties are those that are derived from attention to the work's perceptual properties and those that are relevant to them (such as, perhaps, the representational, expressive, and literary properties).[12] Every tangible object possesses properties, some version of which forms the basis of how we know that object.[13] For Balanchine's 1967 ballet *Jewels*, we attend to the dancer's movements, the themes and rhythms of those movements, the shapes their bodies make—individually and corporately with other dancers—how they occupy and fill the stage, and so forth. With Pollock's *Lavender Mist* (more formally titled *Number 1, 1950*—presumably a point made to emphasize the absence of representational properties), we attend to the lines of paint, the colors of the applied paint, the textures of the paint as it sits on the canvas and as it overlaps with other lines of paint, and so forth. It is from these perceptual properties that formal aesthetic properties are derived. The "formality" of "formal" aesthetic properties is connected to the fact that they focus on arrangements and interrelations of objective properties. A particular perceptual property may underwrite an aesthetic property attribution—a single color may carry the associated aesthetic property of, say, "bright," "garish," or "calming"—but more often aesthetic properties are about relationships among sets of objective properties, the boundaries of those sets being whatever strikes the subject as relevantly connected to the attribution.

Aesthetic properties carry positive and negative connotations, even though these connotations may vary or shift depending on the context of the analysis, culture, the taste of the person citing them, and the account one offers of a particular work of art—which is to say that a property that is positive in one work may be negative in another. But, in general, terms like "arresting, balanced, daring, elegant, harmonious, moving, poignant, sublime, and vibrant" are positive aesthetic attributions, and terms like "busy, cold, derivative, discordant, frenzied, garish, loud, muddled, and stiff" are negative. We know aesthetic properties

---

[12] We mean to follow the work of Nick Zangwill in presenting these definitions: "Formal properties are entirely determined by narrow nonaesthetic properties . . . the word 'narrow' includes both sensory properties, non-relational physical properties, and also any dispositions to provoke responses that might be thought of to be partly constitutive of aesthetic properties." Zangwill, "Feasible Aesthetic Formalism," *Nous* 33:4 (1999), p. 610. See also Zangwill, "In Defence of Moderate Aesthetic Formalism," *Philosophical Quarterly* 50 (2000), pp. 476–493.

[13] We do not mean to wade into metaphysical waters here; we will try to keep our accounts consistent with the metaphysical commitments of all but the most hard-core idealists.

carry positive and negative connotations because we cite them as evidence when we are justifying our aesthetic judgments and judgments about artistic value.[14] If asked why one believes an object to be beautiful, that individual will defend their judgment by citing positive aesthetic features of the object and the interrelations of these properties as they work together to constitute the object's aesthetic form. The same happens in the reverse: an overall negative judgment about the quality of an object is evidenced through citation of negative aesthetic properties or an argument about how the object's aesthetic properties do not cohere.

The manifestation of aesthetic properties requires the attention of a valuing subject. This must be the case if aesthetic properties are bearers of value and if the objective properties of, say, line, color, texture, and so forth, are not. The case of *Lavender Mist* is a good one for this point since two individuals can readily agree on the objective features of this work but disagree when it comes to their summary judgments of the quality of the work—and each will evidence their cases by citing the aesthetic features of the work, which in turn must be evidenced by the objective properties of the object. We understand this process to be interpretative in the sense that the subject moves from the objective to the evaluative; the subject "interprets" the objective properties in ways that lead to the attribution of properties that convey the possession of value.

For some, the story should stop here with appreciation of the object's formal aesthetic properties. The disinterestedness approach described above was formalist. The focus was to be on those properties relevant to aesthetic appreciation for no purpose other than appreciation of them and their relationships to one another. The "art for art's sake" movement, famously championed in the nineteenth century by artists Oscar Wilde and James Whistler, held that any consideration of a work of art that moved beyond focus just on the object's properties and that incorporated anything external to the work is contrary to a truly aesthetic focus. Formalism was advanced in the twentieth century by advocates of New Criticism—including I. A. Richards, T. S. Eliot, Cleanth Brooks, and Robert Penn Warren—and it rested on formalism, prominent proponents of that day including Eduard Hanslick, Clive Bell, and Roger Fry; the most recognized formalist today is Nick Zangwill. For formalists, the aesthetic phenomenon is wholly described once the subject appreciates the aesthetic properties of the object that are derived from the perceptual properties and features closely related to them.

---

[14] This follows the work of Monroe Beardsley. See Beardsley, "What Is an Aesthetic Quality?" *Theoria* 39 (1973), pp. 50–70.

Despite the former popularity of formalism, it is more common today to seek out information about the object and its relationships with other objects. This is more common with art objects than with nonart aesthetic objects, but even with nonart objects we commonly still want to know everything about the object that seems relevant to deepening our experience of it. When it comes to works of art, we seek out information that strikes us as appropriate to the kind of object the object is; this takes us back to the very first step along the path we describe. Relevant information can include

- "genetic factors"—matters pertaining to the artist, to the artist's intentions for the work, and to the conditions of the origin of the work;
- "lifespan factors"—matters pertaining to the existence of the work through time, to the provenance of the work, to the exhibition (performance, etc.) of the work;
- "relationship factors"—matters pertaining to the representative features of the work, the style of the work, relations the work bears to other artworks, and how the work fits into the unfolding histories of art that seem relevant to it; and
- "reception factors"—matters pertaining to the history of the critical review of the work, to how the work has been received by audiences; to trends of audience identification with aspects of the work and to trends of audience cognitive engagement, emotional engagement, and psychological engagement with the work.

Information of this sort rounds out our experience of the object under consideration. As we develop a vision of the object's context, our experience typically deepens, our focus on aspects we find relevant sharpens, and our confidence in our appraisal of the object's worth (in general and as the focus of our particular experience) grows.

One more important point before we explore how the story above relates to appreciation of a garden. There is a difference between "an aesthetic analysis" and "an aesthetic experience."[15] What we have been discussing thus far is aesthetic appreciation as common to all or most people; this is an "aesthetic analysis" where we are analyzing the pieces of the aesthetic phenomenon puzzle. An "aesthetic experience" is particular to an individual, and that experience will include a great deal more than what an "analysis" will include. Aesthetic experiences are events that have personal contexts: if one is tired, bored,

---

[15] David Fenner, "Aesthetic Experience and Aesthetic Analysis," *Journal of Aesthetic Education* 37:1 (2003), pp. 40–53.

frustrated, elated, giddy, hungry, distracted, with friends, alone—these sorts of factors have at times a significant impact on the experience one has. They obviously are not consistent across experiences had by different people or even had by the same person at different times. So while they are noteworthy in terms of accounting for the kind of experience an individual has on a particular occasion, they are not normally (or at all) cited in giving an account (or an analysis) of an aesthetic experience that is meant to be comparable to other such experiences, and these sorts of particularist factors are never cited as evidence for the quality of a work of art or even the quality of the experience of a work of art. Reasons for judgments must function as reasons for everyone who is sufficiently like the person offering the judgment—who perhaps has a similar taste profile, or a similar level of acquaintance with the artform, and so forth. To say that one had a bad experience when attending to a certain work, and then explaining that situation in terms of how hungry one was during the experience, will be dismissed by the hearer of this account as not being about the work but merely about incidental matters related to that person. So, again, while these factors have a real-world impact on aesthetic experience—and so must be accounted for as one reflects on one's experience—they are not typically part of what is relevant about the experience in terms of its character or the judgments that might ensue from that experience.

It is at this point that the path and our story of it ends. The next step, along a next path, is about sharing our experience with others, perhaps recommending that another person seek out a similar experience of that same object, perhaps even writing a critical review. These matters we will take up in a later chapter. For now, we will begin to examine what it means to appreciate gardens as aesthetic objects, in accord and in contrast with the story told above.

Properties of Gardens

We begin with a map. In this chapter we cover a lot of ground, and it will be easier for the reader to follow that progress if the reader knows where we are going and how various matters fit together. Our map is offered in six tables below. They are meant to chart in logical terms—if not in perfectly chronological terms—the life of a garden from inspiration and idea to whether it should be regarded as a work of art. Some of the items/properties in the tables below may make limited sense to the reader at this stage, but each movement will be explained and defended as we proceed. At the end of this conversation of what it means to appreciate a garden aesthetically, we will repeat the six tables as a reminder to the reader of what ground has been traversed.

Genetic Properties

Item or Property	Examples	Derivation
Mission, purpose, categorization, designed functions	"This garden will be a *chahar bagh*."	Choices about whether there is to be a garden, what the garden is to be, and who will design it, are those of the garden's patrons.
Gardenist(s)	"The garden will be designed by Capability Brown."	
Design	Idea, vision, expression, drawing, modeling	The design is derived from the anticipated phenomenal or elemental properties of the forthcoming phenomenal garden.
Initiation	Planning, permitting, installing	The initiation of the garden is driven by the design.

Objective Physical Properties

Item or Property	Examples	Derivation
Site	*Terroir*, topography, climate, hydrology, soils	These items are chosen because of (1) practicalities concerning available land, patronage, resources, etc., and (2) the anticipated phenomenal or elemental properties of the forthcoming phenomenal garden.
Materials	Plants, soil, water, stone, mulch, hardscape, hidden ha-ha walls and irrigation pipes	
Hidden Nature	Unperceived root structures, unperceived shoots of tree limbs above the canopy	These are real aspects of the physical garden that only enter experiences of the garden imaginatively.

Dynamic Process Properties

Item or Property	Examples	Derivation
Nature-based processes of creation	Weather, seasonal change, climate change, ecological activity, plant growth/competition	These are processes that make up nature's contribution to the creation of garden at any given moment.

Item or Property	Examples	Derivation
Human-based processes of creation	Arranging, planting, shaping, pruning, thinning, deadheading, watering, fertilizing, building	These are gardening practices, responsible for human contribution to the garden at any given moment.

Phenomenal/Experiential Properties

Item or Property	Examples	Derivation
Garden Elements	The pink of the rose, the thorniness and height of its stems, the bright yellow of variegated schefflera against the deep purple of a carefully shaped loropetalum hedge, the scent of calamondin blossoms, the sound of a fountain of water	These properties are derived from garden appreciators experiencing states of and perspectives on the four-dimensional, dynamic garden.
Borrowed Scenery	Mountains, skyscrapers, visible trees and architecture outside the garden, sensible sounds and smells outside the garden	

Aesthetic Properties

Item or Property	Examples	Derivation
Formal Aesthetic Properties	Balance, charm, elegance, grace, harmony, peacefulness, grandness, leanness	These properties arise from interpretation of the perceived properties of the phenomenal garden, typically based on the form of the phenomenal garden but occasionally based on individual phenomenal properties.

Item or Property	Examples	Derivation
Contextual Properties	Genetic, lifespan, reception, relationship, and scientific factors—"The New York Botanical Garden is sited on 250 acres in the Bronx." "Jennie Butchart transformed a former limestone quarry into the Butchart gardens we know today." "This garden is made up entirely of invasive plants." "This species only flowers once in its lifetime."	Contextual properties have to do with (1) genetic aspects of the garden (seen at the top of this table), (2) external relations the garden bears to other objects, themes, and persons—historically and currently; (3) responses of audience members, both expressed and not; and (4) those processes discovered through science.
Summative Aesthetic Properties	"This garden is beautiful." "This garden is more beautiful than that garden."	These properties are derived from and evidenced by the formal aesthetic properties of the phenomenal garden and from the "garden form" exhibited by the garden.

Art Properties

Item or Property	Examples	Derivation
Interpretation	"Traversing this garden follows the journey of Aeneas as told by Virgil."	Interpretations are derived from interpretative activities focused on the phenomenal garden, typically with particular focus on "garden form."
Art Status	"This garden is a work of art."	This status is derived from the application of criteria concerning both the four-dimensional physical garden and the phenomenal garden.
Art Judgments	"This garden is a great work of art."	These judgments are derived from the application of criteria concerning both the four-dimensional physical garden and the phenomenal garden.

The Elephant in the Garden: Dynamism and the Objecthood Problem

How are we to condense the changing of the seasons, the calls of migratory birds, the growth and decay of trees, and the movement of the wind into workable aesthetic analysis? Aesthetic appreciation of a garden first and foremost means appreciation of a dynamic object, one that exists in four dimensions essentially and necessarily. There are other dynamic aesthetic objects. All performances—plays, symphonies, operas, concerts, dances—are dynamic. Literature is dynamic as the narrative progresses through time as one reads; film is dynamic in the same way. If one appreciates cloud formations and fireplace/campfires aesthetically, they are dynamic.[16] Movement, progression, and the occupation of time, in and of itself, is commonplace to aesthetic appreciation.

What differentiates these events from the dynamic garden is that in all the aforementioned art cases, the dynamism in question is "closed" in the sense that it has a start, a middle, and an end—designed as part of the object itself.[17] In the case of a closed dynamic system, the aesthetic features of the event are describable in stable ways. Performances may change from performance to performance, as we discussed especially in connection with dance, but (1) aesthetic features, one performance to another, are comparable with each other as the objective features that underwrite them will be similar across performances, and (2) more basically, the aesthetic features of a single performance are stable to that performance. There will be change, but that change is bound in time and once experienced as the single event that was that performance, the aesthetic features of that performance are fixed or permanent.

Gardens, on the other hand, are dynamically "open." As objects, they do not have starts or stops, at least not in ways that typically can be experienced. While some gardens are enclosed in obvious ways by walls or fences, and all gardens are bound spatially in ways that fit human scaling, many gardens do not permit a visitor to take in the whole garden, visually, in a single go. Chinese and Japanese strolling gardens, for example, typically involve irregular paths and obstacles that prohibit a visitor from seeing everything at once.

Appreciating a garden as a dynamic object requires one not only attend to the aesthetic properties as they are commonly attended to—visual, auditory, olfactory, gustatory, tactile, and kinesthetic—but that we also attend to their

---

[16] Certain features of a painting may be more or less pronounced if moved to a new gallery with different lighting, different colored walls, or different paintings in one's peripheral vision; occasionally this can make paintings seem dynamic. At the Rothko Chapel in Houston, the change of sunlight through the octagonal skylight brings a noteworthy aspect of change to the experience.

[17] David Fenner, "Environmental Aesthetics and the Dynamic Object," *Ethics and the Environment* 11:1 (2006), pp. 1–19.

dynamism by noticing how they change through the course of our visit, over successive visits, and over visits throughout the seasons and years. These changes are interesting in their own right. Cognitive appreciation of the processes that account for these changes further enlivens our experience of a garden or of The Garden in general.

Nonetheless, a garden's dynamic character is the core challenge for understanding that garden as an aesthetic object. In the last chapter we referred to this challenge as the "objecthood problem" and began our discussion of this point with Miller's insights on the matter. We need to revisit this challenge now, because the attribution of formal aesthetic properties to a garden is in jeopardy without some resolution.

The objecthood problem, the "framing problem," and one strain of discussion of "the picturesque problem" are all species of the same general problem. We can think of the framing problem in two versions.[18] The first version closely connects to the character of nature being dynamic: we can only see formal aesthetic properties if the perceptual properties on which they depend are static, stable, or fixed. But nature has no such properties. As a garden changes, so its properties change as well. If formal aesthetic properties derive from the objective properties of gardens, then they too will change. If a garden's formal aesthetic properties change, then judgments based on those properties must change as well. This undercuts all activities—comparison, evaluation, interpretation—that require these judgments to be stable.

The second version advances the view that for nature to have formal aesthetic properties, it must be compositionally framed, in the style of looking at nature through a window or through a "Claude glass"—a small mirror, also sometime called a "black mirror," that was used by landscape painters not only to frame a scene (in relation to the dimensions of the mirror) but also to "flatten" the scene into a two-dimensional image that possessed tonal properties capturable in a painting. Formal aesthetic properties are compositional in the sense that formal properties per se are about the relations of objective properties of an area, and so they rely on first delineating the boundaries of that area. One must "frame" an area to see within it compositional properties.

---

[18] Central establishing discussions of the matter of framing nature may be found in the following: (1) George Santayana, *The Sense of Beauty* (New York, NY: Scribner's, 1936); Santayana writes, "A landscape to be seen has to be composed" (p. 101). (2) Ronald Hepburn, "Contemporary Aesthetics and the Neglect of Natural Beauty," in Bernard Williams and Alan Montefiore (eds.), *British Analytical Philosophy* (London, UK: Routledge & Kegan Paul, 1966), pp. 43–62; and "Trivial and Serious in Aesthetic Appreciation of Nature," in Salim Kemal and Ivan Gaskell (eds.), *Landscape, Natural Beauty, and the Arts* (Cambridge, UK: Cambridge University Press, 1995), pp. 65–80. (3) Allen Carlson, "Appreciation and the Natural Environment," *Journal of Aesthetics and Art Criticism* 37:3 (1979), pp. 267–275; "Formal Qualities in the Natural Environment," *Journal of Aesthetic Education* 13:3 (1979), pp. 99–114; and *Aesthetics and the Environment* (New York, NY: Routledge, 2000).

When we view the composition of a painting, we might talk about how that composition exhibits balance, but of course if elements of that composition change, the balance we noticed might be lost. In a garden, the static nature of "a composition"—typified perhaps by taking a mental photograph as we stand still—is illusory in two ways. First, our perspective likely changes as we move through the garden, or it changes as we shift perspectives from our viewing position by looking around or moving about slightly. Second, the items that populate that garden composition change—they grow or fade, and how they are perceived changes as the light moves and as the wind blows.

Nature does not come with visual borders; it extends out indefinitely in every direction around the entire global ecosphere. Nature resists being seen as complete as rendered in any act of framing. Furthermore, application of such a frame can be dangerous in the sense that it holds some landscape area as conceptually separate from the bioregion of which it is a part—or from the full ecosphere with which it is continuous. This is not only anthropocentrically artificial, but if such aesthetic judgments are then used to justify environmental ethical claims concerning preservation, such judgments are corrupted by that artificiality. That is, saving what "looks pretty" to us may be in stark contrast to what has true environmental value—as judged, say, through the lens of ecological science.

The objecthood problem and the framing problem are related to "the picturesque problem."[19] Of relevance to The Garden, "picturesque" can refer to one of the following three:

1. it is the name of a garden style, one popular in England in the late eighteenth and nineteenth centuries; a style associated with the garden designers William Gilpin, Richard Payne Knight, and Uvedale Price;[20] a style that can be regarded as a reaction to Capability Brown's style; one known to include dead trees in landscapes to make them all the more "picturesque";[21]

---

[19] Allen Carlson, "On the Possibility of Quantifying Scenic Beauty," *Landscape Planning* 4 (1977), pp. 131–172; Yuriko Saito, "The Aesthetics of Unscenic Nature," *Journal of Aesthetics and Art Criticism* 56:2 (1998), pp. 101–111; Terry Daniel, "Whither Scenic Beauty? Visual Landscape Quality Assessment in the 21st Century," *Landscape and Urban Planning* 54:1 (2001), pp. 276–281; Russ Parsons and Terry Daniel, "Good Looking: In Defense of Scenic Landscape Aesthetics," *Landscape and Urban Planning* 60:1 (2002), pp. 43–56; Robert Stewart and Roderick Nicholls, "Virtual Worlds, Travel, and the Picturesque Garden," *Philosophy and Geography* 5:1 (2002), pp. 83–99; Donald Crawford, "Scenery and the Aesthetics of Nature," in Arnold Berleant and Allen Carlson (eds.), *The Aesthetics of Natural Environments* (Peterborough, Canada: Broadview Press, 2004), pp. 253–268; Isis Brook, "Wildness in the English Garden Tradition: A Reassessment of the Picturesque from Environmental Philosophy," *Ethics and the Environment* 13:1 (2008), pp. 105–119; Arnold Berleant, "Reconsidering Scenic Beauty," *Environmental Values* 19:3 (2010), pp. 335–350; Roger Paden, "Picturesque Landscape Painting and Environmental Aesthetics," *Journal of Aesthetic Education* 49:2 (2015), pp. 39–61.
[20] Elizabeth Barlow Rogers, *Landscape Design: A Cultural and Architectural History* (New York, NY: Abrams, 2001), pp. 251–254.
[21] Paden writes, "[T]he aesthetics of picturesque gardens is different than that of either *Horti conclusi* or baroque gardens. While *Horti conclusi* emphasize the aesthetics of 'the Beautiful' and

2. it is the name of a formal aesthetic property, one apparently apropos to The Garden;
3. and, finally, it can be taken as an aesthetic term but one that purportedly names a virtue of a natural area—an area of First Nature—that because of its possession of this virtue holds claim to being of either aesthetic or preservation interest (or both).

Roger Paden, in "A Defense of the Picturesque,"[22] lays out the problem, connecting it directly with the framing problem described above.

> The questionable enthusiasm of eighteenth- and nineteenth-century picturesque tourists led Ronald Rees to refer to them as members of a "scenic cult," while Allen Carlson includes these tourists, together with more modern nature tourists and the institutions that serve them, such as the U.S. Forest Service, in what he calls the "landscape cult." . . . Carlson, who has written by far the most cogent critique of the picturesque and the associated landscape cult, argues . . . that while many natural objects such as flowers, crystals, and animals are formally beautiful, landscapes are not, and to perceive landscapes as having formal qualities requires a great deal of creative activity on the part of the observer. . . . Without acts of composition, landscapes are disorganized collections of objects, lacking the order necessary to be judged formally beautiful. As a result, a landscape's beauty, if it has any, is not a quality of the landscape itself, but is instead an artifact of the compositional process. On Carlson's view, this process produces an artificial unity through the use of frames.[23]

These problems together cast a bright light on the conditions that must be achieved to attribute formal aesthetic properties to gardens. If the challenge cannot satisfactorily be met, our intuitive commitment to attributing "compositional" aesthetic properties to gardens will be found to rest on an illusion, and the resulting theoretical commitment we then might be forced to accept is that our *common* practices

- of describing gardens in formal aesthetic terms,
- of comparing gardens one with another on this basis,

baroque gardens emphasize the experience of 'the Sublime', Uvedale Price, an important picturesque gardener and theoretician, argued that 'modern' gardens should emphasize the aesthetics of the Picturesque. On Price's view, the Picturesque was an aesthetic category that was as important as the Beautiful and the Sublime. He defined the Picturesque in terms of 'roughness', 'sudden variation', and 'irregularity.'" Roger Paden, "The Ethical Function of Landscape Architecture," *Environmental Philosophy* 15:2 (2018), p. 152.

[22] Roger Paden, "A Defense of the Picturesque," *Environmental Philosophy* 10:2 (2013), pp. 1–21.
[23] Paden, "A Defense of the Picturesque," pp. 2–5.

- of comparing the aesthetic value one experiences on a first visit to a garden with that of a second visit,
- of evaluating the aesthetic worth of gardens in general, and
- of discovering meaning in gardens through interpretative activity based on the internal formal coherence of the components of a garden

would all be misguided. Were this to come to pass, the case for The Garden being an artform would seem not only a fool's errand but of relatively limited importance in light of the overarching realization that our common notions of aesthetic appreciation of gardens would be derivative or virtual at best.

Acceptance of the problem as unsolvable seems to imply that a garden cannot have formal aesthetic properties. Yet we commonly attribute formal aesthetic properties to gardens. We do it with a firm grasp that such attribution is not only intuitively plausible but appropriate; we do it in such a way that to reject doing it, or to stop doing it, because we cannot develop theory sufficient to justify doing it would almost certainly lead us in the direction of suspicions that we are simply not trying hard enough at our theory-creation efforts. A garden *seems* to possess formal aesthetic properties.

Part of a solution to the problem may lie in recasting the meaning of some formal aesthetic properties of a garden to accommodate its character as a dynamic object. We may no longer understand "balance," for instance, to be a property of a static composition but rather of the overall "balance" of colors, shapes, textures, etc., that populate the whole garden. Or we may take "balance" to refer to the "balance" of natural processes that physically govern a garden, combining or informing our aesthetic view with a cognitivist ecological view advanced by, for instance, Carlson.[24]

> To aesthetically appreciate nature we must have knowledge of the different environments of nature and of the systems and elements within those environments. In the ways in which the art critic and the art historian are well equipped to aesthetically appreciate art, the naturalist and the ecologist are well equipped to aesthetically appreciate nature.[25]

Nothing is lost when recasting aesthetic terms to fit The Garden because, first, aesthetic terms are not fixed or static themselves; they change with personal taste, culture, time, and the evolution of art itself. Second, there exist more

[24] Allen Carlson, *Aesthetics and the Environment: The Appreciation of Nature, Art and Architecture* (New York, NY: Routledge, 2000), particularly his chapter "Nature and Positive Aesthetics."
[25] Allen Carlson, "Appreciating Art and Appreciating Nature," in Salim Kemal and Ivan Gaskell (eds.), *Landscape, Natural Beauty and the Arts* (Cambridge, UK: Cambridge University Press, 1995), p. 50.

dynamic aesthetic objects than simply gardens—dance, performance, music— and so our dynamically altered definitions of aesthetic properties already will have undergone some evolution to accommodate their attribution in these cases.

Appreciating the dynamic properties of a garden and recasting aesthetic terms to accommodate this appreciation takes us a step along the path toward attribution of formal aesthetic properties to a garden, but it does not take us as far as we wish to go. It is likely we will want to attribute a still wider range of formal aesthetic properties to gardens that do not admit of the possibility of this sort of recasting.

Resolving the Objecthood Problem: Phenomenal Gardens

Can a garden have formal aesthetic properties in the same way other aesthetic objects typically do? It might seem, given the common practice of describing gardens using classic aesthetic terminology in ways that seem similar to the ways we use it in other cases, that this is a nonquestion. Ross writes:

> [G]eneral aesthetic principles of gardening, grounded in a philosophical analysis of such concepts as balance and harmony, unity and multiplicity, form and content, nature and art, could be proposed and defended (as in fact they were late in the eighteenth century by Sir Uvedale Price, Richard Payne Knight, and Humphry Repton).[26]

But, given the challenge described above, here we cannot retreat to common practice as the argumentative default.

Paden's solution to the picturesque problem is to redefine what is meant by "the picturesque."

> Rather than implying that we ought to experience nature as if it were a painting . . . "the picturesque" implies that we ought to observe nature as carefully as a painter does. . . . Therefore, criticisms of the picturesque for its supposed assumption of a formalist aesthetic misfire as such criticisms should be aimed instead at the formal gardens that picturesque gardeners rejected. . . . Picturesque gardeners did not seek to impose an arts-based aesthetic theory on nature, but rather they tried to see with a "painter's eye."[27]

---

[26] Stephanie Ross, *What Gardens Mean* (Chicago, IL: University of Chicago Press, 1998), p. 46.
[27] Paden, "A Defense of the Picturesque," pp. 10 and 19.

A similar path is taken by Isis Brook in her defense of the picturesque. She writes:

> What the picturesque allows is the transition to be made from one realm to the other without seeing wilderness or wild places as somehow there for us to tone down and shape to our sensibilities, but to experience as they are. Also we can still value the particular qualities that an ordered garden can bring where human design has the upper hand. The transition zone is where both humans and nature have a hand, but neither is tyrant.[28]

This tact toward redefining "the picturesque" in those terms that would have occurred to those involved in the Picturesque style of garden design, terms that would have been aesthetically focused, not only has historical support but fits common practice. Hunt writes, "Alexander Pope . . . allowed as how 'you may distance things by darkening them and by narrowing the plantation more and more toward the end, in the same manner as they do in painting.' "[29] These painterly insights take the conversation back to its roots, and the emphasis is then not on the "reduction" of nature or the imposition of the human on nature but rather on how humans may come to appreciate nature with greater focus.

Gardens include features gardenists employ to direct our attention. In most gardens, paths direct visitors along routes that are meant to be seen, and that which is not meant to be seen—tools and water hoses, for example—are stored out of view, so that they do not interfere with the composition. Suggestions of frames are offered at sites of benches, entrances to garden rooms, and views from pavilions or windows. Efforts are made to maintain or accentuate the most aesthetically rich compositional views. When a garden designer is commissioned to create a home garden, typically they will express their vision by creating a scale two-dimensional drawing of their garden as viewed from a single point. Gardens differ from unaltered nature in the ways that our views are directed and thereby framed. Miller writes:

> The pattern provided by stopping, resting, and continued movement can also be affected by the gardenist if he or she provides benches, turning points where decisions are required or quotations may be deciphered, pavilions or other vantage points with special views, and so on.[30]

---

[28] Brook, "Wildness in the English Garden Tradition," pp. 116–117.
[29] Hunt, John Dixon, *Greater Perfections: The Practice of Garden Theory* (London, UK: Thames and Hudson, 2000), p. 140.
[30] Mara Miller, *The Garden as an Art* (New York, NY: State University of New York Press, 1993), p. 46.

Hunt describes an interesting phenomenon in this regard:

> When William Gilpin visited Stowe as a young man in 1747 . . . he wrote a Dialogue about the gardens: in it he made one character exclaim in puzzlement about being confronted with a blank hedge and his companion reassure him that this was simply a device to rest the eyes before the next set piece. Gardens were required to behave like perspectival paintings, best viewed from a specific point that could even be marked on the ground in front of them. Out of this kind of experience in gardens came the invention of "stations" for the picturesque tourist in the countryside at large, places where the best authorized view could be taken by a stationary tourist.[31]

These "stations" suggest, in the way a camera captures a moment, so a human may as well—a practice continued in our "lookout points" along "scenic" highways today. That there are best places from which to do this suggests that these places would have provided the bases for valuable aesthetic experiences through the notice of particular visually valuable aesthetic properties. The objective features that underwrote these aesthetic properties were not fixed in time, but the "tourist" could fix them through capturing the moment—and many other moments—in memory. "The moment" is not—or "the moments" are not—the garden but a piece of it, abstracted for the purposes of memory and aesthetic inspection.

Ronald Moore takes a path related to that taken by Paden and Brook by returning, in his critique of critiques of framing, to the important point that while nature may be limitless, humans can only take in parts of it at a time.[32] Our attention cannot be on the whole of nature, and so aesthetic appreciation of it is necessarily only going to consist of appreciation of some piece of it or some particular vista. Moore refers to Jerome Stolnitz, who points out that attention is selective; we can only attend to so much at one time, and so we choose what to attend to and what to ignore, and we do so in accord with our purposes. If our purposes are aesthetic, we choose to focus on what we find aesthetically relevant. Stolnitz writes:

---

[31] Hunt, *Greater Perfections*, p. 137.

[32] Ronald Moore, "The Framing Paradox," *Ethics, Place and Environment* 9:3 (2006), pp. 249–267. We might note that worries about framing are about what Hunt calls First Nature. But as Paden and Brook both point out, gardens are unlike contiguous nature insofar as they are enclosed, at least as history and etymology define a garden. While we find counterexamples to true enclosures being necessary features of gardens, we still advance the related concept that all gardens are "bounded" in various ways. Spatial bounding—within a human scale—is enough to begin to make the point that the framing problem may not apply to gardens as it may to First Nature. Viewing gardens as having such bounds does nothing to diminish the strength of the framing argument; it merely puts gardens out of the scope to which the argument applies.

[A]lthough nature lacks a frame when it simply exists, apart from human perception, this is not true when it is apprehended aesthetically. Then the spectator himself imposes a frame on the spectacle of nature, he selects what he is going to attend to aesthetically and he himself sets boundaries to it. We all do so at one time or another.[33]

The legitimacy of attributing formal aesthetic properties to gardens is based on scaling—and expressing that scaling—as we make such attributions. Gardens and areas of gardens can be and are framed, although the framing is rarely about the final spatial boundaries of the garden (although it can be). It is typically rather about how we bound the garden spatially and temporally as an experiential object. To experience a garden is to experience it at a certain moment in time or over a certain bounded stretch of time, but it is also to experience it as a visual field whose range is set by the subject's focus and attention as guided by the gardenists and gardeners who direct in both subtle and overt ways that focus and attention. In support of this point, Lee writes:

Kant's classification of garden design as a form of painting implies that the imposition of this frame will be a rather straightforward matter, involving little more than the identification of an appropriate garden analog for the customary border around a canvas.... The frame must be recognized as a frame.... Kant's frame must be legible to the garden stroller of his era.[34]

If we couple Lee's observation here with his treatment of Heydenreich's notion of a "mobile frame" that is imposed on parts or aspects of a garden as one strolls through it, we perhaps have a workable garden frame. As we learn from Lee, Heydenreich appreciated the framing problem as early as the time of Kant—as apparently did Kant himself—and Heydenreich's solution, one that informs ours to a certain extent, is to discuss the character of attention itself. As visual attention can only extend out as far as one's visual field may—although in many cases we restrict our attention much more—that visual attention is automatically and necessarily framed. But as we move through a garden, that frame moves. It may be that we experience our attention to the visual aspects of a garden as a film, with the mobile frame working like a succession of dynamically moving pictures—or it may be that we experience it as a series of photographs, where in

[33] Stolnitz, *Aesthetics and Philosophy of Art Criticism*, p. 48. See also Noël Carroll, "On Being Moved by Nature," in Salim Kemal and Ivan Gaskell (eds.), *Landscape, Natural Beauty and the Arts* (Cambridge, UK: Cambridge University Press, 1995), pp. 89–107.

[34] Michael G. Lee, *The German "Mittelweg": Garden Theory and Philosophy in the Time of Kant* (New York, NY: Routledge, 2013), p. 199. David E. Cooper talks about the mobility of framing on p. 30 of his *A Philosophy of Gardens* (Oxford, UK: Oxford University Press, 2006).

succession a series of visual "snapshots" grab our attention. The same type of argument can also be applied to the problem of dynamic nature. Although nature has no stable properties, some properties are more stable than others—erosion, plate tectonics, and the evolution of species occur at time scales so much slower than human perception that we attend to these objects aesthetically as though they were unchanging.

We argued in the last chapter that the objecthood of a garden should not be taken to understand the garden itself—as a physical four-dimensional place—as being "the object" when we discuss gardens in aesthetic terms. Instead, we must understand the "garden object" to be an object of experience, limited in the attention and the focus of the subject, bounded in time and space according to the subject's focus. When we attribute formal aesthetic properties to a garden, as we commonly do, we do not attribute them to the garden as a four-dimensional place but rather to our experiential "phenomenal take" of that garden. This is the only path available if we are to evidence and defend our attributions of formal aesthetic properties on the "objective" properties of the garden. That is, a garden's "objective" properties—when a garden is thought of as a four-dimensional place, being always in flux in a plurality of ways—are not stable enough to function as evidence. So a garden can possess formal aesthetic properties, but only as the object which is that garden redefined in these experiential and phenomenological terms. To hold a garden as an object of experiential consideration is to hold it as an abstraction, something that does not have a one-to-one coextensional correlation with the dynamic physical place that gives rise to that abstraction.

We talk about formal aesthetic properties deriving from an artwork's *objective* properties. "Objective" is a tough term because it conjures up notions of settled stability, the absence of partiality, and the absence of "subjectivity." But, as we mentioned earlier, it is not in this way that we use the term. We use it in the Kantian way to indicate a location. Unfortunately, this does not clear up the matter, because when we think about a location, we intuitively think about a place sited in four dimensions. Ideally, "objective" properties of gardens should be properties of the physical garden in a place and a time, but this will not work in the context of discussing a garden's formal aesthetic properties because properties a garden possesses at this moment may not be properties it possesses in the next. This leads to the complicated notion that for the purposes of aesthetic analysis the "object" which is a garden is located experientially—and so subjectively—and so all its properties are as well. The alternative is to claim that as dynamic objects, gardens cannot possess formal aesthetic properties in the ways that would allow describing them aesthetically in stable ways, comparing them aesthetically one to another, or judging their aesthetic quality. This seems a much greater price to pay than the complexity of locating the "object" which is the garden within the subjective experience of it.

"Phenomenal properties," while properties of an experience and while subjectively located, are not subjective in the sense that they are particular to a given subject. These properties are simply those that are perceived or perceivable, and so they would be common in content (though not exactly alike) across all those who sensed them. "Phenomenal properties" are not "aesthetic properties" but rather the properties from which aesthetic properties are derived; they are meant to be purely descriptive, and commonality across subjects should be readily achieved to the extent that a subject perceives some aspect of the external world as others do. In this way, they serve as a reliable intermediary on which to base further aesthetic interpretation and evaluation.

The temptation to refer to the phenomenal garden as "the aesthetic object" must be avoided because at this stage, the interpretative or translative act that "transforms" objective properties into aesthetic properties has not yet been performed. Once it has, the resulting aesthetic properties bear value and are subjectively informed in such a way that they then cannot be used as evidence of their own existence. "Transformed" is likely not the right word if we were to mean, as Schopenhauer did, that there is an actual ontological difference in the state of either the object or its properties; a better word is "translated" similar to the way the word "gato" can be translated into the word "cat." It takes an act of translation, but nothing of (ontological) substance is different.

It might be best, then, to refer to a garden in two ways: **the "four-dimensional garden" is the physical dynamic garden and the "phenomenal garden" is the focus of an experience of that physical garden.** For the sake of this discussion, when we are talking about a garden in the sense that it possesses stable enough properties to enable a full aesthetic analysis, the garden as an object will mean the phenomenal one, and "phenomenal properties" will mean properties of that phenomenal, experiential object.

Garden Elements as Phenomenal Properties

This "phenomenal garden" path is in tandem to the one Miller takes as she differentiates between a garden's "elements" and its "materials." She writes:

> Elements are always phenomenal, that is, perceived. The materials of the garden are the stones, rock, cement brick, water, earth, air, and vegetation of various kinds. The elements of the garden are colors, textures, shades of light and dark, fragrances, motion and stillness, various tempi and rhythms of plants or water,

relative warmth and coolness (not to be called temperature, because as an element it enters the picture only when it is felt, not measured).[35]

Phenomenal properties are as Miller describes elements: colors, textures, and the rest—and in addition, lines, shapes, symmetries and asymmetries, scents, sounds both endemic and accidental to a garden, and so forth. These are all properties whose existence is phenomenal—as Miller says, in the sense of being perceived (as conscious registrations of the world)—but that we commonly understand to be grounded causally or ontologically in the four-dimensional garden. In offering an account of these properties, we mean merely to be describing what is, and we mean to be doing this in a way that is similar if not identical to the way that anyone with working senses would describe the garden. This commonality of description is important if these properties are meant to serve as evidence for evaluative properties. To the greatest extent possible, these descriptions should be value-free. The degree to which that is possible depends on one's epistemic theory and its axiological implications: even picking out some properties over others may involve, some would claim, an exercise of valuing—so we say "to the greatest degree possible."

Phenomenal properties of gardens include "borrowed properties." Borrowed phenomenal properties are apparently unique to gardens. (That may not be the case in principle but only in practice.) East Asian garden designs and analyses cite as important the "borrowed scenery" that lies beyond the garden's borders. Both mountains and skyscrapers (think Central Park) can constitute this borrowed scenery, and gardenists, and those who appreciate their gardens, believe that these borrowed objects contribute to a garden's overall character. In this way, they become relevant to an accounting of the garden's phenomenal properties. East Asian gardens are not the only gardens that have this aspect; Brownian landscapes involve borrowed scenery as well. Sight lines from the main structure—the manor house, the castle—in Brown's designs stretch out and only end when they come upon a significant sight, usually a work of architecture in an adjoining town. Whatever lies at the end of one of these sightlines is not part of the four-dimensional garden, but as it is relevant to an account of the character of the garden from an aesthetic point of view, it may be said that those beyond-the-garden objects are part of the phenomenal garden.

[35] Miller, *The Garden as an Art*, p. 123.

Process Properties

Phenomenal (or experiential) properties derive both from objective properties (of the physical, four-dimensional garden) and from dynamic properties or "process properties." Process properties may not be perceived directly but are what allow the phenomenal properties to be as they are. Process properties may be divided into two sorts: (1) those that are part of the human-creation processes of a garden, and (2) those that are part of the nature-creation processes of that garden. The former includes all gardening practices—including things like planting, shaping, pruning, thinning, deadheading, watering, fertilizing, building—but it also includes introduction of things like the ha-ha hidden walls that Brown used to extend views and plants placed and arranged with the purpose to screen other scenes from view (maintenance sheds, irrigation equipment, views the gardenist wishes to have sprung upon the garden visitor, and so forth). The latter includes things like weather, seasonal change, climate change, ecological activity, plant growth, plant competition, and the like. These are properties that commonly are known to us through scientific inquiry.

Some process properties are in principle perceptible but are in fact not typically accessible. In those cases, their presence in aesthetic analyses would be rare. Yet, an aesthetic analysis might include reference to them in the same way that to explain visible phenomena in the way a physicist might, we include reference to objects like subatomic particles or even strings. Direct sensory experience of strings is impossible, but if they exist, the proof of that existence is dependent on their place in causal explanatory accounts. In a similar way, it is possible to think about "accidentally invisible" elements of a garden as relevant to an aesthetic description. The same may be the case for historical aspects of a garden. One might see a spent bloom on a rose bush and that might bring to mind a memory of the flower in full bloom that was present just a few days earlier. A weathered rock may bring to mind the imperceptibly slow processes that brought its present state into being.

Process properties describe not only how perceptible properties come into being but also how they change as they constitute aspects of a dynamic object, including the faded rose, the effort to grow plants in a marginal climate or grow tropical plants in a greenhouse, the effort to transplant or prop up large trees, and the general ongoing horticultural work including pruning to specific shapes. For example, *Aloe* plants express different pigments depending on how they are watered or how much sun they receive. The coloration of the plant—the perceived property—is derived from both the material of the plant itself and the process, in this case, of carefully manipulating the amount of water the plant receives. This is also in line with the idea that we do not relate to the microscopic or ultraviolet in our appreciation of gardens, even though these things are highly relevant to the growth and production of the plant. Process properties are important to understanding garden aesthetics, since gardens are always in an "uncompleted" state, involve multiple people and disciplines, and include nature as cocreator. But they are especially important for a full appreciation of the relationship in a garden among objective properties (materials of the four-dimensional garden) giving rise to phenomenal properties (garden "elements" and objects of experience) which in turn give rise to the garden's aesthetic properties that figure into aesthetic analyses, comparisons, and evaluations.

Let us review.

- Formal aesthetic properties of gardens are expected to rely on two things to manifest: objective properties of gardens and a subjective interpretation of those objective properties.
- For a garden's formal aesthetic properties to be stable enough for the purposes of comparison, evaluation, and interpretation, they must rest on objective properties that are stable.
- A four-dimensional garden does not possess unchanging objective properties because it is essentially dynamic and in a plurality of states of constant change.
- Yet it is common practice to appreciate gardens aesthetically by citing their aesthetic features, including their formal aesthetic properties.
- Therefore, when we talk about the objective properties of gardens, on which its formal aesthetic properties rely, we must be talking about a garden-object that is not the four-dimensional ever-changing garden but rather a garden that exists as a phenomenal object, the focus of a subject's experience, one that is a "capture" of the properties of the four-dimensional "real" garden.
- However, if we accept that a garden's objective properties can contain dynamic properties—as would seem necessary to any scientifically informed appreciation of the garden—and if we accept that these dynamic properties can legitimately be cited in accounts of the aesthetic appreciation of gardens (something not uncommon), then we must understand those dynamic properties to be properties of the four-dimensional garden and not of the phenomenal garden.
- Happily, this does not corrupt our account of formal aesthetic properties necessarily resting on the properties of the phenomenal garden. And this is because dynamic properties are relevant to aesthetic analyses of a garden as contextual properties of the phenomenal garden while they are literally objective properties of the four-dimensional garden.
- This is no problem because the four-dimensional garden possesses many properties that are relevant to the aesthetic character of its phenomenal representation, properties readily accepted as contextual properties, properties

like its origins and history, its relationships with other gardens, its relation-ship to its visitors, and so on. None of these properties is strictly thought of as "formal aesthetic properties," so the division is clean.

Formal Aesthetic Properties

Linda Chisholm, in her book *The History of Landscape Design in 100 Gardens*, writes:

> The design principles of contrast and variety are harmoniously woven at Alhambra. . . . Alhambra is a pleasing whole. . . . The Court of Myrtles is one of simple elegance.[36]

> The garden of Vaux-le-Vicomte [designed by André Le Nôtre] appeared peaceful, elegant, and very grand.[37]

> Following in the French tradition, it [Claude Monet's garden at Giverny] is geometric in layout, but planted with a riot of textures, shapes, and especially color. In some places he grew flowers whose colors lie in close proximity on the color wheel, at other places colors that stand opposite. As did impressionistic painters, he used white to brighten and highlight.[38]

> Saihō-ji, one of the oldest gardens in Japan, is lean and subdued.[39]

> The classical and the romantic elements are skillfully arranged at Sissinghurst so that their juxtaposition is never jarring.[40]

> The Garden of Cosmic Speculation dances with charm and wry humor as it presents the newly seen world.[41]

Chisholm's descriptions are embedded in garden descriptions, histories, and relations that make for inspiring reading. And, as we see here, she offers aes-thetic commentary that fuels the inspiration and that feels entirely natural as

---

[36] Linda Chisholm, *The History of Landscape Design in 100 Gardens* (Portland, OR: Timber Press, 2018), pp. 25–26.
[37] Chisholm, p. 81.
[38] Chisholm, p. 321.
[39] Chisholm, p. 325.
[40] Chisholm, p. 388.
[41] Chisholm, p. 476.

observations based on her more "factual" contributions. Without aesthetic observations, her writing might seem incomplete. Attributing aesthetic properties to gardens is important to her work, and just as importantly, seems commonplace to those who read it.

Formal aesthetic properties may not be attributed to a garden, or parts or aspects of a garden, when we think of that garden as a four-dimensionally physical place. (At least those beyond the recasting we described above.) But formal aesthetic properties may be attributed to a garden, or parts or aspects of a garden, when we think of that garden as an object of experience.

Let's examine this claim through an example, and let's use the first quote from Chisholm above: "[The Alhambra's] . . . Court of Myrtles is one of simple elegance." "Elegance" is a classic formal aesthetic property; the attribution and the property in question are both clear. *Myrtus communis* and is various subspecies and cultivars are evergreen trees of scrubs native to Europe, north Africa, and southern Asia, and cultivated in gardens worldwide. These plants have small but showy star-shaped white flowers—that bloom in June and July—leading to dark purple/blue/black round berries in November. Given that it is an evergreen, the Court of Myrtles has an appearance that is relatively stable all year long, but of course when the plant flowers or when it fruits, the appearance changes. The plants in the Court are shaped into hedges, but given the proclivity of the plant to become a small tree, these hedges require regular shaping and pruning. In other words, these myrtles, while not prone to a great deal of change, still, as all plants do, change. The Court is enclosed by relatively high walls, and so the presence of sunlight changes throughout the day. The presence of the rectangular body of water, bordered by plants, adds to the character of the Islamic style of this Alhambra garden.

Chisholm's description of the Court as "elegant" seems very apt, and if the Court were composed simply of the water feature and architectural features, then either discounting the influence of the changing sunlight or suggesting that its influence does not alter the Court's elegance, we might say that the Court's elegance is derived from objective properties possessed by the Court. But the Court includes plants, and plants, even evergreens, change. So considering the Court as a garden, we should say instead something along the lines of "during the course of the day, when the hedges have been freshly cut to display their optimal shapes—despite whether they are flowering, fruiting, or neither—the garden is elegant." If, however, the hedges are at a point when they are not in optimal health or have not been optimally shaped—perhaps a point at which the garden is short-staffed, after a surprisingly hard winter, during a period of drought, or if a gardener is hesitant to prune so that the flowers might better be observed—then the elegance of the Court might wane. Perhaps not, but it is possible.

So the elegance is not derived from and dependent on the objective features of the Court; it is derived from and dependent on a particular state of the objective features, a state that manifests sometimes but not necessarily always. Since this state is one commonly experienced by visitors, Chisholm is entirely justified in saying that the garden is elegant and leaving the matter at that. But were this state only occasionally to manifest, Chisholm would be more correct to qualify her attribution of "simple elegance" by referring to when that state is most likely to be manifested, perhaps something along the lines of "at the point when the hedges display optimal health and shaping."

The water feature and architectural features are objective; they do not change at any perceptible rate. The optimal hedge shaping is perceptual, in the sense that it is a state that must be perceived in order that it may underwrite the attribution of elegance. The elegance of the Court as a garden is ultimately based on the perceived phenomenal properties the garden manifests at particular times and as experienced. It would likely be inappropriate to say of the Court that it is elegant in the absence of this sort of phenomenological qualification because its elegance is not dependent on features of the Court that are unchanging, in the way that the features of a photograph or a painting of the Court are stable and unchanging. The Court's aesthetic properties depend on properties that are only stable to a particular experience of a state of the garden.

The Court of Myrtles is an easy example for us to consider as its features—water, white marble, tightly shaped hedges of evergreen, specific design consideration of the positive impact of sunlight over the course of the day, and its ability to be surveyed from a single vantage point and within a single visual field—are among the most stable of any garden we may find, perhaps almost as stable as an exclusively stone *karesansui* garden. Yet even in the most stable of gardens, change of the caliber endemic to natural dynamism is present, and its presence prevents us from the perfect stability that properties of nonnaturally dynamic aesthetic forms commonly possess, the sort of stability that allows us to derive formal aesthetic properties directly from objective properties rather than from phenomenal ones.

Let's consider a second example, this time of a very small garden, one easily described and conceived. It is a simple residential garden—actually a small garden within a larger residential garden. It has existed for years, but it could have been constructed yesterday. We refer to it as "the Mickey Mouse garden" because of its shape.

Three circles, one larger than the other two, the two smaller contiguous with the larger at the clock-points of two o'clock and ten o'clock. In the middle of the larger one is a twenty-foot-tall mule palm, a cold-hearty hybrid cross between a pindo (*Butia capitata*) and a queen (*Syagrus romanzoffiana*) that looks like a coconut palm. It is surrounded by bush daisies (*Euryops pectinatus*), and they are surrounded by a low hedge of short variegated *Pittosporum* plants. In the middle of the smaller circle at the two o'clock position is a twelve foot pink-flowering crape myrtle (of the genus *Lagerstroemia*). It is surrounded by *Agapanthus* which are in turn surrounded by mature Japanese boxwoods (*Buxus microphylla*). The smaller circle at the ten o'clock position is a field of knock-out roses (a brand of roses introduced to the United States around 2000); in the middle are single orange ones; they are surrounded by double medium-pink ones; and they are surrounded by peach-flowering groundcover roses (from a brand called Drift). The ground is covered in thick cypress mulch.

From this description anyone with a rudimentary knowledge of horticulture could recreate this garden (with the size of the circles scaled to the size of the plantings), and anyone familiar with these plants reading this description likely can easily imagine it.

This garden is charming and colorful. "Colorful" may be a phenomenal property of the garden or an aesthetic property, depending on how it is interpreted. It is a property that does not always manifest—at some points none of the flowers are in bloom (or only a tiny number of roses are)—and so to describe the garden as "colorful" is already to describe only one state of the garden, one that is heavily dependent on season. "Charming" is more decidedly an aesthetic property, one that arises from a variety of phenomenal properties: (1) the garden's colorfulness; (2) the colors being clear and pronounced (green, yellow, blue, pink, orange, cream); (3) the large variety of greens; (4) the large variety of leaf shapes and textures; (5) the large variety of flower shapes, sizes, and population of their plants; (6) the carefully shaped circular hedges; (7) focal trees in the center of two of the circles; (8) the variety of heights, from the top of the mule palm to the top of the very low growing pittosporum hedge; and (9) the simplicity of the garden; it is straightforward and uncomplicated; no plant is particularly exotic.

The attractiveness of the garden is due to the presence of these phenomenal properties and to the fact that these phenomenal properties are frequently experienced. The phenomenal properties enumerated above are what allow visitors and gardeners to experience the aesthetic properties of the garden, and while the phenomenal properties are dependent on the objective properties of the garden containing, for example, *Euryops*, *Pittosporum*, *Lagerstroemia*, *Agapanthus*, and *Buxus*, the inclusion of these plants is not the exclusive cause of the garden's phenomenal properties. We do not mean because objective properties must be "interpreted" for aesthetic properties to manifest; this is true, but what we mean is that these objective properties must be arranged, shaped, tended, and at specific points in their flowering cycles for those phenomenal properties that underwrite the garden's aesthetic properties (of colorfulness and charm) to manifest.

Returning to our first example, *elegance* is what we might term a "midrange aesthetic property" in the sense that elegance is one aesthetic virtue that a garden might have among potentially many others and in the sense that the presence of this virtue might be used as evidence of a garden's overall aesthetic quality. In addition to midrange aesthetic properties, gardens are also described with summative aesthetic properties, properties that are meant to communicate summary or final judgments about a garden as a whole.

"Longwood is magnificent."
"Saihō-ji is sublime."

These properties typically are evidenced by citing a garden's midrange aesthetic properties, and so they normally are a level removed from phenomenal properties. These sorts of culminating properties generally are meant to account for the full range of a garden's aesthetic features, positive and negative, and they are meant to speak not to some aspect of the form of a garden but to that garden's entire form, taken as a whole. In our upcoming discussions concerning the interpretation and evaluation of gardens, we will delve into the various aspects of "garden form" that provide the basis of these summative aesthetic properties.

Before leaving this section, let us offer a few clarifying points. First, it is important to point out that the phenomenal features that underwrite certain formal aesthetic properties in one garden may underwrite different ones in a different garden, and that different phenomenal features will be relevant to a garden of one style over another. This is especially the case when we are dealing with different cultures. The way to know which properties are relevant in a given instance is to understand how properties are cited in patterns of garden designs, patterns that are used to identify particular garden styles. Once this categorization work is done sufficiently well, relevancy reveals itself. This pertains, of course, to what we earlier termed an "aesthetic analysis"—when it comes to a particular aesthetic experience, this may be guided by the categorization (as we would recommend) or it may be entirely free-form, depending only on what the experiencer, during a particular visit to a particular garden, finds engaging and rewarding.

Second, a focus on formal relationships among the properties of a garden is especially important because, in a garden, aesthetic properties must be scaled, and the scale must be expressed, but the scale need not be merely a single visual field. While some gardens are meant to be seen from windows and some from a vantage point like a viewing platform (in the case of *karesansui*) or a terrace (as in the case of baroque European gardens), this is the exception rather that the rule, and when the viewer moves, so does the composition within the frame. Most gardens are designed to be entered, and so the experiencer must bring with them a "mobile frame," one that can move with them as they take "phenomenal snapshots" of a garden.

Beyond the fact of the framelessness of gardens, as three-dimensional places one's appreciation can be scaled very differently depending on nothing more than a slight change of perspective. A garden appreciator can survey a single plant or a small grouping of plants, or they can view a countryside that seems to sweep all the way to the horizon; they can appreciate a small section of a garden, a "garden room" or subgarden, or they can appreciate the full garden in its entirety (evidenced by the fact that no matter how large, we still recommend visits to whole gardens, not to pieces of them).

When voicing an aesthetic property attribution, one should be prepared to express the scale employed in experiencing this property. Without either an expressed mention of the employed perspective or some shared tacit under-standing of it, one may be confused as to what conditions must be met to ex-perience the same property. Expressing scale is common. When we voice an attribution, it is common to include some subtle (or not so subtle) allusion to the scale as a way of quietly inserting an evidential element.

> "When you think about the whole of Kew, you see how the various aspects inte-grate into a single coherent design."
>
> "The white garden at Sissinghurst is incredible."
>
> "You want to visit the rock garden at Ryōan-ji when the sun is coming up or going down so the cast shadows give added dimensionality and weight to the rocks."

A typical Brownian landscape, despite the fact it normally will include exten-sive sightlines and visual corridors, is larger than can be registered in a single visual take. This is certainly true if one's vantage point, as would be common, is from the manor house or castle; one can only look in one direction at a time. The same is true of the gardens at Versailles. A typical East Asian strolling garden—Chinese or Japanese—is meant to have multiple discovery points as one strolls it; it is designed in such a way that it prevents being seen all at once. Botanical gardens and pleasure gardens are all typically far too large to take in at one go. Yet we talk, seemingly intelligently and without controversy, about the aesthetic features of these gardens as whole gardens.

Third, while the analogue we have relied on throughout this discussion may be thought to imply that garden appreciators take phenomenal "snapshots" of single moments in time—and while that may certainly be the case from time to time—a phenomenal "capture" of a garden may be an experience that lasts quite a while. John Dewey talks about "*an* experience" being one that has an internal cohesion or integrity that is marked by having a narrative arc with a discernible beginning, middle, and end. For Dewey, aesthetic experiences are a species of "*an* experience." If we take seriously Dewey's insight, then an aesthetic experience of a garden may be momentary, with that narrative arc occurring just in the moment in which the appreciator registers the experience, or it may last for some time, as would typically be the case in a day-long visit to a pleasure or botanical garden.

When we attend a performance, we typically invest a stretch of time attending to it. The same is true of watching a film or reading a book. These are, as we mentioned, "closed" dynamic aesthetic events, and as closed events, they possess objective properties stable enough to allow for formal aesthetic property attribution. A visit to a garden—say one that lasts three hours—is also a closed event, and while that visit allows for an experience of the garden at a certain season, at a certain time of day, with certain climactic conditions—and so cannot be said to capture that garden in its "four-dimensionally real" entirety—it is stable enough to support an aesthetic analysis of the properties experienced.

Garden visits are more normally like this—spread over a period of time, and temporally scaled to the duration of that visit—but it is also common for a visitor to be struck by a single flower or leaf, a single rock, plant, or honeybee, an elegant parterre curve or symmetry of trees in an orangery or bosquet, or the rich sweep of Versailles almost to the horizon. Admiration of these single elements can be arresting and can easily constitute experiences unto themselves. We all likely have images that are caught like photographs in our minds. And we all likely have particular aromas—a sensory modality frequently employed in the appreciation of gardens—locked away in our hippocampi that, once triggered by a repeat of that aroma, occasion powerful memories.

This third point leads to a fourth: following Dewey, garden appreciation as bounded experientially is the typical way garden visitors tend to describe their aesthetic appreciation of gardens. This is well illustrated in those papers we have discussed that analogize garden appreciation to temporal artforms like music, performance, dance, and film. We experience a garden during a time that has a well-defined beginning, middle, and end. This is the case even with our own residential gardens. We say, "I'm going to work in the garden now," and "I spent three hours working in the garden today." We rarely suggest that our acquaintance even with our own gardens is continuous, and where we do, we express that in conceptual terms.

This brings us to a fifth and final point of clarification. When we make pronouncements about the aesthetic features of gardens that transcend particular experiences of a garden, these observations express patterns that seem well

evidenced by repeat visits. If one wishes to say that a certain garden is particularly colorful, they might qualify this claim by saying "during the spring," or
"during the summer." But where they choose not to offer such a qualification—
where they wish to speak about the garden as a whole, to attribute an aesthetic
feature to a garden regardless of context—that attribution is the citing of a pattern, one found through many concrete experiences with that garden. Patterns,
whether scientific or aesthetic, are conceptual in character because they are
abstractions based on actualities—a finite number of experiences meant to represent an infinite whole. The "white garden" at Sissinghurst Castle is white because in that garden are many white flowers, blooming from many different sorts
of plants. The "whiteness" of that garden is only a property of the garden because
of the many single instances of white flowers. This sort of pattern-drawing is very
common, but it must be taken theoretically as the conceptualization that it is. We
will say more about this below, as we discuss how we identify particular garden
styles by their objective and their phenomenal properties.

Formal Properties of Garden Styles

The act of abstracting a phenomenal object from a four-dimensional object, as
we have been discussing, can be done with an individual garden but, given that
it is an abstraction, it can also be done with a particular garden style. We can
attribute aesthetic features to garden styles, or kinds of gardens, as we can abstract from their instances patterns that not only are used for the purposes of
identifying and classifying those garden styles but also recognizing that garden
styles possess repeated aesthetic patterns. Walter Hipple, in reviewing Franklin
Hazelhurst's book *Jacques Boyceau and the French Formal Garden*, writes:

> Hazelhurst's book shows convincingly that it was Boyceau who, early in the sev-
> enteenth century, elicited from Italian influences and French experiments in
> garden design a distinct aesthetic ideal; who first designed works in this style,
> organically composed in themselves and integrally related to the accompanying
> architecture; who infused the whole with a new monumentality; who elevated
> his art to the rank of a learned profession; and who expounded the ideal of the
> new system in the first major treatise on the French formal garden. . . . I should
> welcome a more explicit aesthetic analysis of styles in gardening. For instance,
> Hazelhurst argues persuasively that Boyceau first developed compositions in
> which the parts are not simply additive but in which they form a true organic
> unity in which "nothing can be subtracted from the design without altering the
> whole." But this unity remains, I think, a classic rather than a baroque unity—a

unity in which each part, though necessary to the whole, has still a life of its own ("may be grasped as a separate entity," as Hazelhurst himself remarks).[42]

This review demonstrates that performing an aesthetic analysis on the oeuvre of a gardenist—as Hazelhurst does with Boyceau—is not only eminently possible but can then be a topic of conversation as another (Hipple) reflects on that aesthetic analysis and points out its highlights and perhaps its limitations. The content of the review and the conversation it represents is interesting, but what is of still greater interest is the fact of the conversation itself. There is no hesitation in engaging in aesthetic discourse about garden styles, about the aesthetic features of those styles, and about how one may describe how another goes about describing those first two.

In the course of two papers—"Visual Perception in Japanese Rock Garden Design" and "Visual Geometry of Classical Japanese Gardens"[43]—Gert van Tonder offers a detailed look at the physical features of these gardens as they underwrite the gardens' aesthetic properties. Van Tonder works from patterns of features of Japanese (or Japanese-style) gardens that are repeated throughout this garden style.

[S]uccessful garden composition partly results from the application of intentional, complex design principles, with anticipated visual effects. . . . Japanese gardeners use comparatively homogeneously textured materials. Saturated colours and high contrast texture markings are avoided, possible since low contrast objects tend not to dominate visual attention. . . . Uniform surface regions with even textures simplify the creation of boundary contours. The process of visual segmentation is thus simplified through the reduction of the number of sub-segments within the interior of each object. . . . Adaptations to spacing between rocks can thus be used to create various moods, such as emptiness or liveliness. . . . Gardeners carefully consider the visual appearance of spatial junctions (boundary contours) between rocks, as well as surface textures of rocks themselves (surface regions). . . . Natural patterns are often self-similar, the outcome of repeated application of the same organizing principle at multiple spatial scales. . . . Duplication, causing visual bias towards any part of the design, is thus avoided. . . . Medial axis transformation is not only useful for compact description of shape regions, but also reveals structure in empty spaces between figure. . . . Structure is embedded in hierarchies of trilateral

---

[42] Walter J. Hipple Jr., "Review of *Jacques Boyceau and the French Formal Garden* by Franklin Hamilton Hazelhurst," *Journal of Aesthetics and Art Criticism* 26:4 (1968), pp. 548–549.

[43] Gert Jakobus Van Tonder, "Visual Geometry of Classical Japanese Gardens," *Axiomathes* 32 (2022), pp. 841–868.

junctions arranged into multi-scale, dichotomously branching patterns in both visual figure and ground, with the main difference being that, while figure converges away from the viewer, the structure of ground converges in the opposite direction. . . . [O]ur work therefore suggests a new link between structure and aesthetic understanding of this type of garden design. . . . The resemblance between the tree structure in Ryōan-ji and natural materials used in other art forms, such as *ikebana*, suggests direct applicability of techniques from one discipline on another.[44]

This long excerpt, taken from various places in the first of the two articles mentioned above, demonstrates both the successful articulation of the physical features typical to Japanese gardens and—as we see in the final two sentences— that the author takes these physical features to underwrite the aesthetic form of the garden design, one that may be repeated in other artforms.

This same sort of project is pursued by Mohammadsharif Shadidi et al., in "A Study on Cultural and Environmental Basics at Formal Elements of Persian Gardens (before & after Islam)."

The Iranian garden is designed in square form. This figure is mostly full square or rectangular. In the square geometry, the quarter is the circle and circle is the depth of universe that hides its essence, From ancient, Iranian know square, as coming from quartering circles, as the basis of their designs in sacred and holy place plans. On the other hand, the structure of Iranian garden is based on circle quartering and Mandela figures with division of water circulation in water canals. . . . In the Iranian garden, there is usually a corridor made by cultivation of cypress tree, plane tree and pine trees in the main axis of the garden. . . . [E]ach section of garden, made in square and rectangular form, is divided into smaller squares. In the peak of each corner of this square shape network, a tree with longer life was cultivated. These squares are then divided into smaller squares and in the pack of each corners, trees with average life and with the same arrangements, trees with shorter life in the peak of smaller squares were cultivated. . . . By finding the precise relationship among those factors in the structure of Iranian garden, each element is placed in its place as its merits and qualification dictates and the Iranian garden comes in harmony with cosmic rules and eternity.[45]

[44] Gert Jakobus Van Tonder and Michael J. Lyons, "Visual Perception in Japanese Rock Garden Design," *Axiomathes* 15:3 (2005), pp. 353–371. Quotes come from pp. 353, 360, 363, 366, 368, and 369.

[45] Mohammadsharif Shadidi, Mohamad Reza Bemanian, Nina Almasifar, and Hanie Okhovat, "A Study on Cultural and Environmental Basics at Formal Elements of Persian Gardens (before & after Islam)," *Asian Culture and History* 2:2 (2010), pp. 133–147; quotes come from pp. 135 and 137. Quotes are original and without benefit of standardization in English.

Motivations for the placement of components in the Iranian garden, as described above, are based on rules developed from metaphysical commitments, but these metaphysical commitments are manifested through the application of design principles derived from them. Patterns among Iranian gardens are easy to pick out as the rules are so clear and prescriptive.

These sorts of approaches are not unique to The Garden. In the dance-categorization example earlier in the chapter, each level of consideration apart from the one directed simply to the particular performance could sustain some aesthetic description of the type. Balanchine ballets have an aesthetic signature; formal ballets have one; ballet as a dance type has a set of characteristics that can be identified with—and, perhaps more importantly, by—a select set of aesthetic features. As all artforms are amenable to various levels of categorization, and as we identify artforms and their various levels by properties that are objective and properties that are aesthetic, it is not only possible but common to abstract from particular instances of these objects patterns that allow for this. The abstraction we do there we also do with gardens, but with gardens—given their dynamism—we must take abstracting all the way down to the "single-instance" level. It should come as no surprise that gardens have sets of aesthetic features that are stable enough to be categorized into garden style. Beginning students of landscape de-sign history are able, in the course of an exam, to identify garden styles from plan drawings or verbal descriptions, and to sketch imagined gardens in a given style.

The descriptions we offered earlier of various garden styles all rely on identifying features that are common, in varying degrees, to all gardens of a par-ticular style. While our focus was like the focus of those scholars mentioned di-rectly above in the sense that we largely focused on objective properties of those gardens, we, like the above-mentioned scholars, take those objective properties to have significance to those gardens because they form the bases of their designs, and decisions that informed those designs were artistic in essence. In other words, while we all may cite objective features to ground our identifications and characterizations, the objective features cited are present because they are driven by aesthetic concerns that drive their gardenists' artistic, expressive designs.

As gardenists design, as they create plans that instantiate their visions for a garden they hope to bring into existence, they work with what Miller describes as "elements." These are composed of sets of phenomenal properties the gardenist wishes to have incorporated in the design and wishes to have present in the resulting garden. A gardenist may approach a design with a particular palette of plants whose colors, textures, and forms will lend themselves to the desired phenomenal properties, or they may begin their designs knowing that their material choices will be limited to certain categories (only native plants, only sun-loving alpine plants, only stones indigenous to the site, and so forth). Still, it is a garden's phenomenal properties that motivate the inclusion of a garden's objective properties. This seems to follow the process of typical garden design—the material components are secondary to which views ought to be maintained, which components ought to be hidden, which routes are needed to control traffic, whether the garden is to be open or enclosed, manicured or naturalistic, peaceful or effluent. Garden designers' choices are motivated by their wish to express a vision through the presentation of elements. And so it is patterns of phenomenal properties that drive design choices; objective properties are what may be manipulated concretely to achieve patterns of comparable experiences of these phenomenal properties, but it is the phenomenal, experiential properties that drive design.

In the next chapter, we move beyond formal aesthetic properties and consider the relevance to gardens of "contextual" aspects, essentially external relations a garden may bear to other gardens, to gardeners, to place and time, and so forth. While aesthetic appreciation, it may be argued, begins with appreciation of an object or event's formal aesthetic properties, contemporary aesthetic appreciation, especially as it is illustrated by the work of art historians and art critics, typically includes consideration of these external relations.

6

The Relevance of Context

In the last chapter, we described what it means to appreciate anything aesthetically, and in that description we said that at the point where one appreciates the
formal aesthetic properties of an object, there the formalist story ends. But we
went on to say that for us, the story continues with appreciation of external relations that simply appreciating through one's senses does not reveal. These are
contextual matters, and for some—likely for the majority of garden enthusiasts—
these matters are highly relevant to appreciating a garden. We want to spend this
chapter exploring the case for the relevance of context and what constitutes the
content of appropriately relevant garden context.

The Limits of Disinterest

Earlier we sketched the history of the disinterestedness tradition as it dominated
"aesthetic attitude" theory. As we saw earlier, Miller takes disinterestedness to
be a major sticking point in how we appreciate gardens, as gardens do not lend
themselves to appreciation that is disinterested. We believe there is a range of
good reasons to reject disinterest as an appropriate attitude for garden appreciation, and we will sketch some below. We might note that a purely formalist appreciation of a garden *may on occasion* lead to a high-quality aesthetic experience
of that garden, but we believe the opposite is more commonly true: an engaged
approach—with "engagement" being the term with which we will contrast "disinterest"—leads to richer and deeper aesthetic experiences of gardens.

Engagement

We begin with a general question: what kind of appreciation is the aesthetic appreciation of gardens? Cooper takes up this question as a central focus of his

Central Park Conservatory Garden, New York, NY
Developed 1937, redesigned by Lynden Miller, 1987
Photo by Samantha Bachert, Spring 2018

The Art and Philosophy of the Garden. David Fenner and Ethan Fenner, Oxford University Press.
© Oxford University Press 2024. DOI: 10.1093/oso/9780197753590.003.0006

book, *A Philosophy of Gardens*. There he explores whether appreciation of The Garden can be understood in terms of how one appreciates art—for instance, paintings—or in terms of how one appreciates nature. Neither he finds to be a sufficient model. For the reasons we explored in an earlier chapter, The Garden is different from many other artforms, and these differences lead to the conclusion that to think of appreciation of The Garden in the same ways we appreciate paintings will not work as an explanation of garden appreciation. The same is true in the other direction. Since The Garden possesses so many characteristics in common with other artforms, to appreciate The Garden simply as nature will not work either. Furthermore, an approach that "factorizes" garden appreciation into art-and-nature appreciation—where we appreciate The Garden as art in terms of those characteristics it shares with art, and we appreciate it as nature in terms of those characteristics it shares with nature—fails as well, because it reduces garden appreciation into those two disparate forms of appreciation and thereby misses that appreciation of The Garden has aspects that are unique and only understandable when appreciation is of a garden as a whole. Among other things, such an approach misses the phenomenon Cooper describes as "garden atmosphere." We need not say that in a garden the whole is greater than the sum of the parts; we need only say that the whole is found only through a sum of the parts.

Typically, we visit gardens to experience them aesthetically, and typically, such a visit is immersive as we stroll through a garden, allowing it not only to envelop us but engage a wide spectrum of our senses.[1] In most cases—though, not all, as we have seen with certain East Asian gardens—we move through a garden as we experience it. Garden experiences present sensory envelopes that contain us. Sensory stimulation comes from all corners and all directions, and many times it comes in all forms. As Ross says, The Garden engages us visually, but as we listen to the wind, the birds, the squirrels, the insects (walk in a treed American Southern garden just after dusk in the summer and try to be heard over the cicadas), and so forth, we are flooded with auditory stimulation. Olfactory stimulation in a garden is potentially just as strong: the scents of flowers, of green foliage, of tree bark, of rocks baking in sunlight, of wildlife scat, of the sulfur present in the well water being sprayed over the plants, and so forth. There are "setting and framing" plants all around Walt Disney World's Magic Kingdom that demonstrate in no uncertain terms that humans enjoy engaging with plants tactilely; they are chewed up from hands feeling them day in and day out. This may be lamentable visually and ecologically, but it might be celebrated as yet another sensory modality engaged as we experience The Garden.

---

[1] Mara Miller says the same on p. 32 of *The Garden as an Art* (New York, NY: State University of New York Press, 1993).

Many love to walk through the grass barefoot or lie on it. And some love to taste a piece—and the gustatory is one more sensory modality engaged. We not only enjoy tasting the purposed produce of vegetable gardens and fruit trees; we occasionally like to taste items in the garden—like grasses, sorrel leaves, or manzanita fruit—that are not commonly thought to be tasteworthy. If we add to this list the physical enjoyment experienced careening down a garden slope, crossing a bridge, following a winding path, making our way up a hill—as we experience a garden kinesthetically—we complete the full list and establish in concrete terms the breadth of the sensory envelope The Garden presents.

In addition to a garden's capacity to engage our senses, as we enter a garden we become a part of that garden and contribute our own traits to it; we become a part of the natural aspects of that garden and the ecological processes of the place. As such intimate participants, it is illegitimate to conceive of ourselves as merely spectators, as being observers only. As we talk, we add to the auditory stimulation; as we bring in food or wear perfumes, to the olfactory; as we wear brightly colored clothing, to the visual. We become garden pests, on a par with aphids and gophers, when we litter, steal, or run through garden beds. Humans are part of nature, and when we stroll the paths of a garden, we are a part of the nature of that garden.

Arnold Berleant follows John Dewey in his celebration of "engagement" as not only a virtue but a defining characteristic of aesthetic experience, especially evident in interaction with nature.[2] An aesthetics of engagement takes to heart the multisensory and enveloping character of nature as well as the fact of the contiguity between nature and humans as parts of nature. It erases the dualism that is implicit in a disinterested approach. Kant's program requires us to hold the art object firmly as merely a phenomenal, experiential focus. In this, Kant's distinction between the object and the subject is not merely conceptual, it is materially dualist: the subject contemplates the object but enters into no other relation with the object but that one. As we enter a garden—even when this entrance is virtual

---

[2] Arnold Berleant, *The Aesthetics of Environment* (Philadelphia, PA: Temple University Press, 1992). See also Noël Carroll, "On Being Moved by Nature: Between Religion and Natural History," in S. Kemal and I. Gaskell (eds.), *Landscape, Natural Beauty and the Arts* (Cambridge, UK: Cambridge University Press, 1993), pp. 244–266; and Emily Brady, "Imagination and the Aesthetic Appreciation of Nature," *Journal of Aesthetics and Art Criticism* 56:2 (1998), pp. 139–147. In writing about Berleant's aesthetics of engagement in relation to appreciation of nature, Carlson classifies Berleant's view as a version of noncognitivism—which is not to mean that the view is anticognitivist but merely that it places emphasis on something other than cognitive engagement as the focus of an aesthetic experience of nature. Allen Carlson, "Environmental Aesthetics," *Stanford Encyclopedia of Philosophy*, online. We take both cognitivist and noncognitivist approaches as essentially antidisinterested, and so both approaches are welcome as arguments for why disinterestedness should be rejected. We only mention Carlson's classification to illustrate the point that a variety of contemporary environmental aesthetic views entail rejection of disinterestedness.

rather than physical—we become a part of that garden, and the dualism between object and observer is erased or at least massively reduced.

Stolnitz makes a special point of saying that disinterest should not eliminate the aesthetic efficacy of an emotional response to the aesthetic object—what he calls "sympathy" with the object. That he felt compelled to add this caveat to his treatment of disinterest suggests that the tradition had been read in ways that discounted emotional connection. This may be the influence of a late addition to the tradition, that of Bullough, who argued that adoption of an attitude of psychical distance entailed that one should hold one's affective engagement with the aesthetic object at a distance (albeit, he said, at the least possible distance). An aesthetics of engagement, on the other hand, emphasizes the importance of an empathetic connection between the subject and the aesthetic object; it makes emotional (affective, psychological) openness and engagement not merely an afterthought but a core principle of what it is to experience aesthetically.

Categorization

In 1970, Kendall Walton wrote an article that has had far-reaching effects on aesthetics, "Categories of Art."[3] In this paper, he demonstrates how art as a field progresses through the relationships that one work of art has with others, how it continues the patterns of art that came before it and how it breaks with them. These patterns are only visible if one first places a work of art within a correct categorization. This same approach works for gardens. To understand a garden is first to understand the kind of garden it is—to understand its purpose or purposes, to understand how it functions as a garden, to see how it is like and unlike other gardens. This is yet another reason why we began this book with a garden inventory; that inventory is a taxonomic approach to this very categorization. To experience or judge a "setting and framing garden" as if it were a "destination garden" is not only to misunderstand it as the garden it is but to hold it to a set of standards that are not applicable. If we expect a constant rotation of bedding plants and massive sweeps of color from every garden we visit, we inevitably will be disappointed by our visits to research-focused botanical gardens. We miss out on the true value of a particular garden if we do not understand it as the kind of garden it is, and this loss is not merely intellectual, it is experientially substantive. An experience that might be wonderful (as an experience), if the object of that experience is misunderstood, may turn sour.

The Kantian advice of not attending to the object of aesthetic judgment under any category, of maintaining a pure focus merely on the presentation of

---

[3] Kendall L. Walton, "Categories of Art," *Philosophical Review* 79 (1970), pp. 334–367.

the object's properties in relation to nothing whatsoever but their own internal interrelations, sets us up for misunderstanding the value of a garden viewed this way. We will judge it incorrectly, and yet this is the very point of Kant's advice: to place us in the correct posture for accurate judgment. Walton's advice is the better of the two, likely for all aesthetic objects but certainly for gardens as gardens come in such remarkable variety and with such a huge range of purposes.

Science

While disinterest *may* pose no danger to appreciating the qualities of works of centrally canonical artforms, and while it is certainly possible to have a good aesthetic experience while only focusing on an object's perceptual properties and the formal aesthetic properties derived from them, the "nature aspects" of a garden cannot be appreciated disinterestedly. Those "nature aspects" of a garden have to do with its physicality, the fact that as nature it obeys natural laws, the fact that it is a "real place" that is existent and bears relationships to other objects, in particular surroundings that are natural. Those nature aspects—alongside the contributions of garden practices—are the "dynamic process properties" that are responsible for a garden containing the elements it does and the phenomenal properties we experience it as having. *Contra* Kant, we must bring the object— the garden under consideration—"under some category" because this is endemic to understanding the garden scientifically as being a part of nature. Our appreciation of the natural processes of a garden are only worthy if they are informed ecologically or biologically.[4] This is not to say that imaginative engagement— seeing a ring of mushrooms as evidence of the presence of fairies—is somehow wrong. But when we are appreciating the natural processes that steer the changes in the garden—either as we plan, maintain, or just enjoy a garden—in scientifically informed ways, we can appreciate the character of these changes as parts of larger patterns and, better still, we can predict their expressions, a delight all garden enthusiasts commonly enjoy.

---

[4] Allen Carlson, "Nature, Aesthetic Judgment, and Objectivity," *Journal of Aesthetics and Art Criticism* 40:1 (1981), pp. 15–27; *Aesthetics and the Environment: The Appreciation of Nature, Art and Architecture* (London: Routledge, 2000); "The Requirements for an Adequate Aesthetics of Nature," *Environmental Philosophy* 4:1 (2007), pp. 1–12. Carlson is joined in a cognitivist approach to the aesthetic appreciation of nature by many: Holmes Ralston, "Does Aesthetic Appreciation of Nature Need to Be Science Based?" *British Journal of Aesthetics* 35:4 (1995), pp. 374–386; Marcia Muelder Eaton, "Fact and Fiction in the Aesthetic Appreciation of Nature," *Journal of Aesthetics and Art Criticism* 56:2 (1998), pp. 149–156; Yuriko Saito, "Appreciating Nature on Its Own Terms," *Environmental Ethics* 20:2 (1998), pp. 135–149; Glenn Parsons, "Nature Appreciation, Science, and Positive Aesthetics," *British Journal of Aesthetics* 42:3 (2002), pp. 279–295, and "Theory, Observation, and the Role of Scientific Understanding in the Aesthetic Appreciation of Nature," *Canadian Journal of Philosophy* 36:2 (2006), pp. 165–186.

A scientific understanding of plants increases our aesthetic appreciation of gardens in a variety of ways.[5] We may give a certain plant special attention if we have the knowledge that it is flowering for the first time in cultivation, or if it is extinct in habitat, or if, in morphological terms, that which towers above our heads is a single leaf. We better appreciate that gardens are ever-changing when we have a basic understanding of plant growth, and we may marvel even more at the clamping of a Venus fly trap if we understand the mechanism by which the leaves move. Every elementary school student understands that flowers and bees have a purpose (pollination), and it is hard to imagine this fact absent from our appreciation of the world. Yet there was a time, fairly recently, when garden visitors would have looked at the springtime dance of flowers and bees as another mystery of the natural world. There are a number of other botanical "mysteries" that have been figured out just in our lifetimes, and this knowledge only adds to our understanding of plants as active participants in the garden world. Even walking on a flat grassy path may be enriched by our knowledge of the fact that the plants are communicating with one another via volatile compounds with each step we take, and that deep underground the roots of many species are sharing resources through webs of fungi. Whether we are looking at the microscopic—biochemical mechanisms and genetic expression—or the macroscopic—evolution and ecological succession—understanding the scientific processes that shape the garden informs and augments our aesthetic experience.[6]

Imagination and Association

One way The Garden invites engagement is imaginatively and associationally. Earlier we introduced the distinction between "aesthetic analysis" and "aesthetic experience." When we engage with a garden experientially, this experience is

---

[5] There are, of course, botanical disciplines—taxonomy, for example—that may not increase our appreciation of plants or gardens. In the past couple decades, the genus *Aloe* has been moved to different taxonomic families based on new genetic evidence. Our relationship to the aesthetic character of these species is likely to be no different if we know that they belong to the family *Asphodelaceae* rather than *Xanthorrhoeaceae* or *Liliaceae*. And there is the possibility that an increased understanding of science could decrease our aesthetic appreciation of gardens—a field of flowering *Ficaria verna* may be one of the showiest displays in the garden, but knowing that this is a noxious weed and invasive plant outside of its native range may cause our admiration of its aesthetic features to sour.

[6] Miller writes, "There is some evidence to suggest that preference for a single type of landscape or biome—the parklike setting with short grass and scattered large trees that approximates the African savanna—is characteristic of our species." Miller, *The Garden as an Art*, pp. 64–65. This is also discussed in her paper "The Garden as Significant Form," *Journal of Speculative Philosophy* 2:4 (1988), pp. 267–287, especially pp. 270–271. If this claim is true, it suggests there may be landscape features—understandable not merely as physical features but also as aesthetic features—that are "hard-wired" as human preferences. This would suggest that there are scientifically discoverable facts not merely about the features of gardens as objects but also about subjects, too.

typified by aspects that are personal and sometimes unique to us as individuals. A certain smell, a certain color, a strong breeze, and we may be reminded of another place or time, another person or significant object. Not only are we connected to that garden as our memory guides, but we may imagine all sorts of fictions that aspects of the garden prompt.[7] Mushrooms and fairies, Aeneas and the underworld, kings and palaces, cats and bugs—The Garden invites all manner of imaginative constructions. As we engage with the "nature aspects" of a garden, characterized by considering a garden as a real place with real relations and following real scientific processes, so we also may engage with the "art aspects" of a garden that encourage flights of fancy and psychological connections that are personal and individual to our particular experiences of a garden here and now. This follows the spirit of Miller's and Ross's focus on The Garden as a virtual place, where illusion is an important component.[8] In this, both Miller and Ross follow the work of Susanne Langer as expressed in her book *Feeling and Form*.[9]

If The Garden is to be thought of as a nature-and-art amalgam—appreciable fully only when it is in its unified, "unfactorized" state—then its artful aspects must be appreciated not only as standing alongside its scientifically focused nature aspects, but also as intermixed with them. This is not to say that science must admit of fantasy but rather that in garden appreciation, one can move seamlessly between the two, and that the scientific appreciation of a garden can complement and inform a fantastical appreciation. Appreciation of the carnivorous nature of a pitcher plant can conjure images of deadly beauty, the suspense of capture, and the horror of a slow devouring demise. The beauty of the plant as fit for the expression of its function is only the start: one can then imagine all manner of fantasy connected to the function itself. This can, not quite as intuitively, work in the other direction as well. The creativity expressed through the instantiation of a garden can direct attention toward greater investment in learning how the garden actually functions. As one is captivated by a plant in a garden whose flowers resemble butterflies or pinwheels or doll's dresses, one will first want to know its common name, then its generic and species name, then how to grow it successfully, then how to tend to its pests, then how to propagate it. One may then want to know where and how it grows in its natural habitat, what it grows with, what types of things pollinate it, what hybrids or cultivars exist, and all else there is to know about the plant. Over time, the hobbyist may join societies devoted

---

[7] While a garden cannot present itself as a fiction, it can provide a context where we conjure fictions imaginatively.

[8] Miller, *The Garden as an Art*, pp. 125–127. Stephanie Ross, *What Gardens Mean* (Chicago, IL: University of Chicago Press, 1998), p. 176.

[9] Susanne Langer, *Feeling and Form* (New York, NY: Charles Scribner's Sons, 1953).

to appreciation of their preferred plants, and, as their informal observations expand year after year, they will become bona fide authorities on the genus.

The gardener gardens over the long term. Few gardeners think of themselves as itinerant.[10] Gardeners are settled in their sites, and this means they commonly spend many seasons with their gardens. They develop historical connections with their gardens, and with these connections come associations to events that took place at various times in the garden's lifespan. Noticing the spring appearance of caladium may bring back memories of Easter egg hunts, and that in turn may take one back years to when their children were little. Seeing a rose bloom from a plant that once occupied a place in one's grandmother's garden reminds one of earlier times. These memories bring to the present experiences that contributed not merely to the life of the garden or the life of the gardener in the garden but to the constitution of a human life itself.

Function

Miller writes:

> Far from being detrimental to the aesthetic status of a work of art (as Kant saw it), our "interest" in it [a garden] and the practical constraints derived from its intended function(s) may well serve as a formal stimulus to creativity.[11]

Miller's focus on a garden's functions strikes at the heart of the disinterestedness tradition as from its inception the core notion was that disinterest meant disinterest in any purpose to which the object might be put. Yet gardens are replete with functions, modest and grand, and even those gardens not specifically designed to serve specified functions still commonly do as places for people to congregate, to play, to socialize, to muse, to problem-solve, to plan, and so forth. To attempt to separate a garden from its various functions—designed and incidental—is to break up appreciation of that garden into constituent parts. This is not only alien to common garden appreciation, but it involves the very factorizing against which Cooper argues.

We said above, "while disinterest *may* pose no danger to appreciating the qualities of works of centrally canonical artforms—," the truth is that in the case of some artforms, disinterest does pose a danger. As we saw earlier, architecture

---

[10] Unlike the case with some agricultural harvesters. The practice of "itinerant farming" encourages a separation between harvester and nature, and this is yet one more reason—added to the horror of the plight within which many itinerant farmers or migrant workers live—to view the practice with shame and a recognition of the injustice being imposed on those trapped in this life.

[11] Miller, *The Garden as an Art*, p. 101.

can only be appreciated as architecture when its functions are taken into account; absent consideration of those functions, architecture is sculpture. Music in film is typically used by directors to encourage in the audience a particular mood, to heighten suspense, to underscore the progression of a particular montage, and so forth—in short, film music is typically functional. John Williams's contribution to the flow of tears at the end of *E.T.*, to the excitement and fun of *Raiders of the Lost Ark*, and to the magic and mystery of *Harry Potter* cannot be dismissed. A film soundtrack (at least one from any Classic Hollywood Cinema film past the silent era) cannot be approached in the same way one would approach a straightforward work of (multipart) music; it must be understood in the context of the film and in terms of its contribution to the aesthetic progression of the film. Something similar is true of other aesthetic forms as well: furniture, cars, boats, appliances, and so forth. While these are not widely accepted as artforms—despite a growing level of discussion focused on "everyday aesthetics"[12]—they are sets of objects where form and function are commingled inextricably.

While a garden need not have any purpose other than to be an object of aesthetic attention,[13] the value of gardens depends on understanding that garden in terms of its purpose. The flower beds of botanical gardens and of pleasure gardens tend to be very different: pleasure gardens tend to have swaths of the same flowering plant, to provide as great a colorful impact as possible. Botanical gardens tend to have single instances or small groups of plants because the offering there is the demonstration of diversity. Pleasure gardens are not "interrupted" by much signage; a botanical garden without signage is not a botanical garden or at least not much of one. This sort of example is readily repeatable; correct garden categorization is essential not only to valuing a garden appropriately but even to experience it appropriately by focusing on those aspects that are relevant to its identity as the kind of garden it is.

Experiencing an object both aesthetically and functionally can at times enhance the aesthetic quality of that experience. The experience of appreciating the aesthetic features of Westminster Abbey while simultaneously engaged in a worship service is a case in point. Some gardens are understood by those who occasionally occupy them as sacred spaces, and the sacredness of the place may be tied, as those persons describe why it is sacred, to its aesthetic features. For these individuals, that garden is at the same time aesthetic and (functionally) sacred, similar to participation in a service at Westminster Abbey. It is common practice to engage with a garden aesthetically while also gardening in it, eating in it,

---

[12] Yuriko Saito, *Everyday Aesthetics* (Oxford, UK: Oxford University Press, 2007); and *Aesthetics of the Familiar* (Oxford, UK: Oxford University Press, 2017).

[13] Miller lists a wide range of purposes a garden may serve on pp. 32–33 of her *The Garden as an Art*.

playing in it, chatting in it, and so forth. The two foci on occasion enhance each other—and time spent in a garden exhibits this phenomenon often and well.

When a garden is designed, it must include access points—routes for gardeners to step into beds for weeding, pruning, maintenance, and the like. The design must conform to the sun exposure of the site, for example, and plants must be chosen so that none will outcompete or cast shade on the others. These are aspects of design that are solely about functionality, but if we were not to include these, the garden would fail. The aesthetic worth of the garden is brought into view only by including aspects of design that both stimulate us from a simple sensory point of view and that serve the function of the garden as a place of growth and continued maintenance.

In response to arguments that hold that the ethical aspects of works of art should not be brought to bear on the aesthetic evaluation of a work, Berys Gaut creates a middle path—"ethicism"—that understands an ethical aspect of a work of art as having aesthetic implications.[14] That is, if it can be shown that an aspect of a work of art that might typically be seen as ethical contributes to decreases the aesthetic value of that work of art, then one is justified in considering that ethical element in their aesthetic judgment of that work. For instance, if a work of art prompts us to respond in a way that is morally compromising—such as adopting an attitude or belief that is morally objectionable—and we do not adopt that morally suspect response, then the work has failed in one of its aims, and this then is an aesthetic failure of the work. And the opposite scenario is true, as well: if a work of art prompts in us a morally appropriate response, and we do indeed adopt that response, this is an aesthetic virtue or success of the work.[15] Gaut's insight demonstrates that matters traditionally deemed external to the set of aesthetic features of an object may be related to this set in ways that do not compromise that primary aesthetic focus. This is an argument against both formalism and disinterest, and it is cleverly constructed to demonstrate the *aesthetic* relevance of these external considerations. The final chapters of this book concern the relationship of gardens to ethical matters, but the point of raising Gaut's argument here is to add to the strength of the general argument that consideration of function can make an aesthetic difference.

---

[14] Berys Gaut, "The Ethical Criticism of Art," in Jerrold Levinson (ed.), *Aesthetics and Ethics: Essays at the Intersection* (Cambridge, UK: Cambridge University Press, 1998), pp. 182–203.

[15] Aspects of this approach are echoed in these two works: Noël Carroll, "Moderate Moralism," *British Journal of Aesthetics* 36:3 (1996), 223–238; and Martha Nussbaum, "Exactly and Responsibly: A Defense of Ethical Criticism," *Philosophy and Literature* 22:2 (1998), pp. 343–365.

Garden Contextuality

While it is possible one might claim gardens, as essentially dynamic, only possess contextual aesthetic properties—as a way to deal with the problem of the requirement to subjectively impose an experiential frame on the garden—and no formal aesthetic properties, this would leave us in the unenviable position of disallowing purely formalist appreciation of a garden.[16] Even if we reject formalism as an appropriate approach for The Garden, there is no reason to limit ourselves unnecessarily if a solution can be found that will open the door widely.[17] In complement, though, contextual aesthetic features commonly fit within aspects that seem quite relevant to experiencing and judging the aesthetic worth of gardens, and so we want to give them due attention. Miller, in discussing whether a garden can be a great work of art, writes, in support of taking contextual matters into interpretative accounts:

> The persistent (and cross-cultural) habit . . . of interpreting gardens in religious, political, social, and psychological terms rather than as purely aesthetic or physical entities, suggests that if gardens should turn out to be art at all, they will be capable of being "great" art.[18]

A garden stands in relation to a great many things, likely more things than most art objects. Some of these will be relevant to aesthetic analyses much of the time, some perhaps only rarely.[19] The so-called butterfly effect suggests that a butterfly flapping its wings might fit within a full account of the causal sequence leading to the path a hurricane takes, but the chances that the butterfly will be credited among the factors of a hurricane's path is a long shot. A garden stands in relation to many things, but most will not get mentioned in an account of what is relevant to the context of aesthetically appreciating that garden. Only matters deemed relevant will, and matters judged to be relevant will have to do with two things: (1) whether they have a sufficiently significant impact on actual aesthetic appreciation and (2) if this impact ranges over many aesthetic appreciators. The second of these criteria has to do with how many subjects' experiences are impacted.

---

[16] For a fuller discussion of this point, see Glenn Parsons and Allen Carlson, "New Formalism and the Aesthetic Appreciation of Nature," *Journal of Aesthetics and Art Criticism* 62:4 (2004), pp. 363–376.

[17] This of course puts us at odds with Carlson on this point. We do not see that a case for garden aesthetics where the aesthetic features are what we are calling purely contextual is a case that has any particular lack of merit; nonetheless we believe, if the purpose of aesthetic theory is to explain common experience, our account that includes provision for a garden having formal aesthetic properties may be preferable.

[18] Miller, *The Garden as an Art*, p. 135.

[19] For a more complete discussion, see David Fenner, "How Should Contextual Matters Figure into Art Evaluations?" *Aesthetic Investigations* 4:1 (2020), pp. 22–35.

(It would be ideal to say "all," but this is practically implausible, so the number must be obliquely described as "many.") A contextual matter is relevant to an aesthetic account when it makes a difference to that account, when the experience of a garden is changed by inclusion of consideration of that matter. The degree of change, the extent of the impact—just like the number of subjects who are actually impacted—cannot practically be quantified in mathematically precise terms. Judging these matters is similar if not the same as judging any evidential matters. At some point, in a given circumstance, the weight of proof shifts the scales and belief is warranted. With the relevancy of contextual matters, at some point enough subjects experience enough of an experiential impact to say of that contextual matter that it is relevant to an aesthetic analysis.

Relevancy, like all other aesthetic matters connected to The Garden, will differ based on place and time. From the perspective of one culture, a matter may be relevant, but from another it may have very limited relevancy. The article written by Shadidi (et al.) noted that many of the elemental inclusions in an Iranian garden, as they described them, are chosen because they express certain cultural or religious metaphysical commitments.[20] Those commitments—that they are expressed, and how they are expressed—are important to Shadidi's description of the elements and the choices that led to their inclusion. But from the point of view of one describing Islamic gardens from outside the culture, whatever motivated the choices of inclusion of elements is less important than that these choices are consistent and form a pattern across gardens identified with the Islamic style. When we look at the English landscape garden from the seventeenth century to the nineteenth, the prevalent style underwent a great deal of change. Baroque gardens—in styles that reflected their Italian and otherwise continental roots— gave way to the naturalistic designs of Capability Brown and then the Brownian style (of which Humphry Repton was the inheritor) gave way to the Picturesque (for which Repton served as a kind of bridge). And as these changes occurred in garden fashion, so the rationales for why some garden elements were eliminated and others chosen changes. As the rationales changed, so also did the relevancy of one explanation over another. "Naturalistic" to Brown meant something different from "naturalistic" to Gilpin, and while Brown thought this was best illustrated through creating clumps of trees, Gilpin might have thought it was best illustrated through the inclusion of fallen trees and overgrown brush.

In addition, the aesthetically relevant aspects of gardens have evolved through history, largely because the relationship of humans to nature evolved. It would not have been uncommon for ancient gardeners to begin to see their enclosed

[20] Mohammadsharif Shadidi, Mohamad Reza Bemanian, Nina Almasifar, and Hanie Okhovat, "A Study on Cultural and Environmental Basics at Formal Elements of Persian Gardens (before & after Islam)," *Asian Culture and History* 2:2 (2010), pp. 133–147.

plantings as having both aesthetic and spiritual dimensions—as we see both in *Gilgamesh* and in ancient Egypt. Today, through centuries of living with and in gardens, our relationships with nature have further evolved. We see ourselves in partnership, in symbioses of various stripes, and we see this in the very way we understand and define gardens. We see ourselves as having obligations, based on these partnerships, that would have been alien to those who saw nature as a thing to be mastered, conquered, and subdued. And while we still take delight—some of us great delight—in the formal gardens typified at Versailles and in more recent versions like Butchart and Longwood, some of us feel just a twinge of guilt at feeling such pleasure, as if such pleasures are out of step with modern garden sensibility that emphasizes greater equality in the relationship between humans and nature.

Like gardens, all these considerations are dynamic. And so while relevancy is itself a matter admitting of opacity and fluidity, a true appreciation of contextual relevancy in The Garden is even more fluid. We need to keep that in mind as we begin to inventory various kinds of contextuality relevant to gardens.

Above, we organized contextual features into four categories:

- genetic factors
- lifespan factors
- relationship factors
- reception factors

Now we would like to explore topics within each of these categories as they relate to gardens. We will attempt to organize matters in accord with these categories to foster the notion that aesthetic appreciation of gardens is not remarkably different—either in common experience or in theory—from aesthetic appreciation of anything.

Genetic Factors

Many aspects concerning the conditions of the origins of objects are relevant to considering their characters and their worth. Let's consider seven.

The Place

Gardens are intentional human place-creations. The Royal Botanic Garden, Kew, is where it is because of its royal origins, but being in greater London affords it prestige, accessibility, and development of horticultural reputation that, were it sited elsewhere, it might not enjoy. Many decades after its founding, Nathaniel Lord Britton visited Kew, and he decided that New York City, to prove itself on equal footing with the cities of Europe, needed a grand botanical garden of its own. The site of the New York Botanical Garden in the Bronx was chosen because of the natural beauty of the site—rock outcrops, changes in topography, and an untouched forest of Hemlock. Why a garden was put in, why it was put in where it was, how its siting interrelates with places that surround it—these are all relevant to aesthetic appreciation of that garden.

The Time

Garden styles that are associated with identities change over time. Some change very little—as we see in the cases of Islamic gardens, baroque gardens, *karesansui*, Chinese strolling gardens, and the like—and some change a great deal—as we see with English landscape design from the seventeenth through nineteenth centuries. The time period in which a garden was designed and instantiated plays a significant role in the determination of the correct categorization of, and so the correct focus on and appraisal of, a garden.

The Gardenist

Gardenists are as important to gardens as painters are to paintings, as choreographers are to dance, and as directors are to films. While gardenists must work within constraints imposed by nature, and so gardens typically are less "plastic" as artistic creations than objects found in other artforms, it is the gardenist who develops a vision for a garden; it is a gardenist who expresses that vision in a design; and it is usually a gardenist who oversees the installation of a garden's initial components.

Many gardenists will have created a plurality of gardens, and for many a signature will be evident throughout their oeuvre that then permits us to associate a unique style of garden design with a particular gardenist. Here we use the term "style" in the classic aesthetic sense. To understand a garden from that oeuvre is to understand it as standing in relation to all others in that set. The American gardens created by Frederick Law Olmsted (and at times Calvert Vaux) may transcend the metropolitan gardens for which they are famous, but to understand New York City's Central Park well is to understand it as an Olmsted garden and in relation to other Olmsted gardens such as Boston's Emerald Necklace and Brooklyn's Prospect Park. Part of this equation is the level of success a gardenist will have achieved both in terms of building a stylistic signature and reputation but also in terms of "landing" garden projects. A gardenist who designs but does not implement will be regarded differently than one who is successful with the manifestation of those designs into actual gardens.

Just as in the case with all artforms, both the intentions of the artist and the conditions under which they created may be relevant to understanding their art. While stains of art criticism in the mid-twentieth century held that these matters are not relevant, it is commonplace at least to want to know what artists thought and how they pursued the expression of their artistic visions. If a gardenist is constrained tightly in design possibilities by the patron or is given free reign and (within reason) unlimited resources—this may be very relevant to appreciation of a garden of their design. If a gardenist is trying to "say something" or make a particular point through their design, this may be very relevant. How a gardenist understands the purpose of their garden—or understands the function of the garden or the functions that will be performed within their garden—these matters likely will be relevant to a fully informed appreciation of their garden.

The Purpose

It is likely that most gardens in the world, in terms of simple numbers, are "setting and framing" gardens or "residential" gardens, and so their purposes tend to be derived from the purposes of the architecture that typically lies at the center. But many gardens have purposes that drive their designs in very strict ways. A botanical garden has a clear purpose, as does a pleasure garden or a "physic" or medicinal garden. These purposes may be read, as one reads words on a page, through observation of the components installed in a garden, but the purposes themselves are not endemically components of those gardens. They are contextual, and yet when it comes to appropriate categorization, knowing these purposes—stated explicitly or tacit, guiding or incidental—is important.

The Patron

As all gardens require resources for their creation and maintenance, sometimes quite significant resources, it will be common that there is someone holding the purse strings. This person may be the one commissioning the garden, and in that case, their influence over what sort of garden it is (and where, when, and how, and so forth) may be great. Or this person may simply be paying the bills, so to speak, for a garden whose impetus and purpose are developed and decided elsewhere, as we might find in the case of municipal gardens, university gardens, botanical gardens, and the like. One of Olmsted's projects was the creation of the garden at the Biltmore Estate in North Carolina. In that case, the patron—George Vanderbilt—was connected intimately with the garden's planning and development. In the case of Central Park, Olmsted worked with a civic commission and was obliged to meet their expectations before a design was implemented. Olmsted also designed the campus of the University of California, Berkeley, and so there his patron was the State of California and its elected representatives. The oldest gardens in Italy are connected in various levels of intimacy with the Medici family, and some of the oldest in China with the emperor. Who the patron of a garden is, the scope of their connection with the project, the depth of their pockets (so to speak)—these matters are all relevant to a full account of a garden.

The Plan

Not every gardenist's plan comes to fruition, and even those that do typically encounter some need for alteration as they are implemented. Nonetheless, access to a garden's original design plans has proven important to those who wish to understand the garden as a product of a certain gardenist and how that garden fits into their oeuvre. Such access has also proven important to the evolution of a garden through time; as the garden changes because of the interplay between nature and gardeners making alterations in response, it is generally considered advantageous—or even essential in some cases—to refer back to original designs to ensure that the garden continues to participate in its status as the art of a particular artist. On occasion, large and wide-ranging alterations to a garden are undertaken to return a garden to its "original state." The American gardenist Beatrix Farrand designed a rose garden for the New York Botanical Garden in her signature axial, French-inspired style. Unfortunately, major structural aspects of the plan could not be completed due to iron shortages in the first world war, and so the plan was shelved and ultimately forgotten. The Peggy Rockefeller Rose Garden opened in 1988 after the plans had been rediscovered; Farrand is credited as the designer even as the planting began several decades after her death.

Siting a Garden

Installing a garden sometimes entails a large change to an area of land that is contained within some political jurisdiction or shares borders with neighbors who are concerned that their interests are not negatively impacted by such an installation. This can happen even in the case of modest residential gardens as "home-owners associations" develop and enforce rules that govern how landscaping may be installed and maintained. This certainly happens when a garden is large enough to require permitting of various sorts from municipal or even state governments. What can be done on a site constrains what may be done, of course, and so a garden's installation may differ from the original plan for that garden. When this happens, it is relevant to a consideration of that garden. Home-owners associations that require certain kinds of grass, a certain uniformity of garden style across all homes in a neighborhood, flower beds to be located in some places but not others, and so forth, exert a very palpable influence on a garden. While such a garden is rarely the focus of sustained attention as a garden, to the extent to which it is appreciated aesthetically, the HOA's influence may be paramount in understanding why it is as it is.

Lifespan Factors

"Lifespan features" refer to those aspects of objects that involve change throughout the life of that object. Let's consider three such features, relevant to gardens.

The Evolution of a Garden

Gardens change as nature changes them, as gardeners react to these changes, and as gardeners redesign aspects of a garden to deal with natural changes, to better suit a garden's purpose, or because the guiding aesthetic of a garden changes. Tracking these changes is important to substantial gardens. A great garden can become a mediocre garden if changes made by gardeners are not principled and well considered. The same can be the case if the resources on which a garden relies dry up, or as different patrons with different priorities assume ownership. On the other hand, a garden well-tended—in both the sense that it is well maintained and that changes are the product of due deliberation—can evolve into being a greater garden than it was when it first began. Trees and plants mature and fill in the spaces originally planned for them; flower production increases; additional plantings to balance color and fill in blank spaces are added; resources originally directed to maintenance that now requires less attention can be used in "capital" ways to enhance the garden. These changes are important to track, not merely to maintain those aspects of the garden we might wish to retain—the original design or what has come to be known as special about the garden—but to ensure that those who return to visit the garden over a succession of seasons or years find the garden that, in some important measure, they remember as that garden. Tracking changes is important for aesthetic experiential continuity, and that is important, as we mentioned, for evaluative, comparative, and interpretative purposes. Most large botanical gardens and many large pleasure gardens devote staff and software to sophisticated plant records departments to track the location, provenance, and lifespan of tens of thousands of plants. Gardens of this status are often hundreds of years old, and their record keeping will include photos, drawings, and descriptions of the landscape as it has grown over time.

Ownership and Patronage

As the influence of a garden's original patronage is important, so it is important to be mindful of the influence of changing patronage over time. Following the claim that The Garden is an artform and that some gardens are works of art, we might describe "changing patronage" as a garden's provenance (which is usually thought of in other arts as ownership and/or the financial source of its original creation). Different patrons may have different ideas as to what a garden should be, what level of resources should be devoted to it (not to mention that different patrons likely will have different levels of resources at hand), how it should be maintained, and even what its purpose should be. In some cases, changes of patronage make little difference—the *karesansui* garden at Ryōan-ji has not undergone any significant *design* change since its creation in the mid-1400s—but in most cases, given that gardens are resource-intensive, patronage changes will mean changes to a garden's properties that in turn impact its aesthetic character. This is often the case when private gardens become public gardens—examples include Chanticleer and Longwood Gardens outside of Philadelphia, Marie Selby Botanical Gardens in Sarasota, The Huntington in Pasadena, and Great Dixter in East Sussex. In all these cases, ownership transferred from one person or family to multiple, and the horticultural aspects of the gardens continued in ways the original owner of the estate may not have envisioned.

Public Access

Not all gardens are open to the public. Some—like Oak Spring in Virginia—are only open by appointment. Some gardens—like Central Park—are always open to the public and, practically speaking, cannot be otherwise. And in yet other cases, whether a garden is open to the public is a matter that, like the garden itself, evolves. The state of access is an obvious factor in terms of the aesthetic appreciation of a garden; if a garden is not accessible, it will not be the focus of aesthetic attention by other than its residents, at least not the typical sort occasioned by being in that garden. This is to speak about a garden being formally accessible, but in those cases where gardens are sited in areas that are remote or perhaps inhospitable, these gardens are likely to fall under the same dynamic as those where access is not allowed: whether a garden is not visited by rule or by practicality, the result in terms of aesthetic appreciation is similar.

Relationship Factors

All objects may be appreciated in terms of relations they bear to other objects. In this, the garden is no different.

The Land

A garden's physical site is aesthetically important in many ways. Attempting to list them all would be too ambitious. However, a garden's topography might be first on the list. Elevation, sight lines, flatness or hilliness—these can all be aesthetically relevant. Occasionally topography can be a defining feature of an individual garden or a garden style. The terraces common in Italian baroque gardens require elevation changes; Japanese gardens tend to include elevation changes to support the "discovery" aspects inherent while strolling through them. In addition, the soil conditions—sand, clay, water-retaining, well-drained, acidic, alkaline—can all make a difference to the growth and expression of plantings. While the soil itself might not be a matter of aesthetic attention, it may still be contextually relevant to explain a garden's living components. Climate and hardiness "zone" are aesthetically relevant, of course. How cold the winter is, how hot the summer, what the average rainfall is, how much sun is normal, whether the wind is typically strong, where there is salt spray or pollutants in the air—these aspects make a huge difference to the kind of garden a garden is.

Wine connoisseurs use a French term—*terroir*—to capture the breadth of these place-focused factors on the resulting grapes grown in a particular area. While a vineyard is likely Second Nature rather than Third—while it is more an agricultural endeavor than a garden—under the right circumstances a vineyard could fulfill all the criteria necessary to be a garden. In any case, *terroir* might be a useful garden term for picking out the very wide-ranging aspects of a site that contribute to the aesthetic features and aesthetic character of a garden placed at that site.

Gardens relate to the land in their design as well as their horticultural characteristics. Some designs will be in concert with their surroundings, others may be in contrast. A Brownian landscape with a ha-ha boundary allows the view to extend from the manor through the pastoral garden and into the sheep pastures beyond. Central Park, when it was first created, had trees planted around its perimeter to shield the park from the sight of the buildings of Manhattan so that the park stood apart from the urban scene outside. This of course was short-lived, and today the skyscrapers of midtown Manhattan are themselves important borrowed scenery for the park's visitors. Considering the character and use of the land around the garden, what views the gardenist has accentuated or hidden, and whether the design wants to blend in or stand out from the surrounding landscape, are important pieces of context in garden appreciation.

The Seasons

Some gardens, such as "winter gardens," are designed to be time-sensitive; they are meant to be at their peak during a particular time. Even in cases where gardens are not designed to be seen in particular seasons, all gardens, even those in the tropics, change in some ways according to the seasons.[21] Some gardens are virtually invisible during the winter months. Some are in full color in the spring, some in full color in the fall. Some are lush and thick in the summer; some are completely dormant. A garden's relationship to the season is an important contextual consideration.

History, Style, and Classification

Categorization of gardens is something about which we have talked at length in this book already, so we will not labor the point here. But, as in the case of the seasonality of a garden, so its correct classification is a centrally important contextual feature.

Relationships with Other Gardens

In further support of the importance of categorization and in support of Walton's general approach to understanding art, Cooper writes:

---

[21] While tropical gardens, depending on the plant choices, may not change at all throughout the year, there will still be differences in hours and direction of sunlight and moonrise, birds and birdsong, butterflies, and so forth.

[S]omeone who honours no constraints—those of appropriate viewpoint and those imposed by categories and functions—when experiencing a garden is not appreciating it *as* a garden, or at any rate not as the garden it is.[22]

The inventory we presented in the second chapter is based on a taxonomy of purposes of gardens. That taxonomy essentially is a representation of one way to illustrate a web of connections among all gardens. Every garden conceptually bears a relationship to every other garden, but most gardens bear some practical relationship with some others. They establish their identity in relation to other gardens that are relevantly similar. We compare gardens that are relevantly similar in judging their relative worth or in understanding, through appreciating their differences, how a garden is special or exceptional. We appreciate how gardens' purposes evolve and spread through cultures and times through comparison one with another. Lee's exploration of the German *mittelweg* is grounded on German efforts at comparing French and English garden styles of a particular time. In this book, we understand all manner of features that all gardens share in common or that some share in common with others through appreciating relationships. The relationship of one garden to another is a core principle of contextualizing aesthetic appreciation of them.

Relationships with Other Works of Art

Gardens have been the foci of other artforms for as long as we have had artforms and gardens. From paintings and carvings to novels: the Sunga/Kusana Empire frieze of a prince reclining in a garden (100 BCE–250 CE), the paintings of gardens in the courtyards of ancient Roman and Pompeiian houses, the Buddhist garden-paradises in the painted caves of Dunhuang, the bas-reliefs of Emperor Sargon II (721–705 BCE) of Assyria, Hieronymus Bosch's *The Garden of Earthly Delights* (1490–1510), John Constable's *Golding Constable's Flower Garden* (1815), Pierre-Auguste Renoir's *Woman with a Parasol in a Garden* (1875), Vincent Van Gogh's *Daubigny's Garden* (1890), Claude Monet's *The Artist's Garden at Giverny* (1900), Paul Cézanne's *The Garden at Les Lauves* (1906), Gustav Klimt's *Country Garden with Sunflowers* (1906), Frances Hodgson Burnett's *The Secret Garden* (1911), John Berendt's *Midnight in the Garden of Good and Evil* (1994), Kate Morton's *The Forgotten Garden* (2008), Tan Twan Eng's *The Garden of Evening Mists* (2012)—the list could go on and on. Miller recounts "The Gardener's Speech" from Shakespeare's *Richard II* and a similar conversation in *Othello*.[23] Hunt

---

[22] David E. Cooper, *A Philosophy of Gardens* (Oxford, UK: Oxford University Press, 2006), p. 39.
[23] Miller, *The Garden as an Art*, pp. 28–29.

devotes a full chapter in his book *Greater Perfections*—"Gardens in Word and Image"—to this theme. He makes special note of how the inclusion of gardeners in paintings of gardens reminds us of the labor that goes into garden maintenance.[24] In his article "Klee's Gardens," Dennis Schmidt cements a relationship between nature and art as he describes Klee's connection of the two:

> I would propose that coming to terms with this movement between nature and art animates all of Klee's work . . . his lines and brushstrokes reenact this movement that itself repeats the movement of life. . . . Klee's point here is that art should not be understood as copying or representing nature; rather, if there is a sense in which art "copies" or "repeats" nature, then it is in this parallel between the coming-to-be of the painting and the life of nature as it is witnessed in the growth of the tree that is rooted in the earth and reaches to the heavens . . . the movement of life that pulses through the tree is the same movement that defines the successful work of art.[25]

Reception Factors

All objects that are designed to be appreciated aesthetically, or simply commonly are appreciated aesthetically, bear relationships to those who appreciate them. How these objects are received by their appreciators is the focus of this section.

Critical Reception

Garden criticism as a discipline or a regular feature commonly found in periodicals is rare, but garden criticism as an activity that is incorporated in celebrations of gardens—of which bookshops are replete—in tour guides and tourist information, in art magazines and periodicals like the *New Yorker* is found easily. There is no clear reason why this dearth should be the case; The Garden as an artform supports critical notice and exploration as well as any other. One could argue that it supports it better as there is, as this long list is meant to demonstrate, an enormity of contextual information that could profitably and interestingly be incorporated in art-critical discussions of gardens.

If a film wins an Academy Award, attention to that film will rise. Some may object that this is disingenuous as the film already possessed, before winning the

---

[24] John Dixon Hunt, *Greater Perfections: The Practice of Garden Theory* (London, UK: Thames and Hudson, 2000), p. 160.
[25] Dennis J. Schmidt, "Klee's Gardens," *Research in Phenomenology* 43:3 (2013), pp. 394–404, quotes come from pp. 395 and 399.

award, all the inherent value that it ever did. But the fact remains that an award-winning film will garner more attention. If a garden is the subject of a film documentary, a painting, a novel, a song—or any literary notice at all—it will benefit from this exposure, too. But beyond this, the content of what someone qualified to talk about a garden says about that garden will form in the minds of those who have the occasion both to read about and to visit that garden a part of the context that informs their experience of the garden. One estimate is that Leonardo's *Mona Lisa* receives 30,000 visitors a day. *A day.* It surely does not receive such attention because of its artistic value, no matter how great that may be judged to be. It likely enjoys such attention because it is a cultural icon at the very center of the art world, and it occupies this position in part because of the attention and regard it has received as a cultural icon (rather than simply a work of art). Critical notice matters.

Audience Reception

Absent critical attention, or absent access to the products of critical attention, gardens enjoy reputations that are formed popularly and passed along visitor-to-visitor until perhaps one day they are written down, codified, and become a part of that garden's history. If a garden widely engages many of its visitors emotionally, this is noteworthy. If they experience a sense of cognitive engagement with the place, perhaps deriving satisfaction from thoughts about or occasioned by a garden, this is noteworthy. Trends in how audiences react—feel, think—are relevant to the general set of aesthetically relevant information about a particular garden.

Personal Associations

We have already, to use a favorite phrase, spilled much ink on the matter of personal associations that one makes with and within a garden. Little more needs to be said about this issue as a contextually relevant matter apart from a reminder that while personal associations do indeed impact one's particular experience of a garden, personal associations do not commonly form a basis for the sort of information that would be useful across subjects about a garden. Personal associations are experientially relevant, perhaps supremely so, but they are not commonly aspects of an aesthetic analysis. We say "commonly" because this rule can have exceptions: if the Queen of England had a personal association with a garden, that could be relevant to royalists eager to experience that garden themselves.

Thematic Associations

It is very common to view a work of art through a particular thematic perspectival lens. We think about issues of gender, of physical and psychical pain, and of Mexican cultural identity when we view the paintings of Frida Kahlo. We think about issues of Mexican political identity, of Marxism, and of revolution when we view the paintings of her husband, Diego Rivera. We think about issues related to racial identity and history when listening to Billie Holiday (or at least when listening to "Strange Fruit") or listening to James Weldon Johnson's anthem "Lift Every Voice and Sing." It would take special effort not to consider the religious context of so much European medieval and renaissance art, paintings and music, or the geometrically infused art from the Islamic world of this same time period.

Cloister gardens in the middle ages may have thematic associations with virginity and chivalrous devotion. Enlightenment Italian gardens may have associations with proportions and man's position in nature. Olmsted parks may have associations with public access to nature and social welfare. And gardens that use native plants may be saying something about national or regional identity. However, gardens on the whole are not utilized in message-driven or didactic ways to the extent many other artforms are. This is in part because of the special character of gardens being as much about nature as they are about art and being less "plastic" than many other artforms. Nature is sometimes read as feminine, but this may only be a holdover from earlier days when a human–nature dualism was normal; apart from obvious analogies with creation and reproduction, nature does not seem necessarily gendered.

On the other hand, if we think broadly contextually—if we think about the garden not merely as a place but a set of relations and a history—we can readily see how the gender of gardenists may be relevant to the aesthetic appreciation of their gardens,[26] how allotment gardens that may be tended more by women than men may influence how we appreciate them, how Versailles and gardens created in styles that expressed a subjugation of nature to the designs of *men* convey particular historical conceptions about masculinity, how the landscapes of Capability Brown demonstrated facts about the class and wealth of those who commissioned and owned them. These sorts of themes are common in the

---

[26] A list of highly influential female gardenists is likely to include, at a minimum, Gertrude Jekyll (1843–1932), Beatrix Farrand (1872–1959), Vita Sackville-West (1892–1962), Bunny Mellon (1910–2014), Lynden Miller, Martha Schwartz, and Edwina von Gal. For an overview of female gardenists, their work, and their impact, see Louise A. Mozingo and Linda Jewell (eds.), *Women in Landscape Architecture: Essays on History and Practice* (Jefferson, NC: McFarland, 2011). Mara Miller reviews the anthology in "The Rest of the Story," *Landscape Architecture Magazine* 103:6 (2013), pp. 120–130.

appreciation of works in other forms, and with many gardens, thinking broadly contextually, they are relevant, too.

Representation

It is rare to find a theory of aesthetic appreciation that does not include representational, expressive, and textual/literary aspects of an aesthetic object among those that are central to its aesthetic character. Earlier we mentioned Stourhead's representation of Virgil's *Aeneid*; Versailles's representation of the human imposition of order on nature; how components of the *karesansui* at Daisen-in represent oceans, rivers, mountains, and boats;[27] how a certain cascade of rocks at Saihō-ji represents a rushing waterfall. Gardens can directly reference historical or mythological people, places, animals, and objects with the inclusion of statues, columns, topiary, or titles. Plants could be used to represent certain themes, such as in Chinese gardens where evergreens represent longevity and the lotus represents purity. We mentioned Cooper's view that The Garden itself represents, either through evocation of the artistry of the gardenist or the intimacy of the relationship between humans and nature displayed in all gardens. So the claim that gardens can represent or have representational features or properties seems well evidenced.[28]

In addition, gardens can represent other gardens and nature in general. We see this as the styles of particular gardenists are emulated, and in the very creation of styles themselves, as aspects are repeated and these patterns take on the role of establishing the typical elements expected to be seen in a particular style. Berthier writes of East Asian gardens, "The garden maker is thus supposed to institute two kinds of movement: one in space, whereby the beauty of famous scenic places is invoked in the specific garden, and another in time, whereby the beauty of famous gardens of the past is emulated in the present site."[29] Lee writes, "Schiller was concerned . . . to point out that the landscape garden . . . is unique among the fine arts in that it represents nature through nature itself."[30]

Hunt devotes a full chapter of *Greater Perfections* to discussion of how The Garden represents. He is careful to point out that when we talk about

[27] Berthier offers an extensive discussion of the representational aspects of Daisen-in on pp. 59–63 of his *Reading Zen in the Rocks: The Japanese Dry Landscape Garden*, ed. and trans. Graham Parkes (Chicago, IL: University of Chicago Press, 2000).

[28] The claim is not that all gardens represent or have representative features; it is not even that these feature heavily in all appreciation of a garden that has these properties—this is discussed a bit in Cooper, *A Philosophy of Gardens*, p. 27—the claim is that where gardens have these features, they can be cited, appropriately and potentially robustly, in the aesthetic analysis of a garden.

[29] Berthier, *Reading Zen in the Rocks*, p. 111.

[30] Michael G. Lee, *The German "Mittelweg": Garden Theory and Philosophy in the Time of Kant* (New York, NY: Routledge, 2013) p. 152.

representation in The Garden, we usually do not mean it in a strict mimetic sense but rather in what Cooper might term an evocative sense.[31] This is surely the case in the examples offered above, especially in the *karesansui* case where rocks, gravel and sand are taken to represent water. Hunt writes:

> [L]andscape architecture is a representative art in the many ways in which the term has been invoked. It re-presents forms and motifs from other natures; it epitomizes the nature and culture of locality, nation, owner, user; it realizes some idea of the particular site, bringing out some of its history.... Landscape architecture also represents the process of its own creation, being necessarily self-conscious because it will (to a greater or lesser degree) need to make its visitors and users conscious of their interactivity with the place it has made.[32]

A garden can represent not merely other objects, other gardens, and nature, but also gardens can also represent ideas, as in the example above about Versailles. One idea that most gardens are especially adept at representing is the safe interface between the human and nature.[33] This representation may go back to our earliest examples and conceptions of gardens as natural but enclosed spaces. We can feel reasonably certain about our safety in our own living room or kitchen, where a sufficiently strong enclosure protects us and our belongings from storms and dangerous creatures. This is not the case in the backcountry of Yellowstone National Park, where humans do not occupy the role of apex predator, and we might feel even less sure about our safety in unnamed, uncharted wilderness. Although there are forces in wild nature that threaten our survival, it is through nature—first through hunting and gathering, next through planting and harvesting—that we derive the elements of our survival. As gardens grow in aesthetic dimensions, so our sense of security grows as the order of beauty fosters comfort, satisfaction, and a sense that all is right. A garden brings the changes and uncertainty of nature into a demarcated, human-sized environment so we may feel at home in the natural world. In this light, gardens represent flourishing on all levels and in all senses. We will return to these issues in the last chapters on the intersection between gardens and ethics.

---

[31] Hunt, *Greater Perfections*, p. 84.
[32] Hunt, pp. 235–236.
[33] Thanks to our colleague A. S. Kimball, Emeritus Professor of English at the University of North Florida, for this insight.

Expression

As gardens have representational properties, so gardens also have expressive properties. "Expressive properties" as a class are difficult in the following way. When a work of art exhibits either an emotional state,[34] a thought (idea, intuition[35]), or a mood or "sense,"[36] we readily imagine that the phenomenal properties of the object cause this emotion to be felt, or this idea to be thought, by the person focusing on that aspect of the object. To be an aesthetic property is to be interpreted as an aesthetic property, so in all cases, aesthetic property manifestation requires subjective interpretation. But what happens when this causal circuit is incomplete and the subject does not experience what the object was supposed to occasion? What if the sad music does not make one sad?

Sometimes, the presence of an expressive property in a causal relationship—where being present means that it occasions a certain reaction in a subject—is not necessary. A work of music can be sad without making listeners sad; we instead simply attribute the expressive property—as we do with all aesthetic properties, formal and contextual—to the object itself. It is the music that is sad. Billie Holiday's "Strange Fruit" is a sad song, even if the reaction of sadness were not occasioned. What is tricky about this is that notes on a sheet or notes played on an instrument cannot feel or think, and attributing the feelings or thoughts to the performers does not help solve the mystery. But if this is not about the occasioning of actual feelings and actual thoughts—which subjects can experience—attribution of expressive properties to objects seems either a bit muddled or metaphorical. Instead, we might find a middle road by saying that expressive properties are aesthetic properties that refer to "being such as typically to produce in properly disposed subjects" expressed feelings and thoughts. If the circuit fails to complete in a particular instance, no foul. But if the circuit never completes, it seems too much a stretch then to say that that expressive property was present (a circuit can be of a type rather than a token; an expressive property can be the sort that causes feelings or thoughts even if that particular property never does).

The example of "Strange Fruit" is overdetermined in its expressive quality of sadness because, first, it is Blues, and second, the content has to do with the depiction of recurrent atrocity. Lyrics, cover notes, or titles are not needed to understand that Chopin's Nocturnes express something different from his Waltzes, and it is often the case in gardens that expression is accomplished without the

---

[34] R. G. Collingwood, *The Principles of Art* (Oxford, UK: Clarendon Press, 1938).

[35] Benedetto Croce, *Aesthetic* (London, UK: Heinemann, 1921).

[36] What Cooper describes as an evocation or intimation, and how he describes a garden's "atmosphere," might fit here well.

use of symbolism or allusion to anything other than the composition itself. Just as the musician has tempo, chord progression, and time signature, gardenists have a full palette of plant form and hardscape elements to compose the desired scene. Chords and plants both come with no intrinsic expressive quality. But just as G-major sounds different from D-minor, a sunny hill of dahlias will occasion a different reaction than a cool, moist fern dell. This is not to say that the quality of sadness is objectively in the recording of a song, or the soundwaves/plants themselves, but rather if the chords and lyrics are the materials, then how they strike receptive ears are the elements/objects of experience, and the aesthetic interpretations in both cases proceed from the phenomenological analysis rather than the physical analysis.

Do gardens express? Most surely. The *karesansui* garden at Ryōan-ji expresses, for many, serenity, physical stability, and transcendence. Butchart Gardens and Longwood Gardens express joy and delight, especially when in full bloom. Versailles expresses grandeur. A Capability Brown landscape expresses pastoral paternalism. Iran's Fin Garden expresses reverence contextualized in a dry Mideastern physical context and an Islamic spiritual one. The Royal Botanic Garden at Kew or the New York Botanical Garden in the Bronx express diversity—but not as an emotional state but rather as an idea alongside the expression of a value for engaging with, supporting, and contributing to that idea. Engagement with these gardens either brings to mind or brings to heart the expressive quality we attribute to that garden, and where it may fail to—because the subject was not properly disposed due to being hungry, rained-on, glued to a cell phone, and so forth—the garden's expressive quality typically or commonly would have.

Francois Berthier writes that "In the realm of art they [Japanese Zen Monks of the Muromachi Period] privileged the garden most of all as a means of expression."[37] This observation not only adds support for the claim that gardens at least once were recognized as central canonical artforms, but it reinforces the view that if art has the capacity for expression, gardens not only have this capacity but have it in abundance. If we push the claim a bit, it might be possible to say that to the extent that art expresses, great art expresses greatly, and so the greater the expression, the greater the art. The logic here does not work out perfectly, but the sentiment might still be intact. Some gardens may be great art because they express greatly. "Express greatly" can mean either or both that they are highly effective in the conveyance of the expression or that they express great feelings or great thoughts. The first seems implicit in Berthier's observation and the observations we may make about gardens we experience (in line with the short set of examples mentioned above). The second seems evidenced at the very

[37] Berthier, *Reading Zen in the Rocks*, p. 3.

least through Kyoto's *karesansui* gardens and Iran's Fin Garden. These gardens are taken to express metaphysical or spiritual truths, at least for those culturally or religiously prepared to read them.

Meaning

We devote an full chapters to the matter of meaning and interpretation, so we should limit what we say here. The only thing that usefully might be said here is that "meaning" belongs in this list of commonly relevant contextual properties. While it might be argued that the meaning of a garden relies exclusively on interpretation of it, and interpretation is an activity that is exclusively subjective, we would disagree. Correct interpretations—we will argue that some interpretations are correct and some are not—are constrained by the phenomenal and objective properties of gardens. And not only that, but we will argue that correct interpretations are dependent on those properties for their substance and content. That is, a garden's meaning is found within the properties, and the relationships among those properties, of the garden itself. And while there is a subjective component, it need be no more a part of the equation than it is in the case of the manifestation of a garden's formal aesthetic properties. All of this is to say that a garden's meaning, when a garden has a form sufficient to support discovery of meaning, is relevant to the experience of that garden as an aesthetic object. Experiences can change radically given different interpretative lenses through which a garden is experienced, and so meaning belongs on this list.

Appreciating Gardens Aesthetically

The general account we offered of aesthetic appreciation, informed by how we modified it to be applicable to garden appreciation, is briefly as follows:

1. One typically begins by categorizing the object at hand as an aesthetic candidate. The first categorization is that of "aesthetic candidate," that the object is worthy of aesthetic attention (with "worthy" cashed out in terms of what motivates the subject to invest attention in the first place). An example of a garden classification might go as follows:
   a. This is a garden; enclosed with walls and chiefly composed of fifteen stones, gravel and moss; located at the Ryōan-ji Temple complex in Kyoto.
   b. This is a garden that exhibits a cultural style (an "identity and style" garden, in our taxonomy).

    c. This is a Japanese garden.

    d. This is a *karesansui* "Zen" "dry landscape" garden.

    e. This is the sort of garden that must be observed, quietly and respectfully, from a viewing platform.

2. Next one may determine the formal aesthetic properties of the object. With a garden, which is endemically and essentially dynamic, one must first create for oneself a stable experiential object, the "phenomenal garden." In a single visit to a (four-dimensional) garden, there may be many such "framed" experiential gardens, each of which may have phenomenal properties that may then be used as a basis for subjective translation into formal aesthetic properties. Formal aesthetic properties may be identified relative to a single framed phenomenal garden or as patterns across phenomenal "captures" of the four-dimensional garden. Formal aesthetic properties can even be based on patterns across a plurality of gardens as one, for instance, identifies garden styles and types.

3. A (four-dimensional) garden may possess aesthetic properties itself, derivable from that garden's objective properties, that are conceptual in character and do not require the stabilization that is necessary to identify formal aesthetic properties of gardens. A good example are process properties discovered through science.

4. Typically, one then turns to collecting information that strikes the subject as relevant to aesthetic appreciation of a garden (relevancy is determined by what might deepen one's experience across all potential subjects). This information may be derivable from the properties of the four-dimensional garden, but the full set normally will include facts that are external to a garden (either the four-dimensional or phenomenal sort).

5. Once all matters relevant to aesthetic appreciation of a garden are taken into account, one then normally turns to (a) overall aesthetic evaluation of the garden and (b) interpretation of the garden, if the garden's form is sufficient to justify interpretative activity. These are the foci of the chapters that follow this one.

Our organization of the aspects that are relevant to aesthetic appreciation of a garden are listed in the six tables below. Dynamic or process properties, as discoverable through science, are appreciated in two different ways: (1) as they stand behind and allow for the derivation of those properties of the garden we can access using our senses (where we tend to use the word "nature" or "natural processes"), and (2) as contextual aspects of the garden as we consider them, through science, not through their perceptual manifestations but as processes themselves. We include "contextual properties" both in the first table that focuses on the development and instantiation of a garden but also in the table focused on aesthetic properties since we believe there is good reason to include consideration of contextual properties in an aesthetic analysis of a garden. Our final table transcends thinking about the garden merely in aesthetic terms to thinking about it in artistic terms; we include this table to capture schematically what we discussed in the earlier chapters.

Genetic Properties

Item or Property	Examples	Derivation
Mission, purpose, categorization, designed functions	"This garden will be a *chahar bagh.*"	Choices about whether there is to be a garden, what the garden is to be, and who will design it, are those of the garden's patrons.
Gardenist(s)	"The garden will be designed by Capability Brown."	
Design	Idea, vision, expression, drawing, modeling	The design is derived from the anticipated phenomenal or elemental properties of the forthcoming phenomenal garden.
Initiation	Planning, permitting, installing	The initiation of the garden is driven by the design.

Objective Physical Properties

Item or Property	Examples	Derivation
Site	*Terroir*, topography, climate, hydrology, soils	These items are chosen because of (1) practicalities concerning available land, patronage, resources, etc., and (2) the anticipated phenomenal or elemental properties of the forthcoming phenomenal garden.
Materials	Plants, soil, water, stone, mulch, hardscape, hidden ha-ha walls and irrigation pipes	
Hidden Nature	Unperceived root structures, unperceived shoots of tree limbs above the canopy	These are real aspects of the physical garden that only enter experiences of the garden imaginatively.

Dynamic Process Properties

Item or Property	Examples	Derivation
Nature-based processes of creation	Weather, seasonal change, climate change, ecological activity, plant growth/competition	These are processes that make up nature's contribution to the creation of garden at any given moment
Human-based processes of creation	Arranging, planting, shaping, pruning, thinning, deadheading, watering, fertilizing, building	These are gardening practices, responsible for human contribution to the garden at any given moment

Phenomenal/Experiential Properties

Item or Property	Examples	Derivation
Garden Elements	The pink of the rose, the thorniness and height of its stems, the bright yellow of variegated schefflera against the deep purple of a carefully shaped loropetalum hedge, the scent of calamondin blossoms, the sound of a fountain of water	These properties are derived from garden appreciators experiencing states of and perspectives on the four-dimensional, dynamic garden.
Borrowed Scenery	Mountains, skyscrapers, visible trees and architecture outside the garden, sensible sounds and smells outside the garden	

Aesthetic Properties

Item or Property	Examples	Derivation
Formal Aesthetic Properties	Balance, charm, elegance, grace, harmony, peacefulness, grandness, leanness	These properties arise from interpretation of the perceived properties of the phenomenal garden, typically based on the form of the phenomenal garden but occasionally based on individual phenomenal properties

Item or Property	Examples	Derivation
Contextual Properties	Genetic, lifespan, reception, relationship, and scientific factors—"The New York Botanical Garden is sited on 250 acres in the Bronx." "Jennie Butchart transformed a former limestone quarry into the Butchart gardens we know today." "This garden is made up entirely of invasive plants." "This species only flowers once in its lifetime"	Contextual properties have to do with (1) genetic aspects of the garden (seen at the top of this table); (2) external relations the garden bears to other objects, themes, and persons—historically and currently; (3) responses of audience members, both expressed and not; and (4) those processes discovered through science
Summative Aesthetic Properties	"This garden is beautiful." "This garden is more beautiful than that garden."	These properties are derived from and evidenced by the formal aesthetic properties of the phenomenal garden and from the "garden form" exhibited by the garden

Art Properties

Item or Property	Examples	Derivation
Interpretation	"Traversing this garden follows the journey of Aeneas as told by Virgil."	Interpretations are derived from interpretative activities focused on the phenomenal garden, typically with particular focus on "garden form."
Art Status	"This garden is a work of art."	This status is derived from the application of criteria concerning both the four-dimensional physical garden and the phenomenal garden.
Art Judgments	"This garden is a great work of art."	These judgments are derived from the application of criteria concerning both the four-dimensional physical garden and the phenomenal garden.

In the next chapters, we explore what it means for a garden to "mean." In other words, can a garden be interpreted, and what does it mean to interpret a garden? (This is the top row of the last table above. The middle row is the subject of chapters three and four—those chapters needed to come early to inform this chapter—and the bottom row is the focus of the last chapters of the book.) In Chapter 7, we examine arguments that hold that gardens are not interpretable or bearers of meaning—and arguments that they are, or at least can be. In Chapter 8, we survey "families" of interpretative approaches to art, focusing on how they apply to gardens, and we spend the bulk of the chapter exploring what it means to "read" a garden. We describe this as a "form-focused" approach to garden interpretation. There we go into detail in explaining what "garden form" consists of, and we give examples of applications of this approach. The chapter ends with a brief discussion of elements of interpretation that are unique or at least special to gardens.

7

Can We Interpret Gardens?

Interpretation is about the discovery of meaning (or sometimes about the creation of meaning).[1] But what "meaning" means can be elusive. We use the term and its variations frequently and effortlessly, but what does "meaning" really mean?

First, meaning seems to be about a relationship between two things, between the thing about which meaning is sought—which throughout this chapter we will call the object, regardless of whether we are talking about a three-dimensional object, an event, an idea, or something else for which we seek meaning—and the meaning itself—which usually comes in the form of an articulation that has the character of being an explanation. If that last part is true—which we take it to be—we can call the object the "explanandum" and the "meaning-articulation" the "explanans."

David Cooper devotes a chapter of his *A Philosophy of Gardens* to "The Meaning of Gardens," but the reality is that his entire book is an exercise in interpretation. He writes:

> Indeed, the "fundamental question" with which the book began, about the significance of gardens, can be rephrased as the question "What is the meaning of The Garden?"[2]

Cooper's ambition is not centrally to explore a mechanism that leads us to what a particular garden may mean. His aim is to explain what "The Garden" means. For those—whom we shortly will meet—who claim that no garden is a bearer of

---

[1] Alan Goldman—in "Interpreting Art and Literature," *Journal of Aesthetics and Art Criticism* 48:3 (1990), pp. 205–214—writes, "Interpreting literature includes paraphrasing or giving the meaning of a phrase in its context; stating the broad theme or historical, moral, political, or religious significance of a work as a whole; explaining the place of a chapter or an episode in a work; showing formal patterns implicit in plot and character relations and developments; analyzing psychological features of characters or their motives as inferred from their actions or thoughts" (p. 205).

[2] David E. Cooper, *A Philosophy of Gardens* (Oxford, UK: Oxford University Press, 2006), p. 123.

Banryu-tei Garden at Kongobu-ji, Koyasan, Japan
Unknown gardenists, 1984
Photo by Adam Dooling, Winter 2017

The Art and Philosophy of the Garden. David Fenner and Ethan Fenner, Oxford University Press.
© Oxford University Press 2024. DOI: 10.1093/oso/9780197753590.003.0007

meaning, Cooper's project stands as a stark counterexample. He not only lays out terms in which any garden can be interpreted; he advances a view about what all gardens—as captured in the category "The Garden"—mean. His thesis is based on following the path of one way in which a garden can mean—that of "exemplification," a term he borrows from the work of Nelson Goodman. "Exemplification" joins a range of other ways that meaning can be understood—Cooper offers a rich list: mereological, instrumental, depictive, allusive, expressive, symptomatic, and associative.[3] What all of these different relationships that describe how a thing can mean have in common, according to Cooper, is that they are all focused (1) on the act of explanation, (2) on relating the interpreted object in appropriate ways to things larger than or beyond itself, and (3) on aspects that constitute a human form of life.[4] It is the first of these that in this chapter interests us most—the act of explanation as a way to think of meaning. By an "explanation," we take Cooper to be describing a final step in interpretation, the conveyance of what has been discovered and what is available for communication to another. His focus otherwise is on the relationship between an articulable meaning—an explanation—and the object that is being interpreted, and such a relationship can be, as we see, characterized in a plurality of ways.

Second, meaning is sought when there is a doubt or a mystery to be addressed and resolved. This mystery is what motivates the investment of cognitive attention necessary to work out the mystery and come to resolution.

Third, meaning derives its existence from a subjective act of interpretation and so is located in the mind. On occasion we may talk about an object "possessing meaning" or "being a bearer of meaning." These expressions refer to one of two things: either (1) to the significance or importance—the "meaningfulness"—of the object or (2) to the aptness of the object for serving as a focus of interpretative activity. In neither case is it the object per se that possesses meaning or meaningfulness; it is rather that one finds the object meaningful or "meaning-ful."

This observation, however, provides no support for claims that gardens cannot be meaningful or bearers of meaning, or that gardens are meaningless. Any theory that argues that (all) gardens are incapable of being the focus of discovered meaning because meaning is always and only subjective are without merit because they would entail that nothing can be the basis of meaning, as no object possesses—as a property of itself—meaning. Yet things abound that are filled with meaning—our lives, our values, our principles, our relationships, and a host of our experiences. And some of those experiences are of gardens, or at least we will argue as much.

---

[3] Cooper, p. 122.

[4] Cooper, pp. 111–112. Cooper's work on meaning in *A Philosophy of Gardens* seems to some degree a distillation and focusing of work from his book *Meaning* (New York, NY: Routledge, 2014, originally published in 2003).

Fourth, while it is possible to imagine that an interpretive act can be done without volition—perhaps in the case of identifying the set of which something is a member, if that can be thought of as "interpretative activity," or "that chair looks uncomfortable"—our focus will be on interpretation as a purposed and purposeful activity. Interpretation involves cognitive engagement, engagement that is deliberate, focused, and goal-oriented.

Fifth, when it comes to art interpretation, meaning has an endemic relationship to expression. A work of art is an expressive instantiation of an artistic vision, and what a work of art expresses will be bound up tightly with understanding what it means. As we interpret a garden, we typically look at a garden's expressive features; this is not a conflation but rather an acknowledgment of the close relationship between meaning and expression.

Sixth, in amplification of the point about expression and of the claim that meaning is subjectively located, when an object's meaning is unclear, we typically turn to the object's creator. Consider what arguably is interpretation's simplest form, that of language translation. Translation from one language to another is a common occurrence, and while there may be important differences between linguistic interpretation and art/aesthetic interpretation, there are salient similarities as well. When one says, "el gato esta en la jardín," one who does not speak Spanish but only French will seek out someone—or nowadays some internet mechanism—who speaks both languages to ask what the phrase "el gato esta en la jardín" means. The translator will reply with "le chat est dans le jardin." Here the meaning of the word "gato" is dependent on the reference, and while there is room for confusion (is "gato" meant only to refer to a male cat?), with some conversation and continued clarification, the meaning becomes clear.

In this case, the meaning is largely dependent on two variables: the rules of the two languages and the meaning of the terms and phrases as purposed by the two speakers. In a sense this mirrors the Aristotelian distinction between form and content: the linguistic rules, including how they manage reference, is like the form of the statement, but the meaning of the phrase—which is what the two persons and the translator purpose to communicate—is the content. And as it is with argument, while there are rules with grammar and reference that are beyond the individual subjectivity of particular speakers speaking particular phrases, meaning is not capturable in the same structured way. For instance, it is possible that the answer the French speaker receives about the meaning of "el gato esta en la jardín" is about the character of wildlife and their affinity for exploration, with cats only interested in exploration within the safety of an enclosure; or perhaps about the character of the proper place for natural creatures being in nature; or perhaps about the significance of the phrase being about a life free of care and worry. The result is that there is a plurality of theories of meaning that might help determine which translation is the correct interpretation of the original speaker's meaning. We will partly structure our discussion of interpretation of a garden around what we might call families of these theories, and to avoid repetition we will discuss the characters of these families at that point.

Seventh, in contrast to the appeal to the object's creator when meaning is unclear, the point above does not automatically require us to be "intentionalists," believing that the meaning of a sentence is exclusively a matter of what the speaker intended—or that a work of art's meaning is exclusively a matter of what an artist intended to express. We can "read" the object as an expression without getting into who expressed or what they intended to express. One who speaks Spanish likely will not seek any clarification when presented with the sentence "el gato esta en la jardín." Each word has a meaning that is known commonly, and the sentence relates the words together to suggest a meaning that seems clear. In this case—and likely in many cases—appeal to the artist, author, or speaker is unnecessary. In complement, the focus on "reading" the object does not automatically require us to be "textualists" or "formalists," believing that meaning is fully readable through merely considering the perceptual or literary/textual properties of the object. We will discuss these matters in greater depth later in the chapter.

Eighth, when a person is attributing aesthetic properties to a garden—either the four-dimensional sort or the phenomenal sort—this attribution requires an act of interpretation where the subject translates the objective or phenomenal properties of a garden into ascriptions of aesthetic properties. We know in part this is an act of interpretation because it necessarily involves the subject, and we know it involves the subject because objective and phenomenal properties as parts of the world external from us are not informed or imbued with value. Subjects imbue objects and states of affairs with value. Since aesthetic properties are value-bearing—which we know because they are used to evidence and justify aesthetic evaluations—they must be products of subjective activities, activities of translating value-free properties into value-bearing ones. This is a kind of interpretative activity: the bridging of the descriptive with the evaluative.

Before engaging the question "how do we interpret gardens?" we first should ask "can we interpret gardens?" This is an important question, not merely for establishing a platform for asking our "how" question, but because if no garden is interpretable in a way that is significant or robust, and all works of art are, then it is unclear whether a garden can be a work of art. If it cannot be interpreted, that not only undercuts the work we undertook earlier, but it may limit the significance of the entire project to which this book is dedicated. That is, if we can take an aesthetic point of view to anything that is capable of being sensed, we might ask "what makes The Garden so special that it deserves books devoted to exploring what it means to take an aesthetic point of view to gardens—that is, if no gardens are works of art and The Garden is not an artform?" Despite the importance of motivating this project, the question of whether gardens can be interpreted is an interesting and important one, one we need to approach with seriousness and openness to where the arguments may lead. While it certainly seems that at least some gardens—like Stourhead—will sustain examination of their meanings, we need to work out, as we did in the case of how gardens possess formal aesthetic properties, whether gardens are truly interpretable or whether our interpretation of them is an activity that cannot be theoretically supported.

If some gardens are works of art, then, following Miller, they will have a form sufficient from which to derive meaning. Whether this is possible with any garden is the focus of this discussion. Different writers on the meanings of gardens have different conceptions of what it means for a garden to have meaning (or, perhaps more precisely, what it means for us to find meaning in a garden). Some writers on the topic have a narrow understanding of meaning. For them, meaning is strictly about the sort of informational messaging we typically find with narrative or story, about what can be immediately read about a garden as capturable in a linguistic form. "Stourhead is about the travels of Aeneas as told in Virgil's *Aeneid*." Very few gardens present meaning in this form and as transparently as Stourhead, so it would seem a shame to restrict the discovery of meaning through garden interpretation merely to this form. We examine this matter in the following section.

The Argument against Interpreting Gardens

It is important we begin with recognition of the point that when talking about interpreting gardens, we are not advancing the view that all gardens can be profitably interpreted. A function-forward kitchen garden behind someone's house is likely not profitably interpretable, neither is a standard American front yard or a standard "setting and framing" garden. In complement, we also do not want to advance the view that *only* gardens that qualify as works of art are interpretable. Even though to be a work of art, a garden should be interpretable, there are gardens that arguably are not works of art that are still interpretable in ways that seem interesting and significant. For instance, while it is likely not every garden designed by Frederick Law Olmsted rises to being a work of art, every Olmsted-designed garden forms part of a rich oeuvre and at least can be interpreted as a contribution to the style that defines that oeuvre.

There are many gardens that, while not works of art and not interpretable in significant ways, are quite beautiful and have earned reputations that make them popular destination gardens. It is likely the case that these gardens were designed in accord with those goals, and in some cases their designers had no intention of creating gardens that would be regarded in terms that transcended beauty and pleasure. We mention these matters before beginning this section's discussion to indicate that just because one particular garden cannot be profitably interpreted is no argument that none can be. And, just as importantly perhaps, just because a garden can be profitably interpreted does not automatically confer on that garden the status of being a work of art. The various boundaries inherent in this discussion do not line up like that.

G. R. F. Ferrari, in an article titled "The Meaninglessness of Gardens," crafts an interesting case for why gardens do not have meanings. It begins with the observation that the elements of gardens are *lives*, for Ferrari believes, as Miller does, that a garden is composed chiefly of living things. Ferrari is unhesitant to point out that in saying that the elements of gardens are lives, he does not mean "living things." Living things are the *materials* of gardens—harkening back to Miller's distinction between elements and materials. No other artform, he says, has as its elements lives. As an assemblage of lives, a garden is like a political state, and like a political state, where some aspect or component of it might have meaning, the whole does not. It simply is. The same, he claims, is the case with gardens.

> Stourhead is undoubtedly a garden whose design aims to get something across to its visitors. That garden historians have offered a wide variety of interpretations of its exact meaning only serves to confirm it as a garden that aims to have at least one. What is more, it is the place as a whole that conveys its meaning, which can only be appreciated from an entire circuit. The symbolic elements do not merely inflect the garden; taken in order, they constitute its point. What has become, then, of my contention that symbolic and pictorial elements function only as meaningful accents in the garden, while the garden as a whole carries no meaning? That contention remains firm; for there is no whole to the garden at Stourhead. Rather, it is a garden that, over its entire extent, has been co-opted in order to produce something other than a work of garden art: a narrative. (Likewise, a garden consisting entirely of topiary, or of symbolic rocks, would be a garden that has been co-opted to produce a depiction.) This narrative does indeed make for an artistic whole; but it is not the artistic whole that a garden alone can be, a whole composed of the lives of plants. Gardens are not, in fact, very good at telling stories or painting pictures, nor should we expect them to be. The strain tends to show when they are co-opted for such purposes. One group of writers on Stourhead astutely compare it to the Disney theme park ride "Pirates of the Caribbean." Here too there is the suggestion of a narrative as our boat whirls from diorama to diorama. The ride is exciting (What will we see and feel next?), but offers none of the involvement in

plot that comes from watching Johnny Depp in the movie. Likewise, Stourhead is a very beautiful place.[5]

There are several items either stated or implicit in Ferrari's claims with which we would take issue. First, he understands meaning in narrow informational terms. This is evident not only in his description of Stourhead's meaning but also when he writes, "A society includes communicators and their instruments of communication but is not itself an example of either. So it is too with the garden. There is nothing for gardens to get across."[6]

The ride "Pirates of the Caribbean" is not the film *Pirates of the Caribbean*. Setting aside for the moment that the ride preceded the film in its original instantiation—a point nonetheless that adds light to the point we are about to make—the "narrative" of the ride and the narrative of the film convey different meanings. The ride's "narrative" is meant to be evocative of general caricaturesque themes of "piratehood" (at least those themes appropriate for general audiences). Each scene evokes another theme, and together they suggest a sort of account of "piratehood" and perhaps suggest a causal sequence of the sort of events that pirates stereotypically get up to (again, for a G-rated audience), but there is very little in the way of a classic story-focused narrative, and there is not supposed to be. To ask of the ride that it convey a narrative of the sort one might find in a film or novel is to misunderstand the purpose of the ride, and so also to misunderstand the meaning of the ride. We do not delight in the ride because of its story; we delight in the enveloping aesthetic as we move through it (by boat, i.e., pirate conveyance) and discover more four-dimensional components set in a space that surrounds and involves us. The pirate theme stitches it together and provides a content focus, but the point is not the telling of a particular story in the sense of conveying a particular set of events in sequence. This is evident from the fact that there is no character development in the ride; there is no unfolding psychological discovery; there is no basis for identification; there is very little that we expect from filmic or literary narratives. This is not to say the ride tells a story badly; this is to say that it does not tell a story *in that sense*.

As the analogy was meant to cast light on Stourhead, we would point out that Stourhead is not meant to be or convey the narrative of the *Aeneid*. Stourhead is derived from the *Aeneid*, and one who had not read the *Aeneid* or at least knew what it contained would not be able to read Stourhead at all—that is, as an account of the story of the *Aeneid*. Stourhead functions as a reminder of the themes of the epic poem. Given the particulars of Stourhead, the reminders are not

---

[5] G. R. F. Ferrari, "The Meaninglessness of Gardens," *Journal of Aesthetics and Art Criticism* 68:1 (2010), pp. 33–45. The quote comes from p. 39.

[6] Ferrari, p. 38.

merely reminders of components of the story in the sense that one person might say to another, "hey, do you recall when Aeneas went to the Underworld? That was some wild time." The reminders are grand, and they are placed in a grand setting, reachable only by undertaking a journey around the garden's central lake. The grandeur is not a reminder of narrative events simply as events; they instead remind us of the epic nature of those events. The meaning of Stourhead is not the story of Aeneas per se; the meaning is about the expression of the epic character of the journey and its events. In this way, Stourhead succeeds beautifully.

To take Stourhead as a substitute for the poem as a literary device is to misunderstand Stourhead's value and, more importantly for our purposes here, Stourhead's meaning. The point of talking about "Pirates of the Caribbean" and Stourhead is to highlight the point that if we take meaning merely to consist in the sort of narrative in which literature, plays, and films excel, nothing but literature, plays, and films will ever fully succeed in being regarded as bearers of meaning. Criticism of the meanings of gardens will evolve quickly into criticism of the meanings of paintings, prints, photographs, sculptures, dances, symphonies, and even operas. "There is nothing for gardens to get across" belies a narrow conception of what it is that may be gotten across.

Second, we disagree that a society cannot convey meaning as a whole, as a society. Patriotism is lodged in and derived from its members finding meaning in being members of the society of which they are part. We do not live in community for the sole sake of the creature comforts, with one set of societal particulars capable of being swapped out for another. "Welsh" or "Scottish" does not stand for a geographically bounded collection of lives and nothing more. It stands for a certain conception, likely different for different members, of what it *means* to be a member of *this* particular bounded collection of lives—bounded not merely by geography but by collective-defining principles of governance and shared values, history, and culture. The point is that while the brutely objective collection of lives bounded together geographically has no meaning—and it does not, as a collective object—that is not what a society is. A society is a mental construct that describes, in terms rich with meaning, how the "brutely objective collection" is regarded by its members and others. Ferrari mistakes the object as the locus of meaning, something we complained about at the start of this chapter.

Third, we would disagree with Ferrari's conception of a garden. While we believe that every garden either contains living objects or is contextualized within a larger garden space that contains living objects, we disagree that the distinguishing element of a garden are *lives*. Ferrari's characterization emphasizes something about the living materials of a garden that is not central to how gardenists and gardeners think of plants. Surely they are alive, and this is obviously an important aspect of them, and surely they are placed and cared for in view of the fact that they are alive and with concern for their longevity, but gardeners do not commonly think about plants as discrete lives the way that a pet owner may think of a dog, cat, or goldfish as a life that exists within the temporal confines of their own. Significant plants, especially long-lived trees, rarities, and plants that are points of emphasis in a garden's design perhaps might be thought of as lives, and gardeners might spend time to grieve if they are stolen, vandalized, or uprooted in a storm. But ten thousand violets are not seen as lives, a couple of boxwoods in a fifty-foot hedge are not seen as lives, even valuable plants in a botanical garden will be culled if they begin to senesce or interfere with other plantings. If gardeners looked at every plant within the garden as a distinct life, others would accuse them of genocide when they systematically eradicate genetically deformed seedlings, apply chemical pesticides to pathway weeds, or clear out whole gardens to make way for new designs.

Gardeners tend to see plants as *items* of life; they tend to aggregate them together as a "life-collective," and so the loss of a single plant typically does not overly concern the mature gardener. The old English proverb reads, "one for the mouse, one for the crow, one to rot, one to grow." It is essential for gardeners to take what some might see as a dispassionate view of their living populations, as gardens are open ecological systems subject to "the circle of life" and plants in high density overcompete for the garden's limited resources. If a certain garden has the space to support six hundred poor plants or three hundred healthy plants, the gardener will prioritize the health of the "life-collective" over the health of the individual "lives." A definition of a garden, or of The Garden, that rests on such an unshared assumption is liable to fail to resonate with those intimate with gardens and those whose conceptions of gardens admit of gardens as bearers of meaning.

Ferrari discusses Ryōan-ji's *karesansui* but (at least initially) rejects it as appropriately falling under the term "garden." As was evident in Chapter 1, we include it on the grounds that many who have witnessed it and discuss it refer to it as a garden—and if asked, they would probably say, "if not a garden, then what?"—and our view is that theory should follow practice. We would note also that the fifteen rocks that compose this garden are probably more closely viewed as distinct lives by the creators and appreciators of this garden than even the biologically living plants that surround the temple complex—evidence of this is clear in the Japanese gardening manual, the *Sakuteiki*—and the same is true for the *taifu* rocks of Chinese gardens. In complement, if the living objects of a garden are not widely seen as lives but rather as parts of a life-collective, then the observations upon which Ferrari's argument for the meaningless of gardens rest are suspect as are the conclusions he draws from these observations.

Fourth, we would disagree with Ferrari's implicit assertion that The Garden can be an artform in the fullest measure without those gardens that are works of art capable of carrying meaning (or, more precisely, capable of being interpreted for the sake of discovering meaning). Ferrari argues that The Garden is indeed an artform:[7]

> I have justified the garden's artistic status by appeal to its pursuit of what all aesthetic arts pursue, beauty. . . . Gardens, when they are art, aim at beauty.[8]

While it may be true that "aesthetic arts" pursue the achievement of positive aesthetic features, beauty chiefly included among the members of that set, the arts in general, since at least Picasso, have not been exclusively driven by this goal. Indeed, the history of modern art is much more concerned with meaning—if we take "meaning" to be the natural result of cognitive engagement—than with beauty. The juxtaposition of "beauty" and "meaning"—one that sets up a dualism between the two—strikes us as dangerous.

> There is something refreshing about a type of art that does not need to tease you with meaning in order to absorb you by its beauty. It is refreshing because it is unusual, and it is refreshing because it offers ease of seeking to the beauty-seeking mind. This is what I meant when I wrote that having no meaning is part of the garden's charm.[9]

Not only may this relegate The Garden to a kind of second-class citizenship within the art world, but there is a lingering feeling that this "second-class citizenship" really is second-class if The Garden is not capable of profitably engaging audiences cognitively. (May we suggest that the juxtaposition of "beauty" with "cognitively engaging" has led to some unfortunate consequences not merely in the world of art but in the world in general?) If arguments for The Garden being a bona fide artform are to be successful, we believe that to the greatest degree possible, the artistic activities that are possible and common with all works in all canonical artforms should be available in some appropriate measure with The Garden as well. A garden that is merely beautiful, and no more, feeds immediately into the arguments that Kant and Hegel made concerning the artistry of The Garden; these are arguments we addressed earlier and will revisit immediately below. Again, it is important to note that not all gardens are profitably interpretable, and for some the pleasure of beauty is all they aspire to; there's nothing problematic with that at all. But there seem to be clear examples of gardens that are in fact interpretable in significant ways, and it is those gardens that are at issue here. Ferrari paints with too large a brush in dismissing all gardens as meaningless.

---

[7] Ferrari, pp. 34, 35, 37.
[8] Ferrari, pp. 40–41.
[9] Ferrari, p. 40.

Jane Gillette, in an article titled "Can Gardens Mean?,"[10] comes to a similar conclusion to Ferrari's, that because gardens are particularly good at, and, for Gillette, defined in terms of the achievement and expression of beauty, they need not be bearers of meaning. In fact, Gillette says, because of the latter, they do not express meaning.[11] She writes:

> My speculative origin of the garden tries, then, to account for a salient feature of all gardens, from a row of flowers along a fence in Grandmother's backyard to Henry Hoare's Stourhead. This salient feature is the one essential function of the garden: to give pleasure of a certain mindless sort.[12]
>
> Every medium champions itself; so that if people make gardens to express ideas, we need to ask what idea requires the garden for its full and best expression, an expression that cannot be adequately achieved by some other medium—poetry, say, or the philosophical treatise, the play, the landscape painting.[13]

Gillette agrees that gardens are particularly adept at illustrating the unity of natural processes: "the unity of all phenomena, spiritual and material, human and nonhuman. This is, conceivably, the only meaning of all gardens."[14] But she goes on to say, "The landscape is identical with the process it professes to express."[15] And she believes this identity relationship negates the possibility of representation or expression: the illustration of unity The Garden offers is not the basis for the derivation of meaning per se; it is simply the instantiation of unity. She claims that illustration of a thing by that thing is not representative or expressive; it is simply showing the thing itself. This point connects to her insights about how meaning relationships seem to manifest—through a process centrally involving "the pathetic fallacy," the anthropocentric attribution of meaning onto objects that are not themselves imbued with that meaning but rather are merely canvasses (our word) on which the meaning is projected.

> Gardens, artifacts, undesigned landscapes, and so forth do not tell, desire, or express anything. Only humans can do that. . . . Personification of the artifact is part and parcel of our emotional desire to be at one with the physical universe (or nature), a symptom of our desire to move the creation of meaning away

---

[10] Jane Gillette, "Can Gardens Mean?" *Landscape Journal* 24:1 (2005), pp. 85–97. Reprinted in Marc Treib's *Meaning in Landscape Architecture and Gardens* (New York, NY: Routledge, 2011).

[11] Gillette, "Can Gardens Mean?," p. 85.

[12] Gillette, p. 86.

[13] Gillette, p. 87.

[14] Gillette, p. 88.

[15] Gillette, p. 89.

from our own self-conscious minds and make it an intrinsic aspect of the physical universe[16]

In an article titled "Gardens Can Mean," Susan Herrington responds to Gillette's
argument. She writes:

> Gillette provides three main reasons why gardens cannot communicate, evident
> in the way a) gardens are described by writers (particularly in their use of the
> "pathetic fallacy"); b) the influences of postmodern culture on approaches to
> landscape design; and c) how the physical medium of gardens make attributing
> meaning to them difficult.[17]

> [a)] The pathetic fallacy is the act of attributing emotions, thoughts, and
> aspirations to objects that could not possibly have these states of conscious-
> ness. . . . [I]f the pathetic fallacy is employed to get friendlier with a notion of
> nature, this does not mean the garden is bereft of meaning. In fact, the pathetic
> fallacy is *not* a way of connecting to external nature at all, but to people.[18]

> [b)] As a lead-in to her suspicions about postmodern culture, Gillette uses
> Treib's critique of postmodern landscapes. . . .[19] Treib cautions that in their
> quest to give meaning to landscapes, designers are overlooking the importance
> of pleasure and human comfort. . . . Yet, Treib finds this emphasis on people's
> use and pleasure blatantly absent from many of the postmodern landscapes he
> discusses. . . . Yet these are not mutually exclusive properties. Interpretations of
> landscapes can be pleasurable, but Treib argues that landscape architects should
> be less concerned with ascribing meaning to their landscapes, and more con-
> cerned with creating landscapes that are pleasurable to all the senses. I think
> designers can intend their gardens to communicate, but they can also give great
> pleasure to the users as part of their meaning.[20]

> [c)] According to Gillette, rocks in a garden cannot express because they have
> not been sculpted into something else, or wildflowers cannot mean because
> they are in a garden rather than a painting. This contention not only ignores the
> role of art in garden design, but architecture, sculpture, ceramics, weaving, not

---

[16] Gillette, pp. 92–93.
[17] Susan Herrington, "Gardens Can Mean," *Landscape Journal* 26:2 (2007), pp. 302–317. Reprinted
in Marc Treib's *Meaning in Landscape Architecture and Gardens* (New York, NY: Routledge, 2011).
[18] Herrington, "Gardens Can Mean," p. 303.
[19] Marc Treib, "Must Landscapes Mean? Approaches to Significance in Recent Landscape
Architecture," *Landscape Journal* 14:1 (1995), pp. 47–62. Reprinted in Marc Treib's *Meaning in
Landscape Architecture and Gardens* (New York, NY: Routledge, 2011).
[20] Herrington, "Gardens Can Mean," pp. 304–305.

to mention the last century of art theory. The notion that art is an imitation of real things in the world (called mimesis), and thus it cannot also be these real things in the world has been challenged by a century of artists such as Marcel Duchamp and Cindy Sherman, and philosophers from John Dewey to Arthur Danto.[21]

In describing the Ryōan-ji *karesansui*, Herrington writes:

> If we believed Gillette and if the rocks were moved and placed in a line rather than in specific clusters, the garden would still be the same. . . . Likewise, if the rocks themselves were manipulated, for instance scrubbed clean of moss, it would be no different because the garden's rocks are rocks and cannot mean anything other than a massing of hard consolidated mineral matter. Yet, because these manipulations would significantly change the way this garden communicates the intention of its designers, the rocks—their placement and their treatment—must be contributing something to how and what Ryōan-ji means.[22]

Herrington describes Gillette's conception of meaning as "insufficient," and we would certainly agree, as may be surmised from our discussion of Ferrari's observations. To reduce the meaning of an aesthetic or art object to what may be fully expressed in another medium—like poetry, the treatise, a play, or a painting—is to misunderstand artistic expression itself.

Consider the first category in this list, poetry; consider this 1891 poem by José Martí:

> Cultivo una rosa blanca,
> En julio como en enero,
> Para el amigo sincero
> Que me da su mano franca.
> Y para el cruel que me arranca
> El corazón con que vivo,
> Cardo ni oruga cultivo:
> Cultivo la rosa blanca.

And now an English translation (achieved by putting the poem into "Google translator"):

---

[21] Herrington, p. 309.
[22] Herrington, p. 310.

> I grow a white rose,
> In July as in January,
> For the sincere friend
> Who gives me his frank hand.
> And for the cruel one who tears me away
> The heart with which I live,
> Thistle or caterpillar culture:
> I grow the white rose.

The informative content is the same. And yet no one would mistake the latter for the former. The former is a great poem; the latter (arguably) is not, despite the fact that the "expression of ideas" in the reductive sense is the same. Martí's poem was written in Spanish, and his artistic expression is tied to the language in which the poem was originally written.[23] The artistry of his poem—the sounds, the cadence, the rhythm, and the beauty with which the content of the poem are expressed give the poem its artistic strength and character.

Consider the description one might offer a friend of a film they recently saw. There is no description, no matter how detailed, that will equate to that friend actually experiencing the film for themselves. This is complicated further because as we offer filmic descriptions, we tend to be fixated simply on the narrative—the plot or the story. Few film describers will include comments about cinematography, editing, costuming, and the like—and where they do, simply out of practicality they cannot offer a sufficient description of the formal aspects of every shot, much less every frame. If the meaning of a film is derived from the full experience of the film, the film's meaning will never be such that it can be captured in another form, especially one that reduces meaning to the film's narrative. Meaning is not merely informational or narrative content; it is more than that. If it were, we would gather the same meaning reading the sheet music as we would attending a concert.

The point of these exercises is to illustrate the irreducibility of aesthetic objects to descriptions of their informational components. Any theory of artistic meaning that reduces meaning to its informational components is bound to be insufficient. The artistic expression which is a garden cannot be reduced to a poem, treatise, play, or painting, just as no work within any of these artforms can be reduced to, or captured by, another work in the same or a different artform. Works of art are singular in their expressive capabilities of artistic visions, and so the same is the case with meaning. Meaning derived from a work of art—whether *the* meaning or *a* meaning—is not derivable in any other way. While a single meaning-account may be the result of interpreting a plurality of art objects, even across a plurality of artforms, this is incidental and does not work in the opposite direction: one cannot

---

[23] Our thanks to Bertha Danon for this example.

set out with an informational content, attempt to manifest or instantiate that in a variety of artworks, and then expect that each resulting work will be interpreted to discover the informational content that drove those creative efforts.

This focus on the meaning of meaning is our chief complaint about those who criticize The Garden as being meaningless. To complete the point, consider the implications of such views as Ferrari's and Gillette's in the cases of other artforms. (Herrington picks up this thread.[24]) Every artform that handles the expression of ideas narratively is going to be prima facie more successful with the expression of ideas: novels, films, and plays will be more successful at communication of ideas than poems, dances, and operas and presumably much more successful than paintings, sculptures, and architecture. Gardens are manifestly more complex as bearers of meaning than most of these other artforms, not only because gardens are inherently very complex objects, not only because they are dynamic and enveloping, but because they are so common as contexts of multiple aspects of our lives. But this is no reason to give up the search for the (or *a*) meaning of a garden; it simply means a bit more attention must be paid and a bit more investment in terms of cognitive reflection must be afforded.

The case against gardens—any gardens—being profitably interpretable includes the following:

1. Gardens are assemblages or collectives of separate entities, with the implication that the interrelationships among the separate entities do not constitute what is necessary to exhibit the *form* (in the aesthetic sense) necessary to bear meaning.
2. Gardens are not good at conveying narrative, narrative (or informational messaging) being central to meaning.
3. A garden can be a work of art without needing to be interpretable, at least in part because gardens are "aesthetic-forward" and the delivery of occasions of experiencing beauty is sufficient for being art.
4. To interpret a gardens is to inappropriately overintellectualize it; those gardens that encourage such intellectual responses typically suffer from a lack of beauty and occasion for provision of pleasure.
5. While objects in gardens may represent other things, gardens themselves cannot represent; if a garden could represent, it would represent nature, but as it is nature, this is not a case of representation.
6. Gardens are not good at expressing, as a different medium likely would express whatever a garden might express better (more effectively, more communicatively) than a garden could.
7. Interpretation of gardens relies on the anthropogenic imposition of meaning rather than on what the garden brings to the equation.

[24] Herrington, "Gardens Can Mean," p. 306.

None of these reasons we find compelling. Most, we believe, turn on a mistaken notion of what meaning may consist of. We likely have said enough in rebuttal of these points, so we will not repeat our objections but rather move now to talk about the positive case for gardens being interpretable and profitably so.

The Case for Garden Interpretation

A case for the claim that some gardens are profitably interpretable likely should begin with observation of the fact that some gardens have been interpreted in ways that seem insightful and explanatorily revealing. At the start of the chapter, we mentioned that Cooper both offers a collection of characterizations for how gardens can mean and ultimately offers a case for the meaning of The Garden as a kind. Charles Moore, William Mitchell, and William Turnbull, in their book *The Poetics of Gardens*, offer a wide range of insights about gardens that largely focus on characterizations of meaning in the gardens they discuss.[25] An exploration of the meaning of gardens is even more explicitly the focus of Mark Francis and Randolph Hester's collection of essays, *The Meaning of Gardens*. Their organization of the essays exhibits a taxonomy of ways in which gardens are commonly thought to mean: "faith, power, ordering, cultural expression, personal expression, and healing."[26] This sort of focus on taxonomizing ways of garden meaning is echoed in Cooper's work, as it is echoed in the work of Laurie Olin, who writes:

> The subject matter or meanings that I believe are being dealt with in the most thoughtful landscape designs today—beyond the programmatic and instrumental—are the following: (1) ideas of order, (2) ideas of nature including a critique of past views as provoked by knowledge of ecology, (3) ideas about the arrangement of cities and thereby society and its desires (as well as needs), (4) ideas about the medium as an expressive one (the landscape as medium) revealing something about our methods and its processes, and (5) considerations about the history of art and landscape design and the history of places—their archeology.[27]

and in the work of Marc Treib where he says:

---

[25] Charles W. Moore, William J. Mitchell, and William Turnbull Jr., *The Poetics of Gardens* (Cambridge, MA: MIT Press, 1988).

[26] Mark Francis and Randolph T. Hester Jr. (eds.), *The Meaning of Gardens: Idea, Place, and Action* (Cambridge, MA: MIT Press, 1990).

[27] Laurie Olin, "Form, Meaning, and Expression in Landscape Architecture," *Landscape Journal* 7:2 (1988), pp. 149–168. Reprinted in Marc Treib (ed.), *Meaning in Landscape Architecture and Gardens* (New York, NY: Routledge, 2011); quote comes from page 68.

During the 1980s, declarations of meaning began to accompany the published photos and drawings of landscape designs. At conferences, landscape architects described their intentions, their sources, and what they believed their designs signified. . . . The emerging generation of designers displayed a new interest in making form; and many of them claimed that these new forms would be meaningful. Landscape architecture from these two decades might be assigned to one or more of five roughly framed approaches and by extension, to a striving for significance: the Neo-Archaic, the Genius of the Place, the Zeitgeist, the Vernacular Language, and the Didactic.[28]

Ultimately, Treib's exploration of meaning in gardens takes a path different from the one we take. Ours focuses on a garden's form, and Treib says that meaning comes through time, knowledge, and relationships of people with places. But whether we agree on the mechanism by which a garden acquires meaning is less important than the fact that we both believe that gardens can indeed mean. In fact, that we disagree on "how" entails that the earlier question of "whether" is settled, at least for us.

Perhaps nowhere is the interpretation of gardens more evident than in Stephanie Ross's book *What Gardens Mean*,[29] where—as she explores the relationship of four gardens to poetry in their capacity to mean—she examines Alexander Pope's garden Twickenham; Stowe Gardens, designed over time by Charles Bridgeman, William Kent, and Capability Brown; Henry Hoare's Stourhead; and Sir Francis Dashwood's West Wycombe Park. Ross does not simply talk about how gardens may mean; she actually delves into the interpretation of the four gardens. She writes:

> [H]ow can gardens deal with great human subjects and represent significant human actions? In what follows, I shall show that some eighteenth-century gardens did indeed accomplish such tasks. Gardens such as Stowe and Stourhead contained complex iconographical programs [about literature, politics, morality, and religion] that visitors could "read" as they strolled the grounds . . . a garden can often convey the same content as a poem. . . . In working out the iconographic program of a garden like Stowe or Stourhead, we are clearly solving an intellectual puzzle.[30]

[28] Marc Treib, "Must Landscapes Mean? Approaches to Significance in Recent Landscape Architecture," reprinted in his *Meaning in Landscape Architecture and Gardens* (New York, NY: Routledge, 2011), pp. 88–89.

[29] Stephanie Ross, *What Gardens Mean* (Chicago, IL: University of Chicago Press, 1998).

[30] Ross, pp. 50–51, 54, 175.

To argue that gardens cannot be bearers of meaning is to argue that her work is ill-conceived, but given that the interpretations she offers serve her case that gardens, like poetry, can indeed mean—that her interpretations are evidence for her claims, and some would say compellingly so—it is difficult to take seriously the claim that what she in fact offers is impossible to offer.

Reflecting on Ross's work, Herrington observes:

> Drawing together the aesthetic theories of Susanne Langer, Richard Wollheim, and Arthur Danto, Ross distinguishes between the physical world of gardens as pieces of land and the virtual world of gardens that are "the sensory experiences of (triggered by) its physical base." For Ross, the perception of the physical and the virtual aspects of a garden are simultaneous. Garden elements are shaped and arranged by the designer, but their presence and design triggers our interpretation of them. . . . Humans express ideas to other humans through the physical world, whether ink and paper, paint and canvas, or mud and stone. When gardens are designed by humans as a medium to communicate ideas and emotions, they become the conveyance of expression.[31]

If some gardens are works of art, as we said earlier, then they are expressions of their artists. This is the case by definition. The artist develops a vision, designs a plan for the implementation of that vision—the *expression* of that vision—and then moves to instantiation. The re-creation of that process in the mind of one reflecting on a garden is the simplest way that garden interpretation occurs, and yet this happens commonly and frequently in the case of reflection on works of art. We ask "why," and we look for explanations that sometimes take the form of recreating the deliberative and decisional path the artist took. To deny that this can be the case with gardens is ultimately to deny not only that they are parts of a class of items that together constitute an artform, but it is to deny that they are the products of human deliberation and decision. This is to think of gardens as First Nature, indistinguishable from nature that has not undergone partnership with the imposition of human artifactuality. Recognizing this point puts the burden of proof on those who claim The Garden cannot mean, but if our review of Ferrari's claims and Herrington's review of Gillette's claims hold water, at least those two accounts do not succeed. Yet even if we take the burden of proof on ourselves, and even if we say that gardens do not participate—as artistic expressions—in what all other works of art participate, we are still left with the brute empirical reality that many of us commonly find reward, sometimes great reward, in cognitive engagement with gardens.

---

[31] Herrington, "Gardens Can Mean," p. 309. Focus on the "virtuality" in gardens is something Ross shares in common with Miller.

Herrington expands her notion of meaning to include emotional engage-ment, and we might—following this tact and Cooper's notion of associational meaning—include psychological engagement as well. These additions do no harm to the notion that meaning is discovered through cognitive engagement; they simply broaden the case. Below we will claim that interpretation is a cogni-tive activity—the result of cognitive engagement—as all explanation is a matter of mental reflection and mental creation. But we have no argument against in-cluding emotional and psychological engagement—which we described earlier as contextually important to appreciating the full range of aesthetic features of a garden—as bases for those who experience gardens to find meaning within them. Some gardens as a matter of fact do communicate expressions of feeling and connect with us associationally and through our identification with parts or wholes of gardens. If we take all communication to constitute discovery of meaning, the communication of noncognitive expressions should be included as well. The matter turns, of course, on how one understands "meaning." Reductive accounts that regard meaning as simply the immediate communication of infor-mation fail, and they fail in large measure because they do not fit our experiences of gardens and how we might describe those experiences. While rejecting one end of the spectrum is necessary, there is no reason to reject the other end, where all manner of experienced meaning is included.

The account of how meaning is derivable that we offer below is a cognitive one, but in truth the cognition involved is one step removed from the aesthetic features—including the full range of relevant contextual features—of gardens. The aesthetic features are the "explanandum," the objects the presence of which are the focus of the explanation. We ask "why" these aesthetic features are present, the answer is the explanation, and the reception of the explanation is the discovery of meaning. These features will include expressions of emotion and psychological associations and identifications. Our account is cognitive not in the sense that it stands in opposition to these (so-called) noncognitive aspects, but rather it is cognitive in the sense that it is about explaining the purpose or point of inclusion of these noncognitive aspects in the garden—or being experi-enced in appreciation of the garden.

These sorts of explanatory activities, again, are commonplace in garden appre-ciation, and they stand as the principal evidence that gardens—some gardens—may indeed be profitably interpreted as places of discovered meaning. While thus far our talk about interpretation has focused on the artistic aspects of gar-dens, on their artifactual character and the gardenists who were responsible for the choices they made in expression of their visions for their gardens, the reason we have done this is that such descriptions find easy analogues in the art world. But gardens, as we have seen, are broader than this, and so an exclusive focus on the artifactual characters of gardens is unnecessarily and perhaps harmfully limiting. (Following Cooper, we take gardens to be wholistic amalgams, where strict factorization of garden appreciation into art-aspects and natural-aspect is unhelpful.) In other words, "why" questions can be answered not merely by appeal to the deliberations and decisions of gardenists but also to those of gardeners and—perhaps more importantly—to the processes of nature itself. Gardens are not thoroughly "plastic," and so the answers to questions about why a particular plant or set of plants (and so forth) are as they are—and contribute to the aesthetic features of the garden as they do—must occasionally (and perhaps always in the case of living components) include reference to nature. The gardenist or gardener will reference the placement and care of plants as compositional elements within functions of the processes of nature: "This groundcover will fill the area in a few years," "I cut this down to the ground in winter so that the new growth will be visible in the spring," "These were planted together because they bloom at the same time," "This was meant to be a formal allée before the windstorm knocked those trees over," "We kept this seedling because it really balances the bed." Accounts of the meanings of gardens serve not only to explain why the aesthetic features are as they are; such accounts also explain how they got to be there. This is not meant to imply teleology; it is simply meant to say that there is a plurality of causal processes at work, and all of them are fair game as "explanans" in the interpretation of a garden. As we will see in the next section, the inclusion of audience reception in the interpretation dynamic has a place as well—on some views, as we will see, the premier place.

To sum up, we take (some) gardens to be interpretable, to be bearers of meaning—when the garden in question has a form sufficient to be profitably interpreted—because of the following:

1. The claim that no garden can be interpreted requires that the person advancing this claim explain why the interpretations of gardens offered by Ross, Cooper, and others—as well as the taxonomies of garden interpretations offered by Olin, Treib, and others—are fundamentally in error, that is, in error in their supposition that at least the gardens they interpret are interpretable.

2. The activity focused on explaining why a garden has the aesthetic features it has (including features like emotional expression), and how they work together to support a quality aesthetic experience of that garden, is commonplace as we (cognitively) reflect on a garden's character.

3. We commonly come away from experiencing great gardens believing that the garden communicated to us some idea; made us feel some feeling; connected us to other people, other places, other times, other species, and so forth. Our experiences in these cases are not only "meaningful" in the sense of being significant, memorable, and valuable; they are "*meaning-ful*"

in the sense that we come away believing or feeling that we now possess something we did not possess before our experience of that garden—we have discovered something.

4. If a garden is a work of art, as a work of art and to the extent that it is a work of art (as part of the amalgam of art-and-nature), it is the expression of a (human) designer (or designers) whose creative processes were, in cooperation with natural processes, responsible for the presence of that garden. Those processes, in various ways (to be reviewed in the next chapter), can be read, recreated, or imagined, and that re-creative/imaginative effort can lead to discovery of meaning.

8

How Can We Interpret Gardens?

Let us accept, at least ex hypothesi, that some gardens may profitably be interpreted, that engaging in the sort of cognitive activity that typifies considering the meaning of this subset of gardens will deliver reward to those so engaged. If this is the case, the next question must surely be: how do we do this? What is the best way or ways to go about interpreting gardens (or, again, that subset we think profitably interpretable)? This chapter focuses on that question. We begin, as has been our trend in this book, with thinking about an answer by considering past answers.

Theories of Art Interpretation and Their Application to Gardens

In the history of aesthetics and the philosophy of art, many theories concerning how meaning should be discovered have been advanced. Instead of reviewing each we consider relevant to garden interpretation, we review these in terms of "families" of interpretative approaches.

We begin with the viewpoint, counterintuitive as it may be given the arguments we have offered thus far, that art should not be interpreted—not simply gardens but all art. This is the view of Susan Sontag.[1] She believes that interpreting art is privileging a cognitive approach that has the result of overintellectualizing artistic experience. As the world of modern art has seen the production of many works that are primarily vehicles for cognitive engagement, and as the role of the critic has risen, partly because of the nature of modern art, we have moved toward interpretation as the primary means by which we engage with art in general. This is lamentable, says Sontag, because it has the effect of devaluing other ways in which we might engage. Emotional engagement, spiritual engagement,

---

[1] Susan Sontag, "Against Interpretation," in *Against Interpretation* (New York, NY: Farrar, Straus and Giroux, 1966), pp. 3–14.

Waterlily House, Royal Botanic Gardens, Kew, Richmond, Surrey, UK
Decimus Burton, 1852
Photo by Ethan Fenner, Summer 2011

The Art and Philosophy of the Garden. David Fenner and Ethan Fenner, Oxford University Press.
© Oxford University Press 2024. DOI: 10.1093/oso/9780197753590.003.0008

inspiration, and what we might call pure aesthetic engagement—in the sense that we seek to focus our experience on the object's formal aesthetic properties, certainly including its beauty—these all take a back seat to cognitive engagement when interpretation, and the cognitive activity it occasions and represents, becomes as focal as it has. If Sontag is correct, then positions like Ferrari's and Gillette's, which celebrate focus on the beauty of gardens, may be seen in a more favorable light. Herrington's discussion of Gillette's worries about postmodernism may fit within an account that is like Sontag's. That is, if postmodern art engagement is intellectually based, and this in turn limits other forms of engagement—such as with a garden's beauty—then we should re-evaluate our position on what it means to engage with a garden. This is not Gillette's stated view, not in her words and not in Herrington's, but such a case would seem consonant with Gillette's general worries.

Our position is that Sontag's concerns are not concerns when it comes to garden appreciation. Despite that there are some fascinating gardens like the Garden of Cosmic Speculation in Scotland, or like Martha Schwartz's Bagel Garden in Boston, in the main there has been no overly heavy push in garden design toward exclusively modernist or postmodernist themes. This is perhaps a consequence of the space, time, and money that garden creation and maintenance requires. Avant-garde gardens do not carry the same curb appeal as traditional designs with centuries of precedent, and relatively few private or public entities are prepared to invest their time, money, and land toward permanent gardens designed to shock their neighbors or give their audiences cognitive pause. As a result, gardens have been the focus of precious little interpretative activity, far less than works in other more-canonical artforms. In fact, enhanced interpretative attention likely would benefit garden appreciation not simply in terms of the claim of The Garden to be a bona fide artform but in terms of expanding the range of ways we commonly approach and appreciate gardens. Expansion of *kinds* of garden appreciation leads to expansion of garden appreciation itself.

Intentionalist Theories

Earlier we mentioned that one way to approach interpretation of works of art is to attempt to reconstruct the motivations, deliberations, and decisions of artists as they developed their visions and expressed them through instantiation. This is a very common way to think about interpretation, and it is a way that is inspired and supported as we think about the relationship between art interpretation and linguistic translation. When we have doubts about a linguistic translation, we commonly turn to the speaker or author and ask, "what did you mean?" The intuitive character of this approach has fueled theoretic approaches that follow suit. Perhaps the most historically prominent theorist who advocates interpretation focused on artist intention is E. D. Hirsch.[2] He argues that when the rules that govern linguistic reference and meaning do not deliver a clear and uncontroversial meaning, we do and should turn to the author (in the case of literary interpretation). Where there is doubt, the only arbiters are the artists themselves. They know what they meant; they know what they purposed to express. So, Hirsch argues, it is to them that we should turn, and it is to them that we should defer when unclarity presents itself.

Intentionalist theories were widely criticized in the mid-twentieth century, with the most well-known attack coming, in 1946, from William Wimsatt and Monroe Beardsley in their essay "The Intentional Fallacy."[3] Not only is there a problem with the potential absence of the artist—due to unavailability or due to the fact they may have died—there is the possibility that the artists themselves may not be in the best position to offer the best interpretation of their own work. If we take seriously the view that works of art are irreducible artistic expressions, an account from the artist may end up as an attempt to reduce their own expression to some description of it. While this is not the way we think about interpretation—an explanation is not a description—privileging the artist may have the effect that we must take whatever the artist says as gospel, even when the artist conflates explanation with description. In the next subsection, we will review what Wimsatt and Beardsley—and others—offered instead of an intentionalist approach and what motivated their approach.

Gardenists are no more available than any other artists, of course, and it may even be that gardenists, as opposed to other sorts of artists, are less inclined to leave written records of their thoughts about their designs. This was clearly not the case in seventeenth-, eighteenth-, and nineteenth-century Britain, and it was not the case throughout Japan's history with the garden; indeed, the *Sakuteiki*[4] dates to the eleventh century, and many of the extant gardens of Japan are interpreted with this manual in mind. But these are, given the breadth of the history of gardens back to Sumer and Egypt, perhaps anomalies. This is a practical problem rather than a theoretic one, so we should take it on that basis. In other

---

[2] E. D. Hirsch, *Validity in Interpretation* (New Haven, CT: Yale University Press, 1967). See also Wayne Booth, *The Rhetoric of Fiction*, 2nd ed. (Chicago, IL: University of Chicago Press, 1983); and Jenefer Robinson, "Style and Personality in the Literary Work," *Philosophical Review* 94:2 (1985), pp. 227–247.

[3] William K. Wimsatt and Monroe C. Beardsley, "The Intentional Fallacy," *Sewanee Review* 54 (1946), pp. 468–488.

[4] Lucius Junius Moderatus Columella (4–70 CE) was a prominent Roman writer on agriculture and gardens. His twelve-volume *De Re Rustica* (particularly Book X) may qualify as the oldest work on the theory of garden design (although a distinction between gardens and agriculture is hard to establish in his work). Some believe the Roman Vitruvius—writing in the first century before the common era—deserves this honor, but Vitruvius's work was really on architecture rather than on landscape architecture.

words, if such records were to exist in abundance, it is still unclear that this would solve the matter of whether we should prefer intentionalist approaches.

The other criticism commonly leveled at intentionalist approaches is that the artist may not be the best person to interpret their own work. In the case of The Garden, this concern amplifies as (1) the gardenist might be only one of a set of designers; (2) implementation of garden plans may be altered as topography, hydrology, and so forth are accommodated in the installation; (3) subsequent gardeners make alterations, following their best lights (with regard to purpose, design, and the partnership with nature), all of which may be viewed as legitimate and all of which may lead to a garden that deviates from the original design; and (4) perhaps most importantly of all, nature's effect as "cocreator" cannot be overlooked. A gardenist who is capable of offering a full explanation for the aesthetic features of their garden must know a decent amount of science in order to be able to articulate the effects of nature as "cocreator." Any gardenist who approaches their work entirely as a matter of art and artifice will, first, likely produce an unsustainable garden or a garden whose aesthetic qualities rather quickly erode as nature's effects take hold, and, second, will not truly understand their own garden, its essential character, and its future. Gardens, as we have seen repeatedly, are not fixed in time, and a full interpretation of a garden likely will include among the full range of its aesthetic features some note of its dynamic qualities. Unlike the case with formal aesthetic properties, an interpretation can and should take into account the full aesthetic gamut, and parts of that necessarily will reference aspects that are really only known by someone who understands the relevant science.

Suppose that the gardenist is available, that there were only one, that the garden is relatively new and the original gardenist's design is easy to read, that the gardenist were responsible for the installation—or directing the installation—of the garden, that the gardenist is not only in full possession of the relevant science but also in full command of the ability to offer an explanation of the coherence of all the aesthetic features of the garden. Given all these caveats, should we defer to the gardenist to provide a correct interpretation of a garden? This is still unclear, and that is because the gardenist will most likely be inclined to provide a single interpretation, a single explanation, and leave matters there. If a visitor to the garden, perhaps one similarly equipped with the knowledge that gardenist has, offers a competing interpretation, what should we do? If authority rests exclusively with the gardenist, that visitor's interpretation should be ignored. Yet if that visitor's interpretation leads to an experience of that garden that is as valuable as, or perhaps more valuable than the experience of the garden as it is informed by the gardenist's interpretation—and perhaps valuable for others as well who adopt the visitor's perspective—on what grounds do we continue to defend the gardenist's view as authoritative and final? That is, if interpretation is aesthetically purposeful, then it will have some influence on the value either of the garden or of experience of that garden (or both), and if experiential value is what drives us to engage with a garden in an interpretative way, what is our argument for rejecting the visitor's view? While deferring to the gardenist's interpretation is the most intuitively plausible avenue when it comes to answering the question "why did the gardenist include what they did?," problems both practical and theoretic might encourage us to consider other approaches to interpretation.

Formalist/Textualist Theories

Art criticism was experiencing its heyday in the mid-twentieth century. The effects of that are still with us today and, incidentally, why Sontag's critique of overintellectualization is still relevant. The most prevalent theory of criticism operant at that time is known as New Criticism. The theory was developed by many, and Wimsatt and Beardsley were among those who had a defining effect. They joined others from the few preceding decades who worked to encourage a formalist approach to criticism—artists such as Oscar Wilde, James Whistler, T. S. Eliot, and Robert Penn Warren, critics such as I. A. Richards and Cleanth Brooks, and philosophers such as Eduard Hanslick, Clive Bell, and Roger Fry. The movement was fueled by several concerns, the initially strong one being an interest to avoid matters that are external to the formal aesthetic character of a work of art coming into play with regard to evaluating that work. That is, there was a push to obviate the intrusion of religious, moral, and other matters external to the work itself playing a role in how that work was received.

New Critics therefore advocated that the focus on a work of art should be purely formal, purely focused on those aspects of the work that are derivable from either literary properties or properties of the work capable of being sensed in an unmediated way. Their thesis was not merely about reception and evaluation but also about interpretation. Interpretation should focus, therefore, on only what is provided by attention directed just to the work itself. As with an intentionalist approach, this view has intuitive appeal. Hirsch himself says that appeal to the artist does not mean that we ignore the properties of the object itself; no matter what Leonardo Da Vinci might have said, the *Mona Lisa*'s representation is that of a woman slightly smiling—this is the case because that is what the sensible properties offer as representational fact. No amount of argumentation to the contrary is likely to convince anyone that the *Mona Lisa* represents something else. As we read literary texts and as we "read" other artworks, we attend to what words and symbols conventionally mean, within the confines of the symbolic system of which they are a part. This governs how these words and symbols should be received. Words have meanings that are generally stable across all their uses—and where they deviate from conventional meanings, that deviation relies on some recognition of that deviation; that is, the conventional meaning remains substantively intact across all contexts. The same is true, although access to meaning may be less obvious, with all works in all artforms, so claim New Critics. Also, as is the case with intentionalist approaches, a formalist approach is very likely to render a single authoritative meaning, the meaning that takes the symbols on face value, on their conventional meanings.

Such an approach might work for parts of a garden or in some cases for whole gardens. Some parts may be seen as having settled meanings that are authoritatively "read" in single ways. It might be difficult to see André Le Nôtre's design work as anything less than formal, and it might be difficult to read his work as anything other than demonstrating the control of nature for the creation of formal beauty. Or perhaps we should state this in the opposite way: that it would be difficult to see Le Nôtre's design as indicative of, say, Buddhist views. While the varieties of presentations of fields of gravel in Daisen-in may be read in different ways, it is common for them to be read as representing areas of water or mountains, depending on material context, and convention suggests that the whole of the *karesansui* there be read as expressing themes that transcend simple materiality. Likewise, certain flowers and certain plants have conventionally been read as representative of things like love, memory, gratitude, glory, purity, steadfastness, and so forth. Among certain groups of people, the meanings of these plants are read in stable form, so much so that lexicons of these representational relationships are readily available.

On the other hand, while every word in a literary text has a meaning, every living and nonliving aspect of a garden, across all gardens, does not have a fixed meaning. Reading gardens as texts comes with many of the same challenges as intentionalist approaches, not least garden dynamism, but the principal problem for a formalist approach to garden interpretation is that gardens typically do not have components or characters that may be read according to any fixed semiotic or symbolic systems of conventional meaning. We will say more about this below.

Value Enhancement Theories

If we adopt an approach that focuses on the purpose of interpretation, understood as resulting in explanations and discoveries that enhance how and the depth to which we engage with gardens, we might then focus our attention on a theory of interpretation that places value-enhancement centrally. Value-maximization is the approach that Alan Goldman describes.[5] Interpretations, then, are developed for and evaluated on their efficacy for increasing the value of the interpreted

---

[5] Goldman described his view in both the aforementioned essay and his book *Philosophy and the Novel* (Oxford, UK: Oxford University Press, 2013). Stephen Davies also defends a version of value maximization in *Philosophical Perspectives on Art* (Oxford, UK: Oxford University Press, 2007).

object or the experience of the interpreted object. Interpretations that enhance our experience of an object by providing an explanatory lens through which to view the object or that lead to us discovering more in the object than simply sensory acquaintance affords are those interpretations we should seek and should value. They are valuable as interpretations because as interpretations they enhance the value of the object or the experience of it. Such an approach is bound to allow and even encourage a plurality of interpretations, each judged worthy for its ability to enhance the depth or richness of an individual's experience. But while such an approach would encourage plurality, it is not an approach that would entail "anything goes." Interpretations that do not have explanatory value—do not aid us in discovery, do not enhance our experiences—are interpretations we would discard. As is the case with the two approaches described above, this one too has intuitive appeal: it utilizes the purpose of interpretation to frame what is appropriate for interpretative activity.

Unlike the two approaches above, this one is more easily seen as useful or applicable in the case of garden interpretation. This approach does not rely on establishing a precise mechanism for developing an interpretation, such as relying on artist intention or textual meaning. Rather this approach allows for all manner of interpretation development, constrained only within the bounds of the facts of the matter (for example, that the *Mona Lisa* represents a woman slightly smiling) and within the goal of experiential value enhancement. This approach jibes well with the mechanism for deriving an interpretation that we propose in the next section.

If there are drawbacks to this approach, they likely rest on the implied connection between interpretation of a garden and evaluation of that garden. If an interpretation is meant to enhance the experience of appreciating a garden, and we adopt a certain interpretation for this purpose, we might inadvertently find ourselves attempting to appreciate a garden that is not worthy of such attention. That is, we may believe that if viewed through a value-enhancing interpretative lens, the garden to which we are attending must be at least as good as the interpretation suggests, but there is the danger that the interpretation may artificially and unjustifiably inflate the true value of a garden. In such a case, we might end up wasting time attending to one garden when a better one might be available for our appreciation. (This concern is addressed by the theorists who advance this theory, but we need not get into that here.)

Response Theories

The reaction to the formalism advocated in the mid-twentieth century in New Criticism came in the form of directing attention away from the artist—as we saw in the intentionalist approach—and away from the object itself—as we saw in the formalist one—and toward audiences. Motivation for response theories may be seen as similar to that for value-enhancing views: the focus is placed on the experience of the object by those who actually experience it and for whom interpretations of the object are useful.

Stanley Fish is likely the best known of those theorists who articulate and advocate for response theories.[6] He works from the premise that objects frequently present themselves to those who experience them as having indeterminate meanings. As we mentioned in the case of the linguistic translation example earlier, it is possible to understand the meaning of "el gato esta en la jardín" to be about merely male cats, the nature of cats, the proper place for natural creatures, or a carefree life. While it may be intuitive to appeal to speaker intentions to solve the mystery—especially in the case of linguistic translation when the speaker, the hearer, and the translator are all present—in the case of art interpretation the artist is not always available and/or not always the most insightful. So Fish looks to those who receive the communication to decipher its meaning. While response theories can be read to privilege the interpretations of individual interpreters—because, after all, the value of their experiences is unique to them as they actually experience the object—it is common to understand the approach as focused on interpretative communities. Such communities—understood as collectives of similar interpretative points of view—have existed in many instantiations and across history. Platonism, Marxist Socialist-Realism, and Maoist Cultural Revolutionism are three social-political schools of thought that also provide the means through which a work of art might be interpreted. Freudian and Jungian psychological theories, Critical Race Theory, Feminism, and Queer Theory all represent interpretive communities. Members of each of these communities—and certainly many more besides—interpret works of art through the lenses supplied by their more general social scientific theories and viewpoints, and so a response theory approach is very likely to generate a significant number of interpretations, each of which is legitimate as it is a product of this approach.

Such an approach, like the value enhancement one, will work in the case of garden interpretation. We have already encountered similar approaches in the articles by Michael Classens and Shane Ralston discussed earlier.[7] Both of these articles view different styles of gardens—urban agricultural gardening and guerilla gardening—through social scientific lenses and develop their insights based on the application of those views to the gardens they discuss. If there is a

---

[6] Stanley Fish, *Is There a Text in This Class?* (Cambridge, MA: Harvard University Press, 1980).

[7] Michael Classens, "The Nature of Urban Gardens: Toward a Political Ecology of Urban Agriculture," *Agriculture and Human Values* 32:2 (2015), pp. 229–239; Shane Ralston, "A Deweyan Defense of Guerrilla Gardening," *The Pluralist* 7:3 (2012), pp. 57–70.

downside to a response approach, it likely rests in the absence of a mechanism by which to sort out good interpretations from less good ones. That is, while it takes seriously the idea that value is located in the audience's (or audience member's) experiential reaction to appreciation of a garden, it does not complete the same circuit as the value-enhancement approach in understanding that experiential value is a means by which to judge the quality of interpretations. Fish's approach has been criticized as an "anything goes" approach, and while that might not be a thoroughly fair critique, it has a kernel of truth to it insofar as the approach is limited in terms of the depth of how it regards the purpose of interpretation.

Reading a Garden Formally

Which of these "families" of art-interpretative approaches do we favor for the interpretation of gardens? The answer is that there are aspects of each we think important, and we think no single approach, at least as they have historically been popular, captures everything we think important. We began discussion of this point in the opening paragraphs of this chapter.

First, it may be important to note that the theories surveyed above are not all of the exact same kind. Some of the theories focus on how we should come to know the meaning of an object, others on the quality of one interpretation as compared to another. In this sense, "value enhancement" approaches are unlike the others because they do not focus primarily on the mechanism by which the interpretation comes into being—as witnessed by the fact that some (such as Alan Goldman's) are more formalist and others (such as Stephen Davies's) more contextualist—as much as they focus on comparison among competing interpretations. As we believe some interpretations are better than others, and such adjudications can be made in terms of the quality of the experiences that result from adopting those interpretations as perspectives on a garden, our view is consistent with value-enhancement approaches. In fact, we would go so far as to say that if any attention is to be paid to capturing the intuition that some interpretations are better than others, some version of the value enhancement thesis likely is true. Interpretation takes time and effort, and activities that take time and effort need to be motivated by some value the "investor" holds. That motivating value likely is connected to the quality of the "felt" aesthetic experience of the interpreter. So it would seem the simplest explanation for this—and the explanation that does not sever the motivating value from the interpretative activity—is that some interpretations are in fact judged better than others when they result in higher quality aesthetic experiences.

Each art-interpretative approach explicitly or implicitly privileges one member of the artistic triad—the art object, the artist, and the audience member—as central to interpretation. We believe quality interpretation takes all three into account, but at ground level, quality interpretation keeps the object central. This is because we are not interpreting the artist or the audience member's response—these are involved in how we go about interpretation—but what we are actually interpreting is the object.

Does this mean that we favor the sort of approach described above as "formalist/textualist"? No—at least not in the form we have seen historically. That family of theories historically adopted commitments about the relative unimportance of the artist and the audience that stole focus from their positive commitment to the reading of the object itself; those commitments narrowed the data set in a way that excluded relevant contextualities having to do with artist, audience, and with relations to other objects. This was an overreaction, so our view is not theirs. The sort of interpretative activity we think best involves a greater investment of aesthetic attention and greater interpretive effort than the reading of meaning described above as using a narrow textual form.

In complement, we do not favor a strict "intentionalist" approach nor an approach where (1) we reconstruct an imagined-artist's intentions from reflecting on the properties of the object, or (2) we construct, from whole cloth, an imagined-artist's intentions through imagining what our own motivations to include this, and not include that, would entail. While we on occasion employ language that seems to suggest we favor this approach, we only mean such a description to capture (1) that artistic expression has a place in interpretation, (2) that artistic expression is the result of deliberation and decision, (3) that deliberation and decision has (or follows) a "logic," and (4) that understanding this "logic"—how arrangements seem to follow a logic—is one key to quality interpretation.

"Response" approaches in general place too little emphasis on the role of the artist's expression and, more importantly, on the centrality of the object itself as the focus of the interpretation. Response approaches do, however, highlight the importance of the Kantian observation that the "rightness" of the arrangement and interrelationships among the characteristics of the object is situated subjectively, in the mind of the person appreciating the "logic" of these interrelationships.

If pressed, we would admit that our approach is most in line with a general formalist approach than the others, but not one that limits interpretative activity as historical formalist approaches have. We would describe our approach as "form-focused." In the end, though, we do not seek to add to the debate about how meaning should be discovered; we seek to show how a garden's elements may be read so that they constitute the data set that then constitutes the explanandum that is the uninterpreted garden.

While there are theories that focus exclusively on the psychological position of the subject to understand what can be known about meaning (such as deconstruction), all the theories surveyed above, even those deeply focused on the experience or response of the audience member, have a place for focus on the object. It is the object—in our case, the garden—that constitutes the content of the interpretative activity. Without an object on which attention is focused, there is nothing to interpret, nothing about which to make discoveries, nothing to explain. The presentation of the object is what occasions the doubt or mystery that motivates interpretative exploration, and it is the object that continues throughout interpretative activity to be what circumscribes that activity. In other words, when we interpret, we interpret an object, and it is the features of that object that provide the primary data set—the explanandum—for our subjectively created explanans.

We believe the same about formal analysis of the object—for the purposes of interpretation—that we believe about formal aesthetic properties. While the conversation is rightly inclusive of matters that transcend the object, it is with the object that we should begin. Even those theories that hold that linguistic or literary translation must appeal to the speaker and their intentions hold that this appeal is necessary only when the words themselves do not allow the hearer to achieve sufficient understanding. As translation begins textually, with the words themselves, so artistic interpretation begins with the object to be interpreted.

What allow it mean to "read" a garden? What features do gardens possess that allows us to "read" them and then on that basis uncover their meaning or meanings? Every garden possesses a set of elements—using "elements" in the technical way we have been—where each element may be understood as belonging to a category of similar elements. By considering these categories of elements, we may achieve a comprehensive overview of a garden's full range of elements. But more importantly, by considering these categories, we come to understand the structure or form of a garden that underlies its possession of particular elements unique to that garden. That is, instead of a granular focus on a garden's elements, we see how all gardens participate in a similar structure—a similar grouping—of elements. This structure is *garden form*. Every artform—or better, "artkind"—has a form. The trick now—if we wish to employ this form-focused approach to understand how to read a garden as a way to develop the basic data set for interpreting it—is to figure out what this form is for gardens.

The Analogue of Film Interpretation

One way to describe our form-focused approach to garden interpretation is to talk first about film interpretation. We think there are lessons to be learned from interpreting films that can be applied usefully to interpreting works of art in a variety of artforms, The Garden being one. Film, from its inception in the late 1800s, has coexisted alongside criticism of film. Indeed, the theoretical attention that film received at the start of the twentieth century was robust, and film theorists adopted positions that were substantial distances from one another. Argument was passionate. Things seemed to cool a bit once American cinema and the classic Hollywood style became not only the dominant source but were commercially motivated. Today film theorizing is robust again, perhaps because, as we hope will be the case with The Garden one day soon, film is approached today more as the artform that it is. (As with gardens, only some films are works of art.) It is common today to talk about a film's "form." Some theorists understand "film form" to refer simply to the style of editing that is employed in a film, but we mean the term more broadly and more in line with thinking about film as an artform. While different theorists articulate this concept in different ways—likely a healthy approach—we attempt to capture "film form" in five categories:

1. A film's narrative. Its textuality; its plot and its story. Not all films are narrative, but the vast majority are.
2. A film's sound. Sounds that are sourced from the world internal to the film are called "diegetic," and most films today include nondiegetic sound such as music to express emotion, to focus attention, and so forth.
3. A film's visual contents. What populates the visual field of the screen: a film's mise en scène which includes everything that may be seen, including sets, props, actors, costumes, makeup, and even the acting itself.
4. A film's photographic aspects. Otherwise known as cinematography; this includes framing, focus, depth of focus, camera placement and angle, camera movement, color, exposure, and so forth.
5. A film's editing style. Shots—components of film that are continuous—are connected through editing. Sometimes a shot can be very long—the classic example is Alfred Hitchcock's 1948 Rope, which has an uninterrupted theatrical style—and sometimes many shots can be edited together in rapid succession, as we see in the "montages" common to the films of Sergei Eisenstein ("montage" has come to be associated with an editing style that involves many shots, but the word denotes simply "editing.")

In the 1940s and '50s, some film theorists began discussing what came to be known as "auteur theory," the idea that certain directors created films that, looked at as a set, exhibited a distinctive style, expressed similar themes across members of the set, and/or had a technical signature that stood out. Theorists advocated that the films of such directors could be viewed as an oeuvre, as was common with painters, sculptors, and so forth. Auteur theory, even to this day, has not been formalized in such a way that there are expressed "necessary and sufficient" conditions for a director being an auteur, and some film theorists reject the idea as unnecessarily elitist in terms of privileging the position of director over that of the many contributors who today make a film what it is. But the idea of the auteur director is still a useful one for the sake of theorizing about film, and the parallel between auteurs and gardenists is rich. As with films, crediting a garden's design and artistry—especially over the course of the life of that garden—to a single person, a single gardenist, may elicit the same reaction as auteur theory does. Still, what may be found in film directors—commonality of style, of theme, of technical expertise, perhaps of artistic vision and expression, perhaps of the quality of their works as judged through time by audiences and critics alike—may be found in gardenists as well.

When we watch a film, we look for internal coherence of the various parts of the film's form. We tend to attribute the presence of such coherence to the influence of a single overarching guiding hand—a sort of artistic "intelligent designer"—who has both the vision and the control necessary for expressing that vision in the resulting film. This is commonly taken to be the director. We expect that a high-quality director will be able to manage all aspects of a film's form—the categories mentioned above as applied to a particular film—to ensure that they all contribute to the expression. Each aspect must work in harmony to focus audience attention and thereby deliver the director's vision for that film. When carefully analyzing a film, one can see the interconnections among the parts, and as one does, one appreciates the artistry and technical care involved. When the film is great, coherence among these interconnected parts becomes visible and appreciable, and the vision of the director comes through as also visible and appreciable. It is then common for those who have done such analyses to attempt to describe that vision in textual terms.

As we articulate how we understand the coherence of those interconnected parts, we are offering an interpretation of the object. As we articulate a statement of how we witness the logic of the internal structure of the object—how the object sets up its own internal rules and then follows them—we thereby articulate the meaning of the object, or, perhaps better stated, our view of the meaning of the object. An example: one could argue that Alfred Hitchcock's 1948 film *Rope* is about the nature of the eroticism of elitism, of one human thinking they are better than another, or of one group thinking they are better than another, along with the context of that elitism resulting from cerebral or academic conjecture and argumentation, and along with the thrill that comes to those who embrace and act upon these arguments. Hitchcock is undoubtedly one of the greatest directors of all, and so we attribute to him the choices that work in evidence of our view: fancy suits, food, drinks, music, and banter for those who see themselves as elite, juxtaposed against the coarseness of a length of common rope used for achieving an event resulting from their self-satisfaction (highlighting the shocking monstrousness of the event); the tension and suspense captured through both cinematic and editorial choices that spotlight the thrill of their "experiment;" the moral denouement of which Hitchcock was famous in the realization that ideas can accomplish the manifestation of actions, sometimes horrific ones; and the implication of the audience as participants in the monstrousness by putting the audience member in the room from beginning to end through the editing. The film is brilliant, and the coherence among the formal components—if the description above holds water—is high and, very important for our purposes, articulable as an interpretation of the meaning of the film.

Formal Analysis of a Garden's Elements

This project is easy to see in film analysis, but it can be applied to any work of art of any artform, gardens included. The key is first to work out what components of a particular art or aesthetic object constitute the salient aspects of its form. The second task is to determine the level of coherence among those aspects. And the third task is to attempt to explain that coherence as an explanation not only of the logic employed in the expression of an artistic vision through the object but also of its quality as an art or aesthetic object. And, of course, its meaning. In this, evaluation and interpretation are conjoined, and rightly so. What we offer, by the end of this chapter, is an account of "garden form," of those categories of elements that structure the garden as an object of artistic or aesthetic appreciation. If we can describe those categories sufficiently well, we then will be able to use that description for the purposes of both interpretation of gardens and the evaluation of them, at least for those gardens, again following Miller, that possess a form sufficient to render interpretation and evaluation profitable endeavors. This approach is supported by the work of Olin, who writes:

> Everything that exists has form. . . . Art, and landscape architecture as a subfield of art, proceeds by using a known body of forms, a vocabulary of shapes, and by applying ideas concerning their use and manipulation. . . . Inevitably this will lead some back to a re-examination of the plant palette, landform, and natural process.[8]

Cooper's approach is partly taxonomic insofar as he maps the categories of relationship between meaning-as-an-explanation and the object that is the focus of

---

[8] Laurie Olin, "Form, Meaning, and Expression in Landscape Architecture," *Landscape Journal* 7:2 (1988), pp. 34–35, 68.

interpretative activity. Our approach, in concert with his first insights, may be described as explanatory in the sense that our focus is on *how* meaning in an art or aesthetic context may be discovered. This is both narrower and broader than Cooper's focus: narrower in the sense that Cooper's thoughts on The Garden transcend thinking about it merely in aesthetic terms; broader in the sense that the particular explanatory mechanism we describe above for interpreting art will almost of necessity render not only separate meanings for separate gardens—a pluralistic path different from Cooper's singular one with regard to his interpretation of The Garden—but very likely a plurality of meanings for any particular garden—which, interestingly, fits in nicely with the plurality that characterizes Cooper's taxonomy of meaning relationships. As we describe our understanding of "garden form," our approach turns on the creation of a taxonomy, but not one of meaning relationships but rather of the elements of a garden, the relationships among which must be explored for the sake of determining *a* meaning of a garden.

What does it mean for a garden to mean, on our "form-focused" proposal?

1. It is for that garden to possess elements that may be understood to stand in formal relation to one another,
2. where we can understand the logic ("logic" in the sense of "nomological or rules-indicative structure") of that form or inherent in that form,
3. where we can see and appreciate the "rightness"—the coherence—of that logic (or the looseness), and
4. where, finally, we can articulate an account—a meaning per se, offered in linguistic terms—that explains those relations as a whole.

We seek, in the case of a garden, to answer the "why" question—to explain (in implied or imagined ways) why the gardenist (or gardenists or subsequent gardeners) included this element and not that, why they placed things as they did, why they took advantage of existing topography in one instance but leveled a patch of ground in another, why they moved in the direction of greater formality or less, and so forth and so on. As these questions are answered, relationships among the answers become visible, and finally the logic of the garden's form is illuminated.

As we mentioned, our approach can be used for the sake of evaluation of a garden—that is partly the question of whether the garden follows its own internal rules consistently—and it can be used to organize one's attributions of formal aesthetic properties to a garden for the sake of creating a summative aesthetic judgment about that garden. But it can also form the basis of discovering *the* or *a* meaning of a garden ("the" or "a" depending on one's preferred theory of interpretation). The meaning becomes manifest as the interpreter articulates an account that describes the gardens' internal logic, in the way we offered above with the film *Rope*. Once we articulate "garden form," we can offer examples of the application of this approach.

Garden Form

"Garden Form" is the comprehensive taxonomy of those categories of elements that make up the structure of each and every garden. These groups of elements are meant to represent the depth of experience a single garden visit can offer; garden form includes only aspects that directly relate to individual experience, not aspects of the gardens that are hidden from our senses (underground, ultraviolet, microscopic, "behind the scenes") or occur on timescales that transcend an individual visit.

Path

The path of a garden can be thought of as the garden visit's narrative arc. As we mentioned in our second chapter, all gardens are human-scaled and capable of being entered, and even those gardens that are not designed to be entered by visitors must still be able to be entered by gardeners. Most gardens contain paths that move through the interior of the garden, and those that do not will still have path-elements such as pavilions or viewing platforms along the perimeter. In each case, the path is where humans belong in a garden—and so when we write about our visual experience of a garden, it is understood that we are writing about our experience of that garden as viewed from a path or platform (not our experience viewed from a tree we climbed, a hole we dug, a planted bed we have leaped into, or a drone image from high above). Successful gardens understand that the path directs our view and frames our vantage points, and the gardener will use their sense of composition from their own perspective, their own path-based mobile frame, to create a visual field that will be similarly experienced by guests walking the same path in the same direction.

The path sets the tempo and rhythm of a garden visit. Some paths are straight lines, others are winding. Some paths are level concrete, others are made of irregular stepping stones that require the visitor to pay attention to their footing. Wave Hill, a garden in the Bronx, uses a curved path over level ground where a straight one would have done just as well—the purpose is to invite the guest to slow down as they approach the view of the Palisades opposite the river, and to see the view from slightly different angles as they enter the garden. Japanese strolling gardens and many Chinese gardens employ zigzag bridges—as we see in the Humble Administrator's Garden in Suzhou—and stepping-stone paths to create a lengthened and more enriching experience through what is a relatively small space.

A path says much about the type of garden through which we are moving. Paths with railings, ropes, or built edges suggest rigidity, whereas a path of loose gravel or pine needles suggests flexibility. The borders may be set with rocks or low walls, and on occasion there is no distinct point where the path stops and the garden begins. Smaller gardens have a limited number of routes to take— perhaps only one loop around the perimeter of a lake—whereas larger gardens have multiple crisscrossing paths that give each garden visitor an individualized journey. In the Covid pandemic era, many highly visited gardens employed one-way markers on their paths to ensure social distance. The experience we have in a garden with set "one-way" paths will differ considerably from our experience of a garden with multiple options of routes where we can move at our personal pace and in the direction we gravitate toward.

Within the element of "path" we include the related elements of benches and gates. These too do much to shape our experience of a garden visit. Both are points with designed views and, in the case of gates, they provide a stable frame through which the garden is meant to be viewed as we approach it (think of the round Moon gates found in Chinese gardens). Benches suggest that the visitor might pause their narrative arc to take in what is assumed to be a composition-ally rich view ahead. Japanese tea gardens employ gates and benches along a path as checkpoints—passing through them as one approaches the tea house marks the stages of our preparation and purification for *chado*.

Structure

"Structure" includes those parts of the garden meant to be the most stable throughout the garden's existence. This comes in a variety of forms: the ground itself, walls around the perimeter of gardens and garden rooms, rocks (as background, accents, or focal points), exterior or interior surfaces of buildings, hedges, parterres, topiary, bed borders, cordons, espaliers, even the trunks of established (or dead) trees. These elements are what some may call the "bones" of the garden—those things that we might experience in a visit to a New England garden in winter. We say "most stable," because (1) all of these elements are subject to change—tree trunks expand, hedges grow, rocks slowly erode, walls crumble over time—and these elements' structures vary in their solidity and permanence relative to one another, and (2) structural elements can be redesigned, redeployed, and, on occasion, may even be designed to change. Nevertheless, these elements are used in the visual composition of a garden as if they will remain in place across a lifetime of garden visits. Structural elements, whether living or not, are maintained to retain their shape and solid character.

Structure is used by gardenists to provide a sense of spatial definition, a contrast or backdrop for planted areas, geometric interest, and a sense of rhythm and movement as these elements are repeated across our visual plane. Often structure is used in contrast with other more mobile or transient garden elements, as a regular drum beat and bass provides the basis upon which the guitar solo can stand out. Parterres and knot gardens are appealing because of their highly structured approach to the planting scheme. Although these plants are growing continuously, the intention is for them to be as structural and crisp as possible, and they require continual work to keep them as structural elements. The regularly spaced cypresses, rectangular pools, and hedges of many Italian gardens create a highly ordered and architectural space. Some trees, like palms, have an intrinsic structure where other trees such as Japanese maples are most prized when their branching structure has been revealed through decades of attentive pruning. Rather than shaping these trees into single shapes, like hedges, structural pruning of trees has almost the opposite goal of separating the branches from one another and opening the canopy, so that each branch stands out to the viewer. With crab apples and cherries, for example, we may remove the suckering crown sprouts from the base of the tree so that our view of the fallen flowers around the trunk is unmuddied by twigs. The *karesansui* at Ryōan-ji is a highly structural garden room, composed of fifteen stones, raked gravel along a level ground, the border around the gravel, and the wall on the opposite side. This and other such dry gardens suggest stability and transcendence when their structural elements are juxtaposed against the constant fluctuations of the clouds above, the movement of the trees just outside the garden wall, or thoughts of the turbulent waters of the Pacific Ocean that the gravel represents.

Artifactual Elements

The artifactual elements of gardens may include buildings, glasshouses, follies, mosaics, sculptures, sundials, fountains, staircases, railings, bridges, flags, pinwheels, orbs, gnomes, and other kinds of ornamentation. This is a distinct category from garden structure because these objects are all products of human design and manufacture, and often they are meant to be the focus of attention rather than a backdrop or compositional complement. At times the artifactual elements of a garden take center stage—a Blessed Mother garden features a statue of the Virgin Mary as its defining point—other times artifactual elements are secondary to the rest of a garden, like a gnome hiding in a corner.

Infrequently, this category of elements of a garden could include living components: Versailles once employed tortoises to walk the garden at night with candles hardened to their shells; Longwood once employed young women to prance about the garden as nymphs; "garden hermits" were once fashionable accessories on the estates of British aristocrats. (We probably would not describe these particular elements as "artifactual," but they are closely related to other members of this set.) An antique plow in the corner of a bed calls to mind the human activity that brought (and continues to bring) the garden into being and ties the landscape to its historical and agrarian past.

Artifactual elements set the tone of the garden, how the patron or owner of the estate looks on their property, and what human associations the gardenist(s) or patron(s) want the viewers to consider. As each artifactual element is a product of design, those responsible for the creation and maintenance of the garden must decide which features look appropriate for the rest of the garden and which look out of place—should the greenhouse be a classic Lord and Burnham style, or something more modern? Should the railing of the bridge blend in or stand out? What font should the wayfinding signs exhibit? Would a pink plastic flamingo add to or detract from the composition? During the winter months when the garden is closed to the public, the horticultural staff at Chanticleer garden outside Philadelphia turn from plants to woodworking, stone carving, painting, and metalworking. The chairs, railings, signs, gates, and other artifactual elements they produce mesh well with the garden rooms in which they are placed, since they arise from a single artistic vision. Larger projects, such as the creation of buildings and site-specific art, will proceed through discussion about designs with the gardenists and patrons. Some public gardens feature exhibits of sculptures, which may be rotating or permanent; in either case, a great deal of thought is required to ensure that the placement of the art cooperates with the garden in which it is placed. The Cummer Museum of Art and Gardens, originally a private residence, features a number of artifactual elements throughout the three main garden rooms: a larger-than-life statue of Diana with her bow pointed to the sky, a wisteria pergola with Doric columns, a three-tiered fountain in the center of the Italian Garden, and a small fountain of a peeing boy in a shady corner. Each of these artifactual elements lends an important piece to the gardens in which they inhabit, and we as visitors interpret the meaning and tone of these garden rooms with these elements in mind.

Space

This is the emptiness that exists between all other garden elements. The openness or closedness of a certain garden relates to our sense of scale, whether we call a garden comfortable or grand, breathable or cozy. The space of a garden is measured as the distance between the viewer and the other garden elements in three dimensions—outward along the ground and upward toward the sky or tree canopy. Often space is manipulated in a garden using structural elements like walls and hedges to create a forced perspective and to accentuate views that extend past the garden as borrowed scenery. Central Park was designed with bottlenecks, areas of narrow space followed by areas of open space as one walks toward the interior of the park. This has the practical purpose of screening the perimeter of the park from the city, but it also allows for the "discovery" of vistas that were obscured earlier in one's walk. Japanese strolling gardens use this hide-and-reveal strategy to great effect to present views from only select angles. Some garden visitors enjoy ducking under tree limbs and easing past overgrowth for the sense of adventure, others look at such gardens as inconvenient or cramped. Some prefer the open air and clear views; others see such gardens as empty.

Water

Water is a component of all gardens, even if they do not contain plants or are sited indoors. Gardens that contain fountains and rills use this element to provide visual and audial interest and to cool the garden's temperature—the four rills of water in the *chahar bagh* garden must have been a welcome element in the Persian desert. It is also common for gardens to contain waterfalls, artificial or natural, to provide a similar effect. Villa d'Este outside of Rome uses massive fountains and waterfalls to create a sense of grandeur, and Longwood Gardens invokes awe with multiple computer-controlled fountains synchronized to music, lights, and sometimes fireworks. A more modest water feature, such as an overflowing vase, a trickling waterfall, or a *shishi-odoshi*,[9] may be more fitting for smaller garden rooms with a more relaxed atmosphere. Even gardens that do not have any water features will have irrigation and ways for water to drain during rain. How the water falls from high to low points is itself a factor in how the garden is created and maintained, what plants can be placed where, and how the visitor moves through the garden's pathways (for example, saturated ground in low places will be mud, dry gravel on slopes will be slick). Traversing a garden with above-ground irrigation may provide excitement to some (especially children) but distress to others. This element also includes the water in the air—humidity—and this too is used to effect within the garden. The

---

[9] A *shishi-odoshi* is commonly a fountain device composed of a piece of bamboo, set on a pivot point so that when it fills with water from above, it tips the water out; such devices have been used to scare animals out of a garden, but they are also used for simple decorative purposes.

evapotranspiration under the canopy of trees will make a grove several degrees colder than an adjacent meadow, and the humidity within a greenhouse helps to simulate the borrowed ecosystem found within. When we enter a conservatory on a cold day, we feel removed from the environment outdoors and transported to another land—the sophisticated mist systems found in modern greenhouses help create the conditions for a unified and multisensory garden experience.

Light

Gardens are lit either by sunlight (or occasionally moonlight) or artificial light. How the light appears influences how the garden is experienced and what type of plants can be grown. A garden could be sited in a desert environment with bright direct light that causes the other visual elements to bleach out. Or a garden could be sited under the canopy of mature trees so that shade-dwelling plants are the only feasible options. Plants in high light will be shorter and squatter whereas plants in shade will stretch and have a more diffuse habit. Gardeners understand that the differences in light quality extend beyond just light and shadow; there is a difference between dappled sunlight, morning sunlight, afternoon sunlight, indirect light, light with red wavelengths, and so forth. Light is obviously essential for plant growth, and the intensity and amount of light affect whether plants are leaning or growing upright, whether they are flowering, what direction the flowers are facing throughout the day, and the pigmentation in the plant's leaves. Light changes throughout the day and seasons, and a visitor may spend several hours in a garden to observe how the light catches certain plants and emphasizes certain features and colors at different times of day. In artificial environments, such as we find within glasshouses, light quality and quantity may be modified and optimized for the plants within, using specific-spectrum LEDs or shade cloth to augment or shield light from the sun. This and other light shifted toward certain wavelengths will affect what we take in visually—red and blue grow lights, for example, are optimally efficient for plant growth but may not be optimal for appreciation of the garden's visual components.

Topography

The ground on which a garden is sited plays a major role in how we move through and experience gardens. Some gardens are mostly flat and lend long lines of vision to distant features on the opposite side of the garden. The Royal Botanic Garden at Kew uses the relative flatness of the area to provide long lines of sight from the Palm House to the Pagoda on one axis and to the River Thames on the other. Gardens like the New York Botanical Garden are hilly, and have relatively few lines of sight, but they offer new views of the same garden if we climb to the tops of local peaks. The University of California Botanical Garden is sited in Strawberry Canyon, and the surrounding hills provide a cradle of trees to the north, south, and east. Topography is often modified in gardens—Central Park is a good example, with Olmsted and Vaux moving massive amounts of earth to create hills in some areas and lakes in others. Brown did the same. Lastly, the topography of a site will determine the amount we must ascend and descend and how water drains through the garden.

Ecology

This is the "nature" aspect of what we encounter in a garden visit, including all the ecological activity of the site—the internal and external biological processes of plants, the microorganisms that are essential to any healthy soil, pollinators such as bees and butterflies, passing migratory birds, insect pests such as mealy bugs and whiteflies, vertebrate pests such as gophers and deer, predators of these pests, pets such as cats or peacocks, and humans. Gardens modify nature not only in its aesthetic components but also in its ecological components. Compared to the land outside the wall, the garden may have more water, more fertile soil, and different species of plants, leading to much different ecological conditions. The result is that the garden is a human-regulated ecosystem, an ecological cell with a semipermeable membrane to the wider ecological world. Some ecological aspects are welcomed and encouraged, other aspects are excluded and removed. It lies with the gardener and the patron to arbitrate what stays and what goes (or at least try to). An enormous amount of time is devoted to killing weeds, but what one gardener considers a weed another might consider a welcome addition. Some weeds are nonnative plants that could overtake the entire garden if left unchecked. Other weeds are indigenous to the ecosystem in which the garden is sited but allowing their spread may either not be good for the health of the garden's plants or interfere with the garden's design and utility. Consider a botanical garden that organizes their collections geographically: anything that is not native to the area of focus is a weed, even if it is a prized plant in the garden on the other side of the path. Similar ecological modifications take place with animal, bacterial, and fungal life: a garden that has too many of the wrong kind of insect might be called "infested," but a garden with no insects at all might be called "sterile;" bacteria that fix nitrogen in the soil are essential, but bacteria that cause disease are forbidden. The last piece in the ecological mix are humans, who may impact the garden's health either positively or negatively.

In recent years, more attention has been placed on the role the garden has within the broader ecological context. In cases where a garden is in a heavily urban site, its land will provide a haven for wildlife that have been excluded from other areas. This makes these gardens a draw for their animal life as well as their plant life, and many visitors spend more time and attention observing birds, pollinating insects, or rare species of newts than other garden elements. There are campaigns to encourage homeowners to keep the leaves that fall on their garden in place to nourish the soil, to withhold water in times of drought, and to leave the dead parts of their herbaceous perennials so that beneficial insects have a place to ride out the winter. Some of this involves a shift in aesthetic values to accommodate our evolved ecological consciousness. The old practices of indiscriminate pesticide applications are now illegal, and we now know about the potential dangers of introducing plant and animal life to new continents.

Ecologically minded gardeners devote a large amount of time to amending misguided practices of their predecessors: clearing out invasive plants imported into the horticultural trade, rebuilding streams from those that had been artificially diverted, and encouraging healthy plants by way of a healthy ecosystem. Professional horticulturists, especially in greenhouses, are increasingly adding beneficial insects to their modified ecosystems to tilt the scales in favor of plant growth and away from pest growth.

Setting

"Setting" is the *terroir* of the garden, parts of the broader world that become a part of our garden interactions. Throughout this book we have talked about how the sunlight, rainfall, soil type, and topography of the site affect how we create and experience gardens, and we also mentioned the uniqueness of the garden element of borrowed scenery (and so we will not repeat ourselves). Here we talk about the exogenous sights, sounds, and smells that affect our garden experience. The gardens at the Battery, a park on the southern tip of Manhattan, is unmistakably a part of New York City, where among its plantings are views to the Statue of Liberty to the south, the noise from the streets to the north, the stream of tourists to and from the ferries, the smell of the hotdog stand, the breeze off the Hudson River, the fluttering trash, the man who invites people to hold his yellow python, the sounds of multinational chatter, buskers, and the merry-go-round. Other gardens make it a point to stand apart from their surroundings; the clean formality of the New York Botanical Garden's paths, lawns, and planted areas are designed to stand in contrast to the surrounding neighborhoods, and it bills itself as an "urban oasis." Within this garden one will find the Thain Family Forest, one of the few old growth forests remaining in New York City, and a reminder of how the setting of this garden has changed over time.

Plant Palette

Within this element are all the designed aspects of the garden's plants, the human contribution to the botanical features of a garden. The palette is determined first by accounting for what plants will be maintained on the property and incorporated into the new design. Gardenists (and gardeners) then take into account practical concerns such as the plant's ability to thrive in the conditions of the site and the availability of this species in the horticultural trade. Beyond this, the gardenist has the full range of botanical forms at their disposal. Some plants will be several stories tall at maturity, others creep along the ground, some move in the breeze, others are able to be cut into a rigid shape, some are chosen for their good smell, or cultural associations, or because the owner has particular memories of a certain plant. Some plants will be chosen for the shade they provide, their year-round color, or their drought tolerance. Each plant on the palette will have vegetative structures such as leaves and stems and reproductive structures such as flowers, and all 350,000 plant species on earth will be at least slightly different from one another. The range of plant possibilities is multiplied several times with hybrids, varieties, and cultivars, which have been bred to exhibit certain traits and are now more common in the horticultural trade than the original species. And so one can choose a multicolored daylily rather than a yellow one, a dwarf hosta instead of one that is full-size, a fastigiate tree that grows straight up instead of the true species that would shade the whole garden. This is where the aesthetic elements—line, shape, texture, form, space, value, color—and the principles of design—unity, contrast, balance, rhythm, movement, emphasis—come into play to the greatest extent, and plant choice and placement is a major contributor to the other formal elements on this list such as space, structure, and light. Each plant is chosen and placed to relate to its immediate neighbors and to the whole of the garden scene. Gardens are complex, and so each of these aesthetic elements is considered not only from different points in space but also different points in time—how the garden's plants will relate to one another season to season and year after year as they mature. Ronda Brands, codesigner of the Heather Garden in Manhattan, describes running up and down the paths of the sloped garden to ensure that a new addition will be planted in the correct location, no matter the angle of approach. Many gardens will have a multilayered palette, with spring-blooming plants being obscured by summer-growing perennials once they begin to fade; the trick is to organize the palette so that it remains a compositionally rich garden any day of the year. The number of species in the design is itself a point of the gardenist's concern; too few species and the design will look sparse, too many and it will look busy. English border gardens, with many species in a relatively small space, will vary height, flower color, and leaf shape—and carefully arrange larger or smaller masses of plants of the same species—so that each species can be distinguished visually. A meadow garden, such as those created by Doug Tallamy, Peit Oudolf, and Jens Jensen will have greater quantities of fewer species designed to visually weave into one another. Roberto Brule Marx used a palette of plants indigenous to Brazil, often grouping the same species together in wide swaths to create a bold visual field that could be read from far away.

Plant Character

The second aspect we consider regarding the garden's plants are their individual characteristics. This can be described as the nondesigned aspects of the garden's plants, those aspects of the garden that are provided by nature, innate to the species, and difficult if not impossible to replicate. How a plant grows over time and reproduces are two of the important aspects of plant character that we consider in the garden. The first aspect of plant character that we encounter in the garden is each plant's care requirements—what kind of soil, water, and light the plant can grow in—that are derived from how the species adapted to its natural habitat. We apply the term "plant character" to unique traits of species but also to individuals within species. Each tree, for example, has a distinct canopy shape and branch structure that can be influenced by gardeners only slightly—a cypress tree with its even, upright conical shape can never be pruned to resemble the wide spreading branches of an oak, even by generations of dedicated pruners, and each individual oak will differ based on its genetics, setting, and history. Mature trees that have been growing in the garden for hundreds of years show their history in their habit—one can see pruning cuts, lightning strikes, how a tree spread in response to wind or competition for sunlight, and the efforts of horticulturists to keep and maintain the tree's structure through bracing, cables, and supports. Often when we prune a tree, we seek to bring out the essential character of the species by cutting at places that seem most natural and directing growth to resemble impressive trees we have seen in nature. Ancient trees in Japanese gardens are adorned with many timber supports to protect their branches from breaking under their own weight. Encountering such practices in the garden (what we have referred to as "process properties") reminds the visitor of the dedication to preserving the unique character of plant individuals, as well as the history of growth and decay that each plant has endured.

We relate to the character of the smallest plants as well as the largest. Mosses are very infrequently cultivated, and despite their cosmopolitan distribution, they are often difficult to relocate and establish in a garden setting. The moss species that grow around the bases of the rocks at Ryōan-ji or among the extensive fountains at the Villa d'Este were not intentionally placed there and tended; rather, a spore found its way inside the garden walls several hundred years ago, and those responsible for the upkeep of the gardens decided to keep it. This moss lends an important attribute to the garden's form, and in the case of the moss at Saihō-ji (also accidental, but now a defining feature), it is delicately weeded and brushed free of leaf litter on a daily basis.

A garden's plant character goes hand in hand with the daily deliberations of the gardener, who interprets the unique growth and reproduction of the individual plants and responds by cutting, culling, or leaving it be. Plants have variable lifespans and propensities to reseed, and most gardeners welcome when a short-lived perennial plant distributes fresh seedlings around the garden when the original plant has aged out. Curating these seedlings is another example of the reaction to nature that gardeners are tasked with—some seedlings are in the wrong spot, but some manage their way into rock crevices and around the bare spots of soil where nothing else will grow. Bunny Mellon, despite her exacting specifications with garden maintenance, encouraged reseeding and creeping of plants over and among paving stones in her courtyard garden at Oak Spring in Virginia. The result is a garden that changes yearly, as annual seeds find new places to go, and a "finished" design that must look very different from how it was conceived on paper.

These eleven aspects make up a structural lens through which to consider a particular garden. As the contents of a particular garden are organized experientially or conceptually into these structuring categories, one develops a full picture of the garden under consideration. This can, as we will see in the next chapter, form a solid basis for developing claims about the quality, or relative/comparative quality, of a garden. But more importantly for our present purposes, appreciating the garden as a structured whole can form the basis of understanding that garden's meaning—as a whole. While gardens have countless aspects that may carry meaning, conventionally and associationally, collectively and individually, a garden taken as a whole can have meaning as well.

Application of Garden Form

At the end of the last section, we promised examples of application of our proposed means of determining the meaning of gardens. This is not an easy promise to keep since a full material exposition of a garden is necessary as a platform for examination of that garden's form, and hence of that garden's meaning. For even modestly sized gardens, writing a full account of garden form that brings in all the variables we have discussed would be long and complicated. But we can start with a simple example, the Waterlily House at the Royal Botanic Gardens, Kew.

- *Path.* This garden has one path, a narrow circle, which is well-defined by the level tile surface and low railings of concrete and black iron. One enters through a door at the center of the greenhouse, passing through a small planted anteroom that further frames the view of the circular pool ahead. Along this circular path are no benches or additional exits.
- *Structure.* The structure of this glasshouse, completed in 1852, is provided primarily by the glasshouse itself. The white iron supports the glass panes of the walls and ceiling, and the circles of white iron inserted in the lofted ceiling echo the shape of the *Victoria* leaves in the pool below. The slightly raised pools give these structures additional definition.
- *Artifactual Elements.* The historical glasshouse that provides the boundary of this garden also provides an interesting piece of human context. It was completed in 1852 specifically for the waterlily *Victoria amazonica* and was at the time the widest single-span glasshouse in the world. Additional human elements take the form of the labels and interpretive signage dotted throughout the garden, sensors to monitor climatic conditions, and railings meant to blend in with the adjacent plants.
- *Space.* The space of this garden is well defined by the walls and ceiling of the glasshouse. Later in the growing season, the space feels more enclosed as the vines expand along the rafters and the tropical plants along the wall grow to their largest size. The space over the waterlily pool is the most open.
- *Water.* This garden features water more centrally than most other gardens, being that it was designed specifically for aquatic plants. The water in the central pool is even and flat, with an organic dye that reduces algal growth and increases its reflectivity. There are also pools on the side of the glasshouse and a relative humidity of seventy-five percent, suitable for the tropical plants within. Although most of the water in this garden is still, one can hear the occasional movement of fish.
- *Light.* This garden is lit by sunlight. The glasshouse contains no shade cloth or supplemental lighting. Most of the sunlight is dappled, filtering through the trees outside and the vines within. In winter, when the light levels in London drop dramatically, the house is closed, and the waterlilies are transplanted back to a nursery with supplemental light for optimal growth.
- *Topography.* This garden is level, and views of the garden are on a single vertical plane.

- *Ecology.* This garden is heated below ground to a temperature of over eighty degrees Fahrenheit to simulate the Amazonian ecosystem in which these plants naturally thrive. Occasionally a bird will fly in through an open door or vent. Within the central pool are fish that provide movement within the water, help to recycle nutrients, and eat water-borne insects. There are pests, such as aphids, that find a happy home among the heat and fast-growing plants and mold that takes advantage of the constantly humid conditions during the growing season. Constant work by horticulturists moderates the spread of these species.

- *Setting.* This is a garden room within the Royal Botanic Gardens, Kew, in London. Its classical iron architecture matches the larger glasshouses nearby and the formality of the garden in which it is sited. From the interior of the glasshouse one can hear children and other visitors making their way around the grounds. In spring and fall, the temperature and humidity inside will be in stark contrast to the conditions outside.

- *Plant Palette.* This garden is replanted extensively each spring, and so certain decisions of plant palette are made each year. But generally, the plants within this glasshouse are from the tropics. Within the pool are *Victoria cruziana* plants, a waterlily with massive round leaves, and smaller aquatic genera like *Nelumbo* and *Nymphaea.* Each of these makes large, striking flowers in a range of colors. The flat round leaves are contrasted by other aquatic plants with narrow, upright leaves such as papyrus and rice. Vines and plants within the hanging baskets are chosen for their pendulous leaves, flowers, or fruits; this includes *Nepenthes*, the tropical pitcher plant, and gourds such as *Luffa.* The greenery around the perimeter of the glasshouse is punctuated by plants such as *Acalypha* species with drooping bright red inflorescences and colorful, variegated leaves. Together, upright and pendulous plant forms draw the eye toward the massive waterlilies at the center of the house. The range of tropical bloom colors/forms, leaf colors/forms, and their placements relative to one another provide a cohesive and visually interesting backdrop to the center pool.

- *Plant Character.* This garden, which is replanted yearly and highly controlled, does not allow for some of the types of character we focused on above (such as unique plant shape and advantageous reseeding). It does, however, invite visitors to consider other aspects of the plants' character, especially the unique growth form of the large waterlilies that awed the local population when they were brought to Kew in the late 1800s. One can also consider the unique characteristics of the other plants, including the carnivorous leaf modifications of the pitcher plants and the thigmotropic[10]

---

[10] "Thigmotropism" is plant movement in response to a touch stimulus.

leaves of certain pea species. Being that this is a botanical garden, each plant is labeled and accessioned, and many visitors are interested to know how old certain of these plants are, how they have been grown and propagated, which are newly discovered, which are extinct in their natural habitat, and which hold records for largest or smallest.

Form-Focused Garden Interpretation

Based on the formal analysis offered above of Kew's Waterlily House, what can we say about its meaning? Let us set aside the questions of whether the Waterlily House is a work of art or whether it can be profitably interpreted; let us suppose it can. We are perhaps warranted in supposing this because formal analyses of gardens that are more clearly works of art are bound to be much more complex, and for the purposes of this discussion, we need to work with an analysis that is manageable.

All interpretative activities begin with an appropriate framework, a context in which the object to be interpreted is appropriately categorized. This must be the case to know the correct "form" of the artkind in question so we can apply that structure to our efforts. When it comes to translation as interpretation—or when it comes to literature—it is the equivalent of knowing which language is the one being used.

The Waterlily House is the equivalent of a garden room, sited next to other garden rooms within the Royal Botanic Garden, Kew. It is therefore important for us to begin with the framework that the Waterlily House is a part of a botanical garden, and so its purposes and goals—which are essential to know not only for interpretation and for evaluation but just to know the garden for the garden it is—are in line with that identification. Since the Waterlily House is part of RBG Kew, we can already know the role that setting plays—within RBG Kew—but also within Greater London, within the United Kingdom, within the jet-stream-fed climate of northern Europe where climatological conditions are very different from the Amazon.

The path and the topography of the Waterlily House are very simple and straightforward, allowing easy but controlled visitation. The entrance of the House is conducive to the drama of coming into what can be a distinctly different climate from outdoors and what is certainly a sight that is distinct from what one would normally encounter in the UK; the dramatic nature of the gigantic waterlilies is heightened by the drama orchestrated by the entrance to the House. The sense of drama is also enhanced by the beautiful iron structure of the Victorian glasshouse—as well as the sunlight entering it—and while it is not the largest glasshouse at Kew (or even close), the space defined by the iron structure is impressive both in how expansive it is—compared, say, to a home conservatory—and how embracive it is, given how the ironwork creates a canopy-like or tent-like feeling around the central pool and those walking around it. This is known as an "even-span" glasshouse, symmetrical along the axis of entry, and the symmetrical arrangement of the structure within makes this room easy to read from a single vantage point. If there is any place in this garden where we are meant to pause and take in the formal attributes, it is the entrance; this is the case because the doorway and the anteroom provide a frame, because the structure within the garden room is oriented symmetrically from the door, and because there are no other benches around the perimeter of the pool. The pool is raised a bit, and this lends itself to the ease of entry by horticulturists whose job it is to maintain the health of the waterlilies and remove what finds a home in the pool that is not meant to. And, more importantly for present purposes, it raises the level of the waterlilies so we can see them all the more clearly. The plant character of the waterlilies is controlled as they are transplanted twice a year, and as waterlilies, their large pads rest upon the water in the pool. The plant palette is dominated by the pale slightly yellow-green of the large round waterlily leaves, brought into focus by the dark water, further emphasized by the upright leaves of the aquatic plants, the downward hanging baskets and pendulous fruit, and the punctuated colors and forms of the tropical plants that make up the perimeter.

We categorized botanical gardens as "preservation and collection" gardens in our taxonomy in Chapter 1, but most if not all botanical gardens are appropriately categorized also as "destination" gardens as most if not all attract visitors and are designed in ways that foster that attraction and that manage visitation. RBG Kew (along with its satellite garden Wakehurst) is one of the most visited gardens on earth, with well over two million visits every year. During the pandemic, visits fell to just over one million a year, but this was enough to secure RBG Kew as the number one tourist destination in the United Kingdom in 2021, bypassing the Tower of London for the first time. Visitors to Kew are not expecting to see merely a diversity of plants; they are expecting to see a diversity of exotic plants[11] and to see them displayed in ways that are "aesthetic-forward." The Waterlily House demonstrates exactly that, perhaps as a microcosm of the whole of RBG Kew.

If one were to come away from the Waterlily house saying that it is a naturalistic garden designed to exemplify a tropical ecosystem, we would say they were mistaken, and we would point to the evidence provided by our account of garden form in doing so. The features are symmetrical and regularly shaped, the plants are labeled and manicured, the paths and structures are made of iron and tile,

---

[11] In the next chapter, we explore some of the morally darker aspects of how Kew's collection of exotics was achieved.

everything is level and organized. To "read" the Waterlily House is to see a garden room that is maximally attractive through its many aesthetic features, through the particular configuration of its aesthetic features, through the classical architecture and historical structure, through the beauty of its plants, through the exotic nature of its plants, through the exotic ecosystem borrowed for the House's climate, and of course through the diversity of the collection presented there. These features are what the House expresses, and it is through appreciation of the coherence of these features that we come to understand the meaning of this simple and elegant house as a quintessential expression of a botanical garden room: diversity, exoticism, drama—all leading to an appreciation, both cognitive and emotional, of the formal variety of the botanical world and the microcosmic glasshouse that supports it.[12]

This simple example is meant to demonstrate the structure of interpreting a garden by developing an explanatory account of why it has the elements it does. But this example can only go so far since it is only a small garden room on the site of a much larger garden, it is frequently replanted, and it may not constitute a work of art (or even a garden at all) by those who suggest that gardens must be out of doors.

Let's now look at other examples of meaning derived from a garden's form, using gardens we have talked about previously.

Andre Le Nôtre's Versailles and Vaux-le-Vicomte

While it might be argued that the various shapes into which hedges are cut—hedges outlining parterres, resembling topiary—accentuate what a natural growth pattern could resemble, the overall effect of many of the garden areas at Versailles and Vaux-le-Vicomte is that of a painting or a tapestry. Plants are highly and tightly controlled to appear as perfect aesthetic forms that seem strongly artifactual and strongly "plastic" in the sense of artistic control. As a consequence, these gardens unambiguously demonstrate the ability of humans to shape nature into geometrically symmetrical aesthetic forms we find highly and immediately beautiful. Their cleanliness and orderliness signal to visitors the power of those who own them to have commissioned them and maintain them. In the case of Versailles especially, the power of the monarch is celebrated not only by the level of control exhibited but by the expanse of the gardens as sight lines reach to the horizon unimpeded. The meaning is majesty.

---

[12] If our consideration of the Waterlily House includes appreciation of contextual features, such consideration may include how the Waterlily House came to possess properties like diversity and exoticism through imperialism and colonialism.

Capability Brown's Stowe, Blenheim, and Highclere

Brown's signature landscapes surrounding great houses or castles demonstrate a paternalism toward nature, how nature may provide a bucolic or pastoral context for the Englishman's life ("English*man*" because of the time period of their popularity). While the landscapes look natural, as if they could have been the result of benign neglect over hundreds of years of pastoral coexistence with farm animals and humans, they signal an anthropomorphic sense of how nature itself may come to respect the dignity of the human who makes his life there, how it responds to being the object of care typified by nature permitted to get on with its work. The result is control without coercion, invisible artifice, and a sense of the mutual respect between the English landscape and the Englishman for the dignity of the other. These features serve as a frame for the life a gentleman—or at least an Edwardian or Georgian gentleman—should lead, an ideal that is a natural metaphor for how one should live with regard to all things.

The Fin Garden

We do not know who created this garden, originally completed in 1590, making it Iran's oldest extant garden. Like other Islamic gardens, this garden illustrates Paradise by representing all the salient and significant features of the Garden of Eden. Its design, as we saw earlier, follows metaphysically derived dictates and is highly formal as the design is so prescriptive. The illustration of Paradise connects the earthly—water, shade, oasis, comfort—with the celestial—reward, perfection, the divine. This connection brings to mind in immediate and tangible ways the realization that the earthly life is only an early part of a greater existence, and what we do in the present life—the fulfillment of obligation and the spirit to do these things joyfully—connects to what happens in the life to come. Here, the meaning is a symbolic mimesis, whereby the organization and conditions of Paradise are condensed into the earthly realm.

Frederick Law Olmsted and Calvert Vaux's Central Park

Central Park was Olmsted's first project, but it clearly exhibits the values that are instantiated and demonstrated in all his designs. Like Le Nôtre and Brown, his signature is very consistent. His gardens demonstrate the affinity humans have for natural spaces even though their lives are contextualized in highly urban settings. This demonstration is enhanced by the easy ways in which the garden is accessed and traversed by city inhabitants. Olmsted offers urban dwellers the oasis they crave, and he imbues this oasis with a rich diversity of purposes to which it may be put. Central Park has ball fields and boat launches, pavilions and meadows, winding paths through wooded lands and arboreta. The form of the garden is meandering, naturalistic, in direct contrast to the right angles and grid streets of central Manhattan. Meaning is found in meeting the craving of the urbanite for the countryside, but within the urban setting; the urbanite can have their cake and eat it too, as the Park is not so much an escape where one can deny the urban context but rather where the urban dweller can bring to that context all the comforts and joys of the garden, writ large. Olin writes, "The central symbol of Central Park . . . is one of healing and purification."[13]

Throughout the book we have talked about many other gardens, often alluding to their meaning as discoverable through the relation of their internal components. Our account of garden form is the theory that follows practice, a rubric for the conversation that we and many others have been having in some form already. There is little doubt that the gardens mentioned and described are not only meaningful for casual visitors—in the sense of being important parts of the lives of those who enjoy them—but are filled with meaning for those who seek them out and contemplate them, spending time thinking about how their various features all work in harmony, representing a process of deliberation (real or imagined) that flows from a unified vision and leads to a conscious inclusion of these features and no others.

Unique Features of Garden Interpretation

There are (at least) three things that make garden interpretation different—that is, significantly different—from interpretative activities focused on objects from other artforms.

Threshold

When one walks into a museum or gallery, they know they will see works of art. They are ready for art experiences and primed for engagement with the objects. The same is the case, of course, when one sits down in a theater. These locations are entered with expectations. This is, however, not the same with gardens. Some gardens are large enough—in size, scale, reputation, age—that one enters them primed for engagement with works of art, but most gardens we visit with any regularity are not like this. Since not all gardens are art, our first interpretative task

---

[13] Olin, "Form, Meaning, and Expression," p. 52.

is to determine if the garden in question has a form significant enough to sustain and reward interpretative/cognitive engagement. This task is only done if one seeks out an opportunity for such engagement.

Unlike the phenomenon of "beauty breaking in" on one, prompting one to attend to the aesthetic aspects of the object experienced immediately as beautiful, there likely are no or few occasions when "cognitive engagement breaks in" on a person. While it might be natural to experience an aesthetic object aesthetically without volitionally adopting a preparatory approach, and while it might be natural to experience such an object as if it were the sort that would reward cognitive engagement if such engagement were occasioned, cognitive engagement itself seems something that requires "turning on." That is, cognitive engagement requires a decision to begin thinking. (For the moment, we have switched from talking about art objects to aesthetic objects because the "breaking in" phenomenon is commonly only outside of typical art contexts.) A decision to seek cognitive engagement with a garden is preceded by a decision about whether such engagement might yield reward. This is what we are calling a "threshold decision." A garden must be such that it seems likely to reward such attention, and since not all gardens are works of art, not all gardens likely would reward such attention. This relationship between a garden likely rewarding cognitive engagement and that garden being a work of art is a relationship that is not only important, but, as we have seen, is one that seems to flow in both directions: works of art commonly reward cognitive engagement of an interpretative, explanatory sort and, in complement, objects that reward interpretative attention seem by virtue of having that capacity to possess, at least prima facie, one important condition for being thought of as works of art.

Complexity

The various ways in which a garden is complex constitutes a challenge for garden interpretation. First, a garden is complex because it is not merely a "plastic" artifactual object. Those aspects of it that are properly thought of as natural complicate the process of interpretation. Intentionalist approaches suffer because nature is not conscious and deliberative—as least on nonsupernatural or teleological models—and if we allow ourselves to anthropomorphize nature for the sake of thinking of natural processes in a "cocreator" role, for the sake of attributing determination of a garden's form in part to these processes, we are then stuck with the challenge of knowing enough science to understand those processes well enough to include them in our explanations for why a garden is as it is.

Second, a garden is complex in the sense that it incorporates many, many parts. Unlike more canonical works of art, the list of items to which attention might be directed in a garden—the list of the items that might figure into a full accounting of the contents of a garden's form—can be astronomically large. Every leaf, every bug, every petal—the list can be large and complex indeed, and while only some of these will engage our attention—the lesson from Stolnitz—a "simple" list of just those items we think relevant can be anything but simple. The description of the Waterlily House above is of a fairly simple garden that is quite small; a description of Longwood could be daunting.

Third, garden interpretation is made complex because gardens are dynamic and always in states of flux. When we account for those contents of a garden's structure, just as it is when we account for the full description of a garden's aesthetic features, some of those contents must be phenomenally stabilized for the sake of our interpretation. Unlike the case with formal aesthetic properties, dynamic properties can certainly be included in our interpretative efforts—because a full aesthetic account of a garden goes beyond merely its formal aesthetic properties—but to the extent that these formal properties play a role, and they certainly should be expected to, they will have to be phenomenally stabilized. This leads to the realization that we may need to index our garden interpretations to a particular time in a particular set of climactic conditions, and so forth; this is no more a problem than we have already discussed, but it does add to the complexity of garden interpretation beyond what is common to other kinds of works of art.

Finally, while most works of art come already scaled for human attention—a painting has a canvas edge or frame, a novel takes (say) fifteen hours to read, an opera takes (say) three and a half hours, as does watching Ingmar Bergman's *Fanny and Alexander* (that is, the short version!)—large gardens do not. Large gardens are typically divided into garden rooms or subgardens, but even a single garden room can spread out beyond what a human can take in during a single course of attention. Even a modestly sized garden, given its complexity, can sustain attention for periods of time that range between the time required to stroll through it and whole days from dawn until dusk, all dependent on what sort of individualized scaling the visitor brings to their experience. So the scaling of both time and space that is necessary for the subject to set is yet one more complexity to be accounted for as one interpretatively connects to a garden.

Distraction

The points that Ferrari and Gillette—and, for that matter, Ross—made concerning how beautiful gardens can be and how appreciation of that beauty can occupy the full attention of someone engaging with a garden are points that should not be overlooked. The reality is that gardens can be so engrossing in so many ways—aesthetic, chief among them—that thinking about whether to think about a garden may not occur to an appreciator. Ferrari's and Gillette's points might have been more secure had they argued that while some gardens may be profitably interpreted, those that reward such attention are very likely those that reward other sorts of attention to a degree that cognitive engagement would be rare. A garden filled with meaning is very likely going to be a garden filled with beauty.

But distraction can come in other forms as well. Gardens have purposes, some of which lie beyond being places possessing aesthetic merits, and so many gardens, as we have seen, provide contexts for all manner of instrumentality: eating, drinking, playing, painting, researching, reading, studying, chatting, sunning, and the list can go on and on. Each purpose and each utility can serve as distractions to engaging directly with a garden, much less engaging cognitively. Interpretation requires not merely the act of thinking but of critical thinking. A case for meaning must be worked through and worked out. This can be intellectually absorbing but also intellectually challenging. Sometimes it is easier and more enjoyable just to "turn off" and watch the path of a butterfly. And there is no case for saying the former is more valuable experientially than the latter. Interpretative attention can be rewarding, and as we have argued is likely important for recognizing a garden to be a work of art, but watching the path of a butterfly can constitute a rich aesthetic experience unto itself, one that is whole, organic, and memorable.

The themes of this chapter, and the arguments for the various positions we took throughout it, seem fairly straightforward. There is likely no need to review them again. But it might be noted as a final word that not only do many of us find meaning in gardens—in parts of gardens and in gardens as wholes—but there are a range of theoretical constructs to explain why and how this might occur, what its implications may be, whether a garden has one meaning or many, how one interpretation can be better than another, and so forth. However it is done in practice, and however it is theoretically explained and justified, it is perhaps of key importance to remember that interpretative attention to gardens is yet one more way we aesthetically connect to them and find in them the experiential depth and richness we do.

In the next two chapters we explore the intersection of The Garden and value. There we examine whether gardens have only positive value, how ethical concerns may be relevant to consideration of a garden, some particular thorny ethical issues, how gardens are evaluated and how they may be compared one to another, what "garden criticism" might amount to, values connected to the practice of gardening, and we offer some thoughts on garden ethics.

9

Gardens and Value

In this chapter and the next, we discuss a wide variety of garden values, including values that specifically have to do with the practice of gardening. We begin with questions focused on the character of value (merit, worth, esteem) gardens possess and how moral considerations concerning gardens are relevant to their overall value. We move from there to questions concerning garden evaluation: on what suppositions it rests, how it can be done, and what "garden criticism" may entail. Then we move to talking about the value of gardening and the values that are involved and fostered by gardening. We conclude with a brief discussion of garden ethics, or in other words, how we choose the right course of action as garden creators, gardeners, and garden visitors. We do not believe we leave aesthetic discourse in talking about these matters; we argue throughout that matters of garden value are relevant to their aesthetic value.

Do Gardens Have Only Positive Value?

In consideration of the value that gardens possess, we might begin with exploration of the claim that gardens do not possess value, that they are neither good nor bad. A garden is a natural place. To say that it is "in part" a natural place—in respect of that fact that all gardens are the products of human design and intervention—is dividing matters inappropriately. The whole of a garden is a designed place, and the whole of a garden is a natural place (absent the obvious artifactual components like sundials and bird baths). This harks back to Cooper's rejection of a factorized approach to understanding garden appreciation. If we attempt to divide a garden into its natural parts and its artifactual parts, such a division would discourage the garden appreciator from noticing and valuing aspects of a garden that are only visible in the amalgamation of both the natural and the designed. As a garden is a natural place, it might be claimed it is neither good nor bad, but it simply is. Worth is a human appellation. Nature is only good

Desert Garden, The Huntington, San Marino, CA
Multiple gardenists starting with William Hertritch, 1908
Photo by Ethan Fenner, Summer 2021

The Art and Philosophy of the Garden. David Fenner and Ethan Fenner, Oxford University Press.
© Oxford University Press 2024. DOI: 10.1093/oso/9780197753590.003.0009

or bad as humans interpret it to be, as it serves our purposes, meets our interests, causes us pleasure, causes us pain, and so forth. So, in this sense, gardens as natural places are value neutral; they are neither good nor bad.[1]

This claim, however, is limited when it comes to gardens because at the same time gardens are designed places. If they were merely natural places—if they were First Nature—such a claim *might* be compelling. But just as they are fully natural places, they are fully designed places as well, and with those designs come purposes, aesthetic considerations, and a host of other aspects that seem indelibly value-bearing.

In contrast to the neutrality claim above, there are reasons to support the claim that gardens are always good, that they possess only positive value.

Aesthetic Engagement

To fulfill the conditions for being a garden, as described in Chapter 2, a garden must be aesthetically engaging. This is not the case with many works of art in other artforms. We attend to works of art because they present themselves to us within a category that prompts our artistic attention; our engagement is a precondition of our accepting a work under the category "art." When the art is in a museum, we assume upon crossing the threshold that what is on the wall or on pedestals is worth our attention. So, as we see in some modern art, there is no obligation for a work of art to reward our engagement positively. This means that a work of art can be revolting or horrible—we think of Piero Manzone's 1961 *Merda d'artista*, Chris Burden's 1974 *Transfixed*, or Damien Hirst's 1990 *A Thousand Years*. These works are found by many as off-putting, yet they are important enough contributions to the world of art that they come to mind quickly.

Gardens are unlike this. A garden must be seen to be a garden to be a garden. To come upon an area where a garden once was, to see perhaps an enclosed space overgrown with weeds—and perhaps even to be told that this is "such-and-such garden"—is not to see a garden. Gardens must be engaging; they do not come "pre-engaged" as work in other artforms might. If a so-called garden is in bad shape—and it is not a garden over which we exercise any control—we turn and leave. Pursuing forward could entail hiking through overgrown bush, over downed trees or washed-out paths, requiring more physical risk than we signed up for. We will not recognize it as a garden, and we will not take the time to try to tease out its positive features from its overarching negative ones.[2]

---

[1] This reasoning follows William James, "The Moral Philosopher and the Moral Life," *International Journal of Ethics* 1:3 (1891), pp. 330–354.

[2] David E. Cooper writes, "For obvious practical reasons, gardens need to be 'reassuring,' 'benign,' usually 'cheerful' places to be in." *A Philosophy of Gardens* (Oxford, UK: Oxford University Press, 2006), p. 106.

Whether this is true of nature beyond gardens is unclear. Allen Carlson and others[3] argue—in a thesis known as "Positive Aesthetics"—that nature (what Hunt would call First Nature) is always aesthetically positive. There are versions of the Positive Aesthetic thesis—Ned Hettinger lays out four[4]—but each contends that nature is aesthetically positive on its own, whether or not aspects of nature can be aesthetically negative (some versions of the thesis are wholistic/systemic and others particularist), and each incorporates some reliance on a positive aesthetic appraisal of nature being based on a proper understanding of nature, scientifically derived or informed and usually with all aspects of nature understood within a context of interrelationships.[5]

It is tempting to employ some version of Positive Aesthetics as support for the claim that gardens must be aesthetically positive, that even when a garden includes aspects we find aesthetically negative, on the whole, understood within the proper context, all gardens are aesthetically positive. But in the end, such a thesis cannot be applied to The Garden. This is not only because The Garden is not simply nature but rather an amalgam of nature and human design—reason enough for Positive Aesthetics not to be relevant here, since all versions of the thesis focus on nature unaltered by humans—but because of the observation that if a garden is untended and it reverts to being First Nature, which according to the Positive Aesthetics thesis is aesthetically positive, this "once-garden" is simultaneously aesthetically positive as First Nature and aesthetically negative as Third Nature. Understood through the lens of the Positive Aesthetics argument,

---

[3] Allen Carlson, "Nature and Positive Aesthetics," *Environmental Ethics* 6:1 (1984), pp. 5–34; Holmes Rolston III, *Philosophy Gone Wild* (Buffalo, NY: Prometheus Books, 1986); Eugene Hargrove, *Foundations of Environmental Ethics* (Englewood Cliffs, NJ: Prentice Hall, 1989); Glenn Parsons, "Nature Appreciation, Science, and Positive Aesthetics," *British Journal of Aesthetics* 42:3 (2002), pp. 279–295. Carlson writes:

> If the natural world seems aesthetically good when perceived in its correct categories, this cannot be simply because they are correct; it must be because of the kind of thing nature is and the kinds of categories that are correct for it. . . . A more correct categorization in science is one that over time makes the natural world seem more intelligible, more comprehensive to those whose science it is. Our science appeals to certain kinds of qualities to accomplish this . . . ones such as order, regularity, harmony, balance, tension, resolution, and so forth. . . . These qualities that make the world seem comprehensible to us are also those that we find aesthetically good. . . . The determination of categories of art and of their correctness are in general prior to and independent of aesthetic considerations, while the determinations of categories of nature and of their correctness are in an important sense dependent upon aesthetic considerations. . . . These categories not only make the natural world appear aesthetically good, but in virtue of being correct determine that it is aesthetically good. (*Aesthetics and the Environment: The Appreciation of Nature, Art and Architecture* [New York, NY: Routledge, 2000], pp. 91 and 93–94)

[4] Ned Hettinger, "Evaluating Positive Aesthetics," *Journal of Aesthetic Education* 51:3 (2017), pp. 26–41.

[5] It is unclear that areas untouched by human involvement continue to exist or exist largely enough to make Positive Aesthetics a compelling thesis, and it is unclear that such land (if it exists or exists on a large enough scale) does not aesthetically benefit from human involvement in the form, for instance, of controlled burns.

an untended garden reclaims aesthetic positivity when it ceases to be a garden and reverts to First Nature.[6] An unfortunate implication of this is that we have no reason, on grounds that prioritize tending gardens for aesthetic reasons, to tend them since their reversion will make them aesthetically positive in a different way. We will set aside this argument as its applicability to The Garden is limited.

Utility

Gardens serve a wide variety of purposes—intentional by virtue of their designs and accidental as visitors employ them in all manner of ways. They may serve as a place for raising food or medicine, for religious worship or contemplation, for the internment and memory of the dead or a historical event or period, for conservation efforts, for educational purposes, or as a setting for a residence, a business, a hospital or school, a collection of sculpture, a soccer field or playground, an amusement park, or even a parking lot or public thoroughfare. A garden's purpose can be simply for aesthetic appreciation—if "purpose" is the right word when the only activity is formal aesthetic engagement and appreciation of beauty (we have been careful about calling this a "purpose" thus far in the book). Many gardens—perhaps most—can be used as venues for relaxation, contemplation, reading, eating and drinking, making phone calls, writing notes, jogging, walking a dog, feeding ducks, playing a game of chess or frisbee—the list is probably endless. While it is possible that some purposes to which a garden may be put may be negative—a matter we turn to below—for the most part the purposes gardens serve enhance our lives and collectively add to the civic good. Incidentally, gardens serve many of these purposes so well, so frequently, and so easily specifically because of their positive aesthetic qualities.

Cultural Value

In cases where nations have gardens that are typical of a style associated with that nation, those gardens enhance and sustain the cultural identity of that nation. In the end, nations are remembered for their cultures and by their cultural products. Gardens clearly contribute to that. We remember the ancient gardens of China's Forbidden City and at Suzhou. We remember the *karesansui* at Kyoto, the English landscapes surrounding Stowe and Blenheim, the extensive gardens

---

[6] "Revert to First Nature" may be a problematic concept, but if Positive Aesthetics only focuses on land that has never experienced human interaction, the thesis ceases being interesting because so little land qualifies.

surrounding Versailles, and the terraced gardens created for the Medici. Gardens of this sort—historically and perhaps artistically significant gardens created in styles intimately associated with particular nations or regions—may be among those cultural products that establish a nation's identity and legacy more than others and in more lasting ways.

Gardening and the Good Life

Gardening is a classic and common component of living a good life. Perhaps no one made this point more evocatively than Voltaire, where at the end of all of Candide's travails, he is advised finally to go and cultivate his garden. While it is possible Candide is being advised to be stoic, or that "cultivate your garden" is meant metaphorically, it is more interesting to think that Voltaire was signaling the special place that gardening holds in the constitution of a good life. The person so engaged is engaged with the creation or sustaining of good: the aesthetic good of the garden, the various purposes the garden may serve, the virtues that the garden encourages, and so forth. Later in this chapter we go into this matter in depth.

In complement, there are, of course, reasons to think that gardens are not always uniformly good, that on occasion they include negative aspects. In the next section we begin to explore the intersection of ethics and garden aesthetics, but even without getting into moral matters we can see examples of the negative aspects of gardens.

Poor Design. Some gardens are poorly designed, and their capacity for aesthetic engagement and as a place people wish to spend time in is diminished because of this.

Poorly Maintained. Some gardens—more than we might like to think—are poorly maintained. They are unintentionally overgrown; dead plants are not removed; plants or parts of them are dying; flower production has waned; areas of hardscape and artifactual elements are deteriorating.

Populated with Invasives. Even if a garden is well designed and well maintained, if it is planted with invasive plants, ones that might contribute to an area's environmental degradation, we could well think of this garden in negative terms.

Can a garden have negative aesthetic features? The answer must surely be yes.[7] A parterre, hedge, or formal bosquet may have missing or dead plants that make them imbalanced, uneven, or shaggy. Colors of flowers of intermingled plants

---

[7] Yuriko Saito, "The Aesthetics of Unscenic Nature," *Journal of Aesthetics and Art Criticism* 56:2 (1998), pp. 101–111; Marcia Mueldner Eaton, "Beauty and Ugliness In and Out of Context,"

may be muddled or jumbled. A plant may be so heavily infested with aphids that it becomes difficult to identify, or a tree may have pruning cuts made so incorrectly that we pray for its recovery. Species needing full sun may be languishing in shade. Lack of weeding or lack of pruning may interfere with appreciation of a garden's positive aesthetic features.

But there is a more fundamental reason why gardens can have negative aesthetic features. We evaluate and compare gardens one with another; we have idealizations of garden styles against which we may compare a given garden; and we have idealizations of what our own gardens should be, and we work them into states that more closely resemble those idealizations. These acts are predicated on the supposition that gardens can have both positive and negative aesthetic features. Those who claim that such an idealization-dynamic requires only positive and neutral aesthetic features is either denying the set of examples we offered above or believes that the presence of neutral aesthetic features is enough to motivate us to spend hours of sometimes backache-inducing work on our gardens to rid them of these value-neutral features. That is unlikely.

If a garden can have negative aesthetic features, is it possible that a garden possesses more of these than positive ones? The answer is no; it cannot. Like a long-neglected "secret garden," once we come upon a garden with more negative aesthetic features than positive, we are inclined toward one of two courses of action. We understand either that we must invest time, talent, and treasure to reclaim the "once-garden" into the garden it was[8] or that the garden is "a wash" and needs to be either eliminated or redesigned from scratch. Both inclinations are grounded on our recognition not that there is a garden before us that needs care but rather that there *was* a garden present at one time—a "once-garden"—that could return in one form or another if we wish it to. In other words, we do not see a garden as a garden if its negative characteristics outweigh its positive ones. A similar situation may obtain in the case of a building deemed a work of great architecture; if the roof caves in due to lack of inspection and maintenance, we categorize it as a ruin rather than a building.

While it is entirely possible that we may call a "once-garden" a "garden," calling it this is (1) to recognize a general form, where objects that possess this form constitute a set identified by a certain label—"garden," (2) to recognize what it once was, or (3) to recognize what it might be again. The recognition of its

in Matthew Kieran (ed.), *Contemporary Debates in Aesthetics and the Philosophy of Art* (Malden, MA: Blackwell, 2006), pp. 47–48.

[8] Occasionally, bringing a "once-garden" back to being a garden requires not the investment of time, talent, and treasure, but rather a withholding of it, as in the case when a gardener has been overzealous in their application of fertilizer, in overwatering or overpruning, in overly fussy (and so ultimately unnatural) design-work, in the addition of too many artifactual elements, and the like.

general form is a linguistic convenience, but it does not carry any ontological weight on its own. To come upon a human corpse, and say, "I found a person," is not actually to have found a person but rather a body; in a situation like this, it is more common to say ,"I found a body." And so while we may come upon a "once-garden" and say, "I found a garden," the application of this label is similarly imprecise because (1) we would not then think of such a place as one where a gardener would take up routine gardening practices (like weeding, fertilizing, or watering) without first taking grander efforts (clearing brush, removing fallen trees, pulling up invasives) to get it to a point where routine gardening practices make sense, and (2) we would not employ such a place as a venue for the many common purposes that gardens serve, such as a venue for relaxation, contemplation, reading, eating, and so forth.

To be a garden, a garden must have certain features intact. Our claim is that to have these features intact is for that place to have more positive features than negative ones, even if the place is in desperate need of weeding or pest management, repairing parterre lines, rebalancing the plant palette, and so forth. Given this claim, and given the list of values gardens serve—aesthetic engagement, utility, cultural value, and contribution to the constitution of "a good life"—we believe that every garden, *qua* garden, is good and has positive value. This is not to say, however, that certain contextual matters related to the existence of a garden do not result in that inherent "garden goodness" being overridden by such concerns, even to the point where we come to the conclusion that a garden should be eliminated. We turn to considering these sorts of contextual matters next.

The Relevance of Ethics to Gardens

As we saw earlier, not everyone agrees that the inclusion of moral considerations in aesthetic judgments is a good thing. Formalists, New Critics, those who believe in "art for art's sake," believe that making such connections can be distracting from the aesthetic or artistic value of a work or can subvert the work into being a tool for a social, political, or religious agenda. Even if we grant the argument that Berys Gaut offered—the argument that an ethical fault in an art object can constitute an aesthetic fault in that object—can this argument be applied to gardens? Gaut's point is that when a work of art recommends either a character trait or a course of action we find morally objectionable, and then on the basis of our finding the recommendation unacceptable we reject the recommendation, this constitutes a failing of meeting an aim of that object and that failing is relevant to judging the aesthetic worth of that object. But can gardens, especially ones that do not "front" those properties that would be most apt in developing interpretations of gardens—representative, expressive, narrative properties and the like—have as their aims the recommendation of character traits or courses of action we find morally objectionable?

The answer is yes. Gardens—or, perhaps we should better say, contextualist aspects of gardens—can and do recommend a range of ethical responses and nurture in those who tend them and those who visit them a range of ethical attitudes and dispositions. They may not do this in the same way novels, films, and operas might, but they do so nonetheless. And they do this in ways that may surpass the ways works in other artforms do. If aspects of a garden—particular gardening practices; decisions about where, when, and how to establish a garden; decisions about what kind of garden to establish and what it might take to do so; decisions about what purposes a garden may serve—recommend or endorse ethical positions, both positive and negative, then they can be subsumed as relevant under Gaut's argument. The presence of a negative ethical consideration relevant to a garden may alter the aesthetic reaction we have to that garden; we may find such a garden less aesthetically valuable because it includes among its contextual properties some ethically suspect matter.

Gaut's argument may be taken to demonstrate the relevance of moral considerations to those gardens where the moral matter is relevant to the aesthetic character of a garden, but what about those cases where the moral matter is not relevant to the garden's aesthetic character? There are two ways to answer this question. One way is to suggest that on occasions when moral matters do not appear at first glance to be relevant to the aesthetic character of a garden, they really are.

- Gardens that require less resource consumption may be more admirable aesthetically because their aesthetically positive features do not incur more cost than other such features might. This may come in the form of water-wise plantings, organic treatment of the soil, or abstention from pesticides that are harmful to pollinating insects or birds. In other words, efficiency in accord with natural processes may be an integral part of an aesthetic account of a garden.
- Gardens with more paths may allow gardeners to maintain the integrity of those gardens with more ease—and we might regard that ease as indicative of attending to the maintenance of life and health, concerns we regard as moral—but more paths may also mean more opportunities for aesthetic appreciation of the garden.
- Gardens involved in conservation work or in scientific study may be found more aesthetically engaging than gardens whose aims are purely visual pleasure. To see plants thriving in a garden is a welcome but common experience, but to know that such a plant is endangered, extinct in habitat, confiscated from poachers, or used to develop medicines brings special

attention to that plant's features. This could account for why, when given the choice of a visit to a pleasure garden or a botanical one, some will choose the botanical garden even though its visual attractions may not be as immediately striking as the pleasure garden.

- Gardens that are wildlife havens may be regarded as morally better than they would be without serving such a purpose, and that they are wildlife havens renders such gardens all the more engaging and beautiful because of the presence of that wildlife.
- Gardens that are especially accessible to those with mobility challenges may be regarded as morally better than otherwise, and it may be that such accommodations prompt in all of us who experience that garden a greater sense of satisfaction through knowing that larger audiences of appreciators of the garden—aesthetic appreciators—are possible.
- Gardens open to the public, with public access and nondiscriminatory pricing or policies, will be better gardens than those that exclude members of certain classes or races or imposes arbitrary rules that make groups of people feel unwelcome. Happy and well-treated visitors are more likely to engage aesthetically with a garden.
- In complement, a garden may be experienced aesthetically negatively when, for example, it has become toxic due to the overuse of pesticides, or dangerous because of untended trees, or consumptive of too many resources or of resources that have deleterious environmental effects. Ethical matters that are negative can impact aesthetic experience as much as those that are positive; we say a bit more about this below.

Such examples demonstrate that, even when not obvious, moral considerations can in a practical way be relevant to cases for the aesthetic merits of a garden, not to mention relevant to particular aesthetic experiences of that garden.

The second way to answer the question is the one where the ethical matter makes no difference to the aesthetic character of a garden. Here one likely should say that where relevancy cannot be established, the proper path is the one consonant with formalism. The warning by those who cautioned against including considerations when they inappropriately interfere with aesthetic appreciation of an object is still a good warning. But, we believe, such cases are rare. Relevancy of some degree is not hard to establish in many cases.

The inclusion of moral scrutiny in considering the worth of a garden focuses not on the garden itself but rather on contextualities having to do with the motivation for installing a garden at a particular place or time, for installing a certain kind of garden, maintaining it in a certain way, and so forth. In an important sense, these morally relevant matters have to do with *gardening* rather than with the garden per se. This gets a bit tricky, however, when thinking about the intersection of gardening motivations and garden purposes, as a garden's purpose is part of its identity as a garden and a large part of the basis upon which we correctly classify it. Suppose a person installs a garden with a nefarious purpose.

- One may install a garden to make their neighbors feel bad that they do not have the resources to install such a garden.
- One may install a garden that encroaches on the property of a neighbor as either a slight or some sort of localized "manifest destiny" expansion.
- One may install a garden to make those who admire that garden forget that the person who installed it has flaws that, absent the garden, might become more focal.
- One may install a garden to make it harder for a municipality of which they are part to take "eminent domain" kinds of action.
- One may install a garden to dissuade others from trespassing on their property, for example to stop the local school children from congregating on their lawn as they wait for the bus.

The value of a garden depends in part on understanding that garden in terms of its purpose, so purposes—even nefarious ones—are relevant to a garden's identity. Are these examples of "negative gardens" or "bad gardens"?

Consider the example of keeping children off one's property. Suppose a homeowner decides the most efficient way to accomplish this purpose is with a plant palette of *Opuntia*, *Euphorbia*, *Acacia*, *Toxicodendron*, and *Heracleum*—plants that can sting, burn, scrape, and potentially even kill trespassing children. There are examples found in botanic gardens, pleasure gardens, and even personal gardens where such plants are used, and sometimes they are used with the intent to deter human/animal encroachment. But in all cases, these exist in the context of larger compositions. While they may be used for the purpose of deterrence, the garden of which they are a part is not designed simply for the purpose of deterrence. If it were, we would argue that it is not a garden. A wall of thorns and poisons planted around the property is exactly that—a wall—and so it is better categorized as Second Nature rather than Third. Now, if a garden were designed to showcase such plants, it could well be a garden, but then its purpose as a garden is not about deterrence—the idea that a garden would be created simply for the purpose of keeping people out seems amusing if not bizarre. We mentioned in an earlier chapter the Alnwick Garden in Northumberland that contains a Poison Garden whose cast iron gates have "THESE PLANTS CAN KILL" written alongside skull and crossbones. This is a bona fide garden, of course, and while its purpose is to showcase deadly plants, the purpose is not to employ the plants to kill visitors or to deter visitors from entering.

Consider another example. Suppose for Halloween a designer of haunted houses creates a "haunted garden." Such a thing could be fun, but while it might be called a "garden," whether it really is one depends on the overarching purpose driving its design and the practical implications of the design. If the purpose is to scare visitors, and the "garden" is composed of dead plants and/or plants that are temporarily assembled (say, left in pots) to create a haunting atmosphere, this is clearly not a garden. In the same way that one might set out tombstones on one's front lawn for Halloween, no one will mistake a residential front yard for a cemetery. On the other hand, if one creates a garden—one that fulfills in all respects the definition of a garden we offered in Chapter 2, even if it is temporary (like the Chelsea Flower Show)—and, perhaps like the Alnwick Garden, populates and arranges it with plants that are meant to contribute to a scary atmosphere, this could be a bona fide garden. As Sissinghurst has a White Garden, so a Halloween Garden, a Scary Garden, or a Black-and-Purple Garden could certainly be a garden.

In rare cases a garden, either in its coming-to-be or its continuation, will entail some problematic dimension. It is possible that this problematic dimension could be great enough to overwhelm the goodness of the garden as a garden. In that case, it is possible that the correct judgment—moral, political, social—is that the garden should be eliminated. Suppose a municipal garden became the location for uncontrollable rampant crime.[9] Despite that the garden might be perfectly charming as a garden and that there is nothing objectionable about the garden itself, it might still be judged that the garden should be eliminated for the sake of addressing the crime taking place there. Or suppose a wealthy individual bequeaths their extensive garden to a municipality. Suppose the municipality is a modest one, and suppose that to maintain the garden as it is, or even in a reduced state, would require resources that would entail hardships on the municipality's populace. Suppose there is no one who could help with private support or by purchasing the property. In this case, the correct judgment might be to eliminate the garden. None of this requires us to judge the garden itself as "not-good." People may love the garden. But sometimes external considerations can take precedence over aesthetic ones.

Since some gardens may be criticized as being brought into existence or maintained in existence in ways that are morally suspect, we can criticize those

---

[9] An interesting case-in-point is People's Park in Berkeley, CA. The University of California, which owns the land, has since the 1970s been trying to turn the park into a parking lot, a soccer field, a volleyball court, and, currently, student housing. Each attempt at development is marked by intense and often violent protests. Some see People's Park as a site of crime and an eyesore, valuable space that can be put to better purposes; others see it as a historical site and an embodiment of the free speech movement and community activism that led to its founding. People's Park, as of this writing, remains in a constant state of "degardening," by the university and "regardening" by the park's devoted residence and advocacy groups.

aspects of such a garden on such grounds. Even if we grant the notion that all gardens are always good, they are good in varying degrees, and so comparison of them is certainly possible. These comparisons can and do include as data points morally relevant facts about a garden. And it is possible that these morally relevant facts are great enough to cause us to consider whether it would be better for a particular garden not to exist. We might be tempted to say in these cases that the garden in question is a "bad" garden, but this description would not be entirely accurate. It is not the garden per se that is bad; it is some contextuality that is at issue.

The parallel with works of art from the late twentieth century and twenty-first century is apt. There have been many works of art whose existence is morally questionable, and in some cases it seems not only appropriate but obligatory to censor the exhibition of those works, at least for certain populations such as children. While all censorship should be viewed with a critical eye, as censorship can be employed in ways that are inappropriate and can have the effect of curtailing artistic expression and cultural development, to say that no work of art should ever be censored is to invite problems. This can be the case even when the work itself is artistically important or meritorious. A work of art that incorporates shooting a gun at an airplane filled with passengers, providing illegal drugs to addicts to incentivize them to engage in prescribed actions, photographing nude children, or promoting racism[10] should be subject to critical ethical reflection. Whether or not these works should be censored can only be decided through consideration of their individual cases, and our descriptions of them are extreme for the purpose of making our point, but it seems clear these and cases like them should be the subject of reflection. Gardens, both that are works of art and that are not, are not immune from such reflection when ethical questions are raised.

Ethical Problems Associated with Gardens

The Colonization Problem

Let's begin by considering an issue raised by Isis Brook[11] and by Michael Moss:[12] the installation of gardens as a significant part of one country's colonization

---

[10] Each of these descriptions refers to either an actual work of art or a body of works of art. These works are all available to be viewed on the internet, and so it might be concluded none were deemed illegal. But whether they each pass the test of ethical reflection is a matter left open.

[11] Isis Brook, "Making Here Like There: Place Attachment, Displacement, and the Urge to Garden," *Ethics, Place, and Environment* 6:3 (2003), pp. 227–236.

[12] Michael Moss, "Brussels Sprouts and Empire: Putting Down Roots," in Dan O'Brien (ed.), *Gardening—Philosophy for Everyone: Cultivating Wisdom* (Chichester, UK: Wiley-Blackwell, 2010), pp. 79–92.

efforts of another. Installing a garden in the style of one from home, populated with plants with which one was familiar at home, was a way not only to make the colonizer feel more comfortable in their new surroundings but at the same time allow that newfound comfort to increase their sense of being settled in the new colony. In complement, colonization historically meant that many species of plants were making their way back to Europe for botanical collections. Botanical collections in Britain especially grew as exotic plants were taken from British colonies circling the globe and as corporations sought to find new commercial opportunities in new species of plants. Increasingly over time we find the gravity of the ill effects of colonization, and while the populating of botanical gardens and the installing of familiar gardens do not seem on their own moral travesties, it is still noteworthy to recognize the role that plants and gardens have played in colonialization.

The effects of botanical colonization are felt today. Some plants "discovered" by European colonists were given common names and Latin binomials that imply false information about the plant or include derivations of racial slurs. At some botanical gardens, the names of certain plants on their labels refer not to terminology that was/is indigenous to the region but rather to the names given plants by colonizers. For instance, instead of indigenous names of southern African plants based in Zulu or Xhosa, names of many South African plants derive from Dutch Afrikaans. This perpetuates the misplaced representation of the plant as a result of colonization.

In addition to the societal tolls of botanical colonization, there are environmental tolls as well. It was the mindset of early botanists that plants found in a new land were up for grabs, and so they were taken wholesale from their native ecosystems and brought to foreign soil. Today, all respected botanical institutions adhere to strict rules and permitting processes to ensure that the collecting process is sanctioned by local governments and limited to certain species, certain areas, and does not interfere with the plant population's health in situ. However, the colonizer mindset is still a very real problem in plant conservation, and poaching from native environments by collectors is now a major cause of species decline. Second, the intercontinental movement of species comes with certain dangers—soil can be brought in with nonnative pathogens, or nonnative plants may take hold in an ecosystem and outcompete native plants, in some cases endangering populations of native birds or other wildlife. In the United States, the most damaging tree blights were brought over accidentally in shipments of other materials, but our most invasive weeds were brought over on purpose—for railroad ties (*Eucalyptus*), erosion control (*Carpobrotus*), cattle forage (torpedo grass), or, frequently, for horticulture (Japanese knotweed, Amur cork tree, kudzu). Of course, there are many other nonnative plants in our gardens that do not cause environmental damage—but it is the responsibility of the garden to monitor, control, and remove plants that pose legitimate dangers to the wider ecosystem.

Land Ownership Problems

There are a variety of ethical dilemmas associated with the ownership of land on which gardens are placed.

- Some private estates occupy many acres, with (1) gardens that are only enjoyed by the owners and those few they invite to visit their gardens, (2) land that might be purposed in ways that might benefit more people, and (3) land that may have been acquired through processes akin to gentrification, where former inhabitants are displaced.
- Some gardens occupy unceded land. Brooklyn Botanic Garden acknowledges that it occupies part of the unceded ancestral homeland of the Lenape people, for instance.
- Some gardens evolve from being free and open to the community to being institutions that erect fencing and charge entry fees, potentially hampering local populations from regular visitation. As an example, the Bronx-sited New York Botanical Garden was founded in 1891 and, because of dwindling municipal support, started charging for admission in 1994.
- Some gardens may have been built through the use of slave labor.

While an example of an agricultural moral matter—so, Second Nature rather than Third—the Indigo Revolt of 1859 reveals the ethical depths land ownership problems can reach. During the British Raj in India, some Bengalis were forced to grow indigo crops that displaced the food crops they would have grown. This threatened their survival. The interests of the farmer, it could be argued easily, were significantly more important than the interests of the owners of the land who wished to grow crops merely for wealth acquisition. Those who organized the Indigo Revolt may have been morally justified, just as those who boycotted the picking and the consumption of crops in California under the direction of Cesar Chavez and the United Farm Workers may have been justified in their actions. In like fashion, would one be justified in stopping the installation of cultivated fields—of crops or for cattle—where such installation involves destroying Amazon rainforest and/or encroaching upon the lands of a particular indigenous Amazonian society? It used to be common to hear of environmental activists chained to trees, occupying positions in tree canopies, and interfering with machinery used to clear land.

Are there situations, like the agricultural ones mentioned above, where the morally correct course of action is taken to be the eradication of a garden on land not belonging to the gardenist? Above we sketched some reasons for why the existence or maintenance of some gardens may be wrong, given that their existence or maintenance entailed the violation of ethical principles or aims that were of greater normative import than the good of the presence of that garden. In such cases, the ethical action may be to eliminate the garden. Presumably this would be done by—and only by—those who owned the garden. But one could imagine someone other than the owners undertaking this elimination—perhaps a "Robin Hood-esque" character whose aim is to destroy private estate gardens and redistribute plants to public gardens to be preserved and enjoyed by many. In those cases, is the "eliminator" morally justified? If the situation is a legal one, there is little in the way of a puzzle: the state (apparently) has the right to enforce the law even if the garden owner disagrees (after, of course, due process of law). But what if the matter is moral and not legal? Is a party who does not own a garden morally justified in acting to eliminate a garden that is (to use a shorthand description) immoral?

Guerilla Gardening

In complement to the puzzle of eliminating a garden on someone else's land, we have the puzzle of creating a garden on land owned by another. "Guerilla gardening" occurs when one engages in garden practices on land not one's own. Common occurrences of guerilla gardening include the occupation of disused municipal lots for community allotments, the planting and tending of medians and roundabouts, and "seed bombing" where (commonly) wildflower seeds are sown on someone else's land ("seed bombing" may not actually produce a garden). Sidewalk and pavement gardens, sneaking into a colleague's office to water and prune ignored plants, and planting plants or encouraging the spreading of plants on one's neighbor's land may qualify as guerilla gardening.

Shane Ralston offers a justification for engaging in guerilla gardening.[13] But his vision of guerilla gardening focuses on reclaiming unused urban land, with the implication that the land is publicly owned. Although he argues that property rights are not ultimately inviolable, the sort of guerilla gardening he discusses is not the sort that would typically involve the violation of property rights in opposition to one wishing to retain those rights in an intact and unviolated state.

---

[13] Shane Ralston, "A Deweyan Defense of Guerrilla Gardening," *The Pluralist* 7:3 (2012), pp. 57–70.

Guerilla gardening, in that case, only becomes an ethical dilemma when the infringement on property rights is taken to be a violation or taken to incur harm to the owner of that property. Such a situation might obtain if "seed bombs" were launched over agricultural land to interfere with the growth of crops or if "seed bombs" were deployed in garden situations so that weeds were sown and that "purposed plants" in the garden were threatened. There are also plenty who believe that plants in public botanic gardens, whose operations are often subsidized by city resources, are just as public as anything else, and so they have no problem lopping off a portion for their own collections. Yet most guerilla gardening seems more along the lines that Ralston sketches: cases where publicly owned lands are unused and abandoned, and where citizens take it upon themselves to use that land for gardening. In these cases, Ralston argues, enough good—of various sorts—comes from the "take-over" of the land to justify the use. In contrast, if a guerilla gardener were to interfere with rights based on the significant material interests of the owners of a plot of land, justification for such interference would need to be forthcoming.

In the 1971 Hal Ashby film *Harold and Maude*, the almost-octogenarian Maude finds a small tree that she claims is suffocating in its urban setting. She insists that she and Harold remove it and take it to the forest. When Harold objects and says they cannot touch it because it is "public property," Maude looks at him and says, "well, exactly!"

In addition to ethical aspects that have to do with particular gardens, gardens as a whole have been the focus of some ethical criticism. We begin with two topics on which Cooper focuses[14] and then move to a third.

The Dominion Problem

The dominion charge holds that every imposition of human will on nature is morally suspect because it proceeds from the view that nature is appropriately subject to our control. The charge holds that this anthropocentric perspective and impetus for action places the conferral of value exclusively with human endeavor and products, and it does not appropriately account for the value of nature on its own. Cooper's answer to this charge is this:

> That a gardener "uses" living things as pot plants or for shade no more entails a denial of their intrinsic worth than his "using" human beings as plumbers or for delivering letters entails a denial of *their* intrinsic worth.[15]

[14]   Cooper, *A Philosophy of Gardens*, p. 99.
[15]   Cooper, p. 102.

This is a lesson we might have learned from Kant's ethics as he articulates the second form of the Categorical Imperative in a way that does not preclude our "use" of other persons but rather prohibits our "merely using" other persons.[16] We rely on other persons—as we rely on nature—to survive and flourish, but this does not and should not entail that this reliance circumscribes their total worth. The gardenist does not seek to alter all of nature and in altering a small part does not seek to thwart the processes of nature or ignore the contributions of nature. The successful gardenist designs in respectful cooperation with nature. If they do not, their garden is short-lived.

We might add, however, that Cooper's observation likely should apply in varying degrees to various gardenists and gardeners. Gardens that rely heavily on annual color—either for bedding plants or displays such as orchid shows— import thousands of plants to be used and discarded within a season. There is a widely shared view that the gardens designed by Le Nôtre at Versailles were meant to demonstrate the French king's imposition of will and dominance over all, nature included. This view apparently figured into the advent of the more naturalistic designs of the English gardenists of the successive centuries (Brown and the Picturesque gardenists), and it apparently figured into the German "mittelweg" orientation to find a middle ground between the naturalism of the English and the formality of the French. The strength of Cooper's observation may differ with different garden styles.

The Deception Problem

The deception charge holds that gardens are deceptive representations of nature, that their creation is motivated through the aim of the gardenist to present nature in a way other than it truly is. No plant grows in as precise a geometric way as a to-piary suggests.[17] No group of trees grows in the perfect symmetry of the bosquet or the *chahar bagh*. In the case of gardens meant to be more naturalistic—English cottage gardens, native plant gardens, meadow gardens—where the aim is to present plants that look as if they had received no tending whatsoever, one could claim that this is the greatest deception of all, since they require a great deal of tending to maintain their perfectly natural appearance. Certain Japanese gardens are like this in their representation of idealized landscapes, pruning that

---

[16] Kant would be unhappy with this allusion; for him, persons are worthy of respect in themselves, but no other living thing is. In addition, when we "use" persons, we can compensate them for that use; the parallel apparently does not hold with plants. (Although, Michael Pollan, *The Botany of Desire* (New York, NY: Random House, 2001), makes the claim that our use of plants has in fact helped those species tremendously.)

[17] For a thorough discussion of topiary, see Isis Brook and Emily Brady, "Topiary: Ethics and Aesthetics," *Ethics and the Environment* 8:1 (2003), pp. 128–141.

follows the tree's natural form, and invitation of certain elements that give a natural patina, like moss. Maintaining these gardens requires many hours of dedicated work by gardeners who have trained for years, if not decades, to know what to remove and what to keep. Nature abhors uniformity, or so the expression goes, and so land left unto itself will include a teeming variety of plants. Yet gardeners spend perhaps the majority of their working hours weeding, extracting from their gardens those plants they do not wish to be there to maintain the illusion.

Cooper's response is that the representational aspects of a garden are never meant to be passed off as completely formed representations of the items or processes of nature. They represent only to a limited degree, and the person who mistakes that representation for portrayal of the full truth of nature makes a mistake. Gardenists who seek to inform their gardens with representational aspects of nature do so knowing that those aspects tell only part of the story, yet it is a part they deem worthy to tell. So a botanical garden that has an alpine area does not seek to demonstrate the full breadth of all plants that grow in alpine settings, or even a representative patch of alpine ground. The gardenist seeks to represent the diversity through offering a selective subset, often from alpine regions around the world, chosen for the breadth of their representation but clearly not offering the full and complete range of alpine plants. Even a naturalistic garden that includes only plants native to a certain habitat of a certain province of a certain country does not intend to portray these plants as they exist in the wild. The idea is to show these plants at their best, away from pests and grazing animals, with as much water as they want, often cleared of dead branches and spent flowers. At the absolute end of the spectrum are tracts of undeveloped, unplanted land within sites of larger botanic gardens—but even here the site is maintained in a way that controls processes of ecological succession, periodic fires, prolonged drought, and encroachment of invasive plants. Here too the goal is not to mislead the visitor into thinking this is unmediated nature but rather to present an example of an idealized ecosystem.

Cooper's final insight on the matter is this:

> [T]he dominion and deception charges share a common premiss—the premiss, I suspect, of much contemporary environmental ethics. This is to the effect that our practical dealings with natural places, including those we are tempted to turn into gardens, should be guided, above all, by some ideal of how we should relate to "wild" nature.[18]

Cooper turns the tables on those who offer the charges, exposing the hidden anthropocentrism that may lie at the argument's heart. Ideals are human constructs.

---

[18] Cooper, *A Philosophy of Gardens*, p. 106.

Guidance on how we should relate to nature based on an ideal of this relationship presumes that such an ideal is achievable in enough precision that it should guide us. No matter how well we understand the workings of nature, no matter how comprehensive our scientific understanding of all the aspects of nature, to claim that this understanding is comprehensive and deep enough to form the basis of a single ideal that guides us appropriately in our relations with nature is a challenging notion. If, as Cooper seems to suspect, this ideal goes beyond what it properly should, it must be held up for scrutiny. Ideals in this context are dangerous things insofar as they suggest, in their monolithic and absolutist versions, that all proper normativity flows from a single vision and a single understanding.

Second, if the logic of such an observation, coupled with the substance of the original charges, entails that we should not interact with nature at all, given the limitations of our understanding and the suspicions of our motivations, this conclusion must be rejected. We must interact with nature to survive. At the most basic level, at least since moving to an agrarian existence, we must alter nature to grow food. How we interact with nature, within the bounds of respectable environmental ethics, is not somehow endemically unnatural—even in the absence of a guiding ideal.

Third, if a statement of the "ideal of how we should relate to 'wild' nature" amounts to justified environmental ethics—if the guiding that is done is appropriately limited in scope (to what we understand, to what we know is injurious, to what we know is beneficial, to only those aspects of nature we as a matter of fact interact with), then this is easily accepted. We do not need a comprehensive ideal in this case; we need only a "limited ideal" fit to our understanding of nature and our impact on it.

Fourth, through the course of his book *Second Nature*, (journalist) Michael Pollan lays out a thorough case for how the dualism of "human-or-nature" is misguided—and how this dualism can lead to the mistaken notion that humans should never interfere with "nature." This is abundantly clear in the case of gardens, but Pollan makes the case that prohibition against human interaction with nature in general is based on conceiving of humans as separate from nature, a claim that does not wash.[19] In aid of his point, he offers a quote from Shakespeare:

[I]n the *Winter's Tale*, to Perdita, who spurns the hybridized rose because it is "unnatural": "This is an art/Which does mend Nature—change it rather; but/ The art itself is nature."[20]

Even once we have recognized the falseness of the dichotomy between nature and culture, it is hard to break its hold on our mind and our language. . . .

[19] Michael Pollan, *Second Nature: A Gardener's Education* (New York, NY: Dell, 1991).
[20] Pollan, p. 49.

Our alienation from nature runs deep. Yet even to speak in terms of compromise between nature and culture is not quite right either, since it implies a distance between the two—implies that we are not part of nature. . . . What we
need to do is confound our metaphors, and the rose can help us do this better
than the swamp can.[21]

[I]ndeed, the whole idea of nature being "out there," a kind of abiding metaphysical absolute against which we can judge messy, contingent culture—is *itself* a
cultural construct.[22]

The Use and Distribution of Wealth Problem

This problem is an amplification of the sorts of problems we saw above, under
the heading "land ownership problems." In the taxonomy of gardens we offered
in Chapter 1, the first taxon under "Identity and Style Gardens" is "Wealth and
Prestige Gardens," and the example offered there is Bunny Mellon's Oak Spring
Garden in Virginia. This was a garden established for the enjoyment of the
Mellon family and, while one may visit the garden by appointment, the garden
is not open to the public. Oak Spring may be a particularly good example of this
category of garden, but examples of such gardens are plentiful. Many such gardens are known among garden enthusiasts because, after all, one cannot display
one's wealth and prestige through their garden by keeping the garden a secret.
But the existence of many of these sorts of gardens is kept quiet because their
chief purpose is not the demonstration of wealth and prestige but rather simply
the enjoyment and pleasure of those who have the wealth to create and maintain
them. In countries or cultures where the achievement of wealth is a sign of success, it may be difficult to mount a compelling argument for the immorality of
installing a garden either as a sign of that wealth or as a source of enjoyment by an
individual or family who has had the good fortune to amass wealth (supposing,
of course, that the wealth was acquired in ethical ways). That is, an argument
against installing a garden is a harder argument to pursue than an argument
against the amassing of wealth itself, especially in those cases where wealth ac-
cumulation involves any degree of exploitation of other humans or (so-called)
natural resources. Nonetheless, gardens that are exhibitions of wealth that has
been achieved in ways that may be to some degree morally suspect are gardens
that are then tinged with these same suspicions—as are, recently, museums of art
financed by wealth acquired in similar ways.[23]

21 Pollan, p. 97.
22 Pollan, p. 167.
23 Michael Classens, in his article "The Nature of Urban Gardens: Toward a Political Ecology of
Urban Agriculture" (*Agriculture and Human Values* 32:2 [2015], pp. 229–239), writes, "On the other

If a garden through its installation or continued existence contributes to unbalancing the equitable distribution of resources in a society and does so in a morally suspect way; to shoring up institutions and traditions that contribute to this; to promoting entitlements that are undue, elitist, or work against inclusion of all members of a society in the freedoms and goods of that society, then ethical concerns such as these may overtake the goodness of the garden as a garden.

A straightforward utilitarian could easily argue that until greater needs of sentient creatures are adequately met, the expenditure of resources for the creation and maintenance of places that are primarily for aesthetic appreciation is unethical. A similar argument could be made for the acquisition of works of art. The $10,000 that went to purchase a painting for one's home could have been given to charity, and the good that might have been done with those funds might dwarf the aesthetic enjoyment of a painting hanging in one's home and, for good measure, only enjoyed by that person and their family. In other words, the promulgation of support for aesthetic objects—all aesthetic objects—is hard to justify on utilitarian grounds when there is suffering that might be eased by channeling that support otherwise. (We will return to this point immediately below.)

Some gardens suffer from another aspect of the wealth distribution problem—namely the allocation of natural resources. The living components of gardens require sunlight, water, and nutrients, and when environmental conditions are not favorable to the garden—say in periods of drought—then the garden must use supplemental irrigation to keep the plants alive. At a certain point in the drought, the city may restrict water use for gardens to ensure that there is enough water for the populace to drink and to grow food crops. This comes at the expense of the garden, but again it is hard to justify prioritization of aesthetic objects over the health of a community. Peat moss, used frequently in potting mixes, is in effect a finite resource, and environmentally conscious botanic gardens are now attempting to reduce their use by employing alternatives such as coconut coir. A garden must decide whether their mulch must come from specifically harvested trees, or if it can come from as a byproduct of tree removal. There is also the question of energy use in the creation of synthetic fertilizers and for the garden's power tools, for example. In all these cases one can imagine a point at which garden maintenance constitutes a waste of energy, natural resources, and emissions.

In those cases where ethical concerns are not hypothetical or speculative but exist as empirical realities, they must be taken into account. In those cases where the concerns are less concrete, it will be down to the honest reflection of

hand, and more recently, a cohort of critical scholars has argued that urban garden projects simply enable the reproduction of contemporary neoliberal policies and subjects—the very conditions urban gardens are meant to address." The quote comes from page 231.

the person in control of the flow of resources where those resources should flow. Arguments that require full attention to the lower levels of Maslow's Hierarchy[24] before any attention is due to higher levels—especially when thought of not in localized ways but rather societally, culturally, or globally—are arguments that do not always land. Beauty is a good, and while hunger is a greater evil than beauty a good, critics may suggest that the situation is not a zero-sum game where the two must be placed in head-to-head competition. Critics may argue that there is room for attending to both concerns, to attending to all levels of Maslow's hierarchy simultaneously and with a balance that seems morally justified. This argument may not be found compelling by Peter Singer–style utilitarians, but then not everyone's moral intuitions and commitments are utilitarian in that way.

Many people value gardens as escapes from the ordinary work-a-day world. They distract us from our problems; they offer a mental oasis. While an argument could be made to suggest that distraction and escape are morally unwarranted— that we should keep our eye on the real and the existentially important—it is difficult to hold simultaneously that the psychology that describes the need of humans for respite and distraction is in error. Escape through aesthetic means may be important to our mental health, individually and collectively, and gardens are a commonly available means of achieving this escape. The good gardens do for us mentally may be, in appropriate balance, equivalent to the good agriculture does for us in terms of filling our stomachs. The trick to getting the ethics right is to know what "appropriate balance" means. We believe this is an effort that must be pursued through individual or particularist moral reflection. Sometimes the garden will win, but sometimes it may not.

Moral considerations are relevant to garden aesthetics because they are relevant to the full case we offer for the aesthetic value of a garden. We judge the aesthetic worth of a garden—if we are judging in ways that are broader than narrowly formalist—by including aspects of that garden that go beyond mere immediate sensory connection. If a garden has a purpose that strikes us as valuable, and if that garden serves that purpose well, and if that purpose has a moral dimension, this may be relevant to how we judge the garden in terms of its overall aesthetic character. Inclusion of consideration of purpose is relevant if we believe that the functional aspects of aesthetic objects are relevant to their aesthetic characters and their worth, and this belief is most easily had when we see that function bound up with our experience of that object as aesthetic.

---

[24] As a reminder, Abraham Maslow famously suggested in 1943 a schema of a hierarchy of human needs with basic needs connected to biological survival at the base, psychological needs above that, and spiritual and self-fulfillment needs at the top.

Can the Value of One Garden Be Compared to That of Another?

Before we begin—in the next chapter—to examine *how* gardens may be evaluated, we first need to ask the more basic question about whether gardens *can* be aesthetically evaluated. It is possible that when one advances a view about the quality of a garden, they are only advancing a "dressed-up" version of their own personal attraction to the object, of their likes and dislikes, preferences and pleasures. Whether a person likes an object or not may be mildly interesting—for the purposes of getting to know that person and, if we find ourselves in the position, developing a basis for contributing to their exposure to opportunities to experience more objects they may enjoy. And frequently when we begin an aesthetically focused conversation, we start with the question "did you like it?" But this report on one's reaction as an application of their preferences should not then be taken as anything more than that. The leap from "I liked it" to "this is an object worthy to be liked" or "this is an object worthy to be liked and those who fail to like it are in error" is a jump unwarranted without further argument. That argument begins with whether there is any path that leads from the first to the second. If one likes an object for a reason that should function as a reason for someone else liking that same object, then this is more interesting than the simple report that one liked the object in question. There are arguments on both sides: that there is such a path and that there is not, and so we should take a moment to explore some of those before focusing more sharply on the "how question" of garden evaluation.

For the sake of creating a stable platform for discussion, we define "realism" in this context as the position that when we attribute aesthetic qualities to an object, that attribution is "real," that there is some fact to the matter that we are capturing. It is the position that aesthetic judgments, including aesthetic property attributions, can be right or wrong—that there is something to be right or wrong about. "Realism" is sometimes conflated with the notion of the "objective," but realism per se is not an ontological position (in this case) but an axiological one. It is about correctness as it is applied to ascription of value. And this, frankly, is the more interesting question as contrasted with questions about the location of aesthetic qualities: can we be right or wrong when advancing an aesthetic judgment about a garden? When we make an aesthetic garden judgment, are we merely giving voice to our likes and dislikes or are we instead saying something that is true or false? We have four arguments "for" and three "against." (In the end, we think the prorealism "for" arguments win.)

Through consideration of the arguments against aesthetic realism, we will include discussion about the nature of taste. We use the word "taste" in aesthetic conversations in four ways: (1) to denote what happens when something is in one's mouth, in contact with one's tongue, and that one typically can smell; (2) the capacity of humans to translate objective, descriptive properties into value-informed aesthetic properties;[25] (3) the strength of one's ability to discern objects of high aesthetic quality from ones of lesser aesthetic quality—this reference occurs when we talk about "good taste versus bad taste" and when we talk about the progress involved in developing and exercising this capacity; and (4) patterns of likes and dislikes within aesthetic categories, such as when we describe one having a taste for minimalism (in opposition to the baroque), for classical music (over rock), or for British New Wave (over Punk). These patterns may be personal in the sense that they can only be described as patterns of a single person, or there may be patterns that a plurality of people share in common. It is this fourth sense that will occupy our attention after we discuss realism "for and against."

Common Usage and Literal Property Ascription

The first argument for realism begins with the observation that garden comparison and evaluation is common, and such conversation typically includes use of aesthetic terms.[26] And, for good measure, these are used with common reference, which is to say that when we say that a parterre arc is elegant, we mean precisely the same thing as when we say that an arced line in a painting is elegant or that the movements of a dancer are elegant. We mean to use the relevant terminology in ways that do not differ from their usages in any other aesthetic contexts. We do not mean their usage to be metaphorical, substitutionary, allusory, or anything other than literal.

While our pretheoretical intentions are to use this same vocabulary with gardens as we would in any other aesthetic context, we could of course be wrong. But this would need to be demonstrated. It is more complex to hold the view that there are two different sets of aesthetic vocabulary—one for "ordinary" aesthetic contexts and one for gardens—and this would shift the burden of argumentation to the person who claims that this more complex description of the situation is the correct one. What would such an argument look like? One would need to show first that terms in the nonliteral vocabulary had different referents or different meanings. One would need to show how those were relevant in garden

---

[25] This is the way Frank Sibley famously uses the term in "Aesthetic Concepts," *Philosophical Review* 68 (1959), pp. 421–450.

[26] The line of argument here was inspired by the work of Eddy Zemach in places such as "Aesthetic Properties, Aesthetic Laws, and Aesthetic Principles," *Journal of Aesthetics and Art Criticism* 46 (1987), pp. 67–73; "Real Beauty," *Midwest Studies in Philosophy* 16 (1991), pp. 249–265; and *Real Beauty* (University Park: Pennsylvania University Press, 1997).

situations as opposed to "ordinary" ones. One would need to give a reason for why two different vocabularies are necessary. And that argument very likely would need to turn on salient differences between The Garden and any other aesthetic form. This last requirement likely would need to refer to the sort of differences we sketched in earlier chapters.

The most salient difference is that gardens are dynamically open. One might argue that given this fact, when a person makes a claim about a garden possessing the property of elegance, that person must initially clarify whether they are describing a static property or a dynamic one. For instance, one could say that a certain parterre arc is elegant and by that be offering an observation about a phenomenal property of the garden, made static in the experience of the attender. And, in complement, one could say that the lifespan of a garden could be described as an arc, and that arc is elegant. Barring arguments that the line of a parterre arc is not a line but a conceptualization of a line, it would not be diffi-cult for one to press the point that the parterre line is an actual line but the line of a lifespan is a metaphor (and so if that line arcs, then that arc is a metaphor). We may think of other properties that are applied to both static and dynamic conditions—"balance" may refer to the garden's (phenomenal) composition or to a garden's dynamic ecological balance—but the arc example works well here because even when the property is described so differently, the aesthetic property used to describe it—here "elegance"—means precisely the same in both cases. In both cases, the aesthetic term refers to the same quality of both lines, despite the fact that one line may be objective (and so literal) and one may be metaphorical. The reference is to a particular quality of a curved line, and the categorization of what kind of line it is makes no difference to the attribution of this quality to it. Accounting for the difference in reference to "line" requires no more than an ex-planation of what line, or what kind of line, we refer to. "Elegance" is the same in all these contexts.

When we communicate with others about whether a certain parterre line is elegant, we have a good chance that our communication will be successful. It might require some clarification of what line we mean, perhaps what "parterre" means, and perhaps what we mean by "elegance" to someone less accustomed to using aesthetic vocabulary, but then our communication is likely to be suc-cessful. If aesthetic terms were unstable, or sometimes metaphorical and some-times not, or in some other way referentially or meaningfully opaque, we would not have the confidence we do in the eventual success of communicating our aesthetic observations. And yet we do. We do because the vocabulary we share with other speakers of our language is stable in the ways it needs to be for com-munication to take place. As words generally have stable enough meanings to allow for easy communication, so aesthetic words do as well. The best explana-tion for this stability of meaning is stability of reference. A word that refers to an object in the world, like the elegance of a parterre line, most likely connects to a set of objective facts about the curvature of the edge of a parterre. Our reference is to things outside ourselves—this is the most likely explanation for the ease of shared reference—and so when judgments are offered that are connected ultimately to these stable objectivities, we may expect our judgments to have just as much basis for stability and for being shared. If a parterre line is straight, the person with whom the observation that the line is elegant is being shared either will correct us or ask for clarification. Straight lines are not elegant (in the typical usage of the term); that is not what "elegance" means. Or, better stated, the straightness of a line is not grounds for correctly attributing to it the property of elegance. This lack of correctness goes all the way through: from the simplest of reference all the way to the most summative of aesthetic judgments, because they all ultimately rely on that simplest of reference.

The Evidentiary and Argumentative Structure of Our Aesthetic Claims

This first argument feeds well into the second. If one claims that a certain parterre arc is elegant, and the person hearing this claim disagrees, the person making the claim will retreat to talking about the curvature of the line and what strikes the claimant as being descriptive and objective about the particular curvature of that line. If the person hearing the claim disagrees, they likely will do the same in terms of evidencing their disagreement by pointing out properties about the line they think are as available and obvious to the person making the claim as to the person disagreeing with the claim. While the two may ultimately disagree about whether the arc is elegant, what they almost certainly will agree on is that the matter should be resolvable by reference to the objective properties of the curved line. They see it—aesthetically—differently perhaps, but they see what is there to be seen, in terms of the objective properties of the curvature of the line, the same.

They must see what is objectively there to be seen the same or else their disagreement would not be worth arguing over. They only argue because they believe the other person is missing something they see as obvious. They believe that if the other person came to see what is plainly objectively there as they do, the other person would no doubt come around and believe as they do. The same structure is the case as we move up the evaluative ladder. Broader or more summative claims are argued on the basis that the object in question possesses the "mid-range" aesthetic qualities that are claimed for it, and claimants expect that if their conversational partner agrees with their observations that the garden possesses the aesthetic features they both agree that it possesses, then surely they will come to support the same summative claim they do.

We argue—in the sense of rationally advancing and defending claims (and not in the psychological or behavioral sense)—expecting arguments are worthwhile. If we realize our argument cannot be successful in terms of potentially persuading the other person of the correctness of our claim, we change gears and either launch a new argument or give up and call it a day. Since it is commonplace to develop and then to share judgments about the aesthetic quality of a garden, especially in cases where one is making comparisons with other gardens, and since it is equally commonplace to argue one's view if others fail to agree, we conclude that these sorts of arguments, at least pretheoretically, are the sorts of arguments people take to be meaningful and, frankly, winnable. Time and effort spent arguing when there is no hope that the argument is meaningful—that it has the chance of persuading another person, rationally and through deliberation, of the correctness of the argued claim—is a waste. If all aesthetic arguments—that we pretheoretically take to be meaningful—are in fact meaningless, investment in such arguments sums up to a lot of wasted time.

The sense of the meaningfulness of arguing about aesthetic judgments relative to gardens is half the matter. The other half is the evidentiary function itself. When we offer arguments, we evidence our claims. This is not only because we think such evidence will be compelling and serve our aim of persuading our audience of the correctness of our claim; it is also because we think that the evidence we offer in support of our claims really is evidence for those claims. The beliefs of rational humans are based upon and compelled by evidence, and as the evidence for our aesthetic garden claims is the same sensory-based evidence that is available to all equally, this evidence should prompt and compel belief. If it fails to, then either the evidence was not as strong as we took it to be—despite the fact that we find it compelling enough upon which to ground our own belief—or our audience is not responding to the evidence in the way rationality seems to require.

Aesthetic Education and the Development of Taste

Thousands of students engage in art education. They sign up for classes; they take instruction; they practice. If we are talking about university education, they likely do this for three or four years, and they spend many thousands of dollars for the privilege of doing so. In each successive course they take, they believe they improve, and for all but a tiny minority of students their professors believe they improve, too. Rigorous professors hold their students to high standards, and commentary about what works and what does not work in terms of art creation is frequent and sometimes blunt. Each person invested in this educative process believes that improvement is possible and that it takes place as course follows course, and creation follows creation. Improvement implies that there are standards of quality, that progress toward reaching higher standards is not only conceptually possible but practical and attainable, and that students are genuinely better when they finish than when they start.

The same situation obtains with students of garden creation and landscape design. Those who wish to become landscape architects engage in education just as any student in an aesthetically based discipline does. When studies are completed, landscape architects are ready to practice, and just as importantly, they are seen to be ready to practice. If American, they may seek out licensure and membership in the American Society of Landscape Architects, a group that encourages and recognizes the acquisition of the credentials necessary to practice. The ASLA offers continuing educational opportunities, and it recognizes significant achievement in the practice through awards. All of this is premised on the belief that there are standards that can be achieved, standards that are real, that are worthy of investment, that once achieved are worthy of recognition, and, finally, that form the basis upon which respect for being prepared to practice ought to be accorded.

Above we mentioned that there are four ways in which we use the term "taste." The third definition was this: *the strength of one's ability to discern objects of high aesthetic quality from ones of lesser aesthetic quality—this reference occurs when we talk about "good taste versus bad taste" and when we talk about the progress involved in developing and exercising this capacity.* Many art and design educators work to expose their students regularly to opportunities to practice honing their skills to see (hear, etc.) and then to articulate what it is about one object that is aesthetically better than another. Through years of continuous exposure to aesthetic objects, and of continuous practice with crafting explanations and defenses of their judgments, aesthetic appreciators grow and their skills improve.

If the matter were simply a case of claiming such growth, we might be inclined to suggest the alternate explanation that they simply have been acculturated to favor what those who advance similar claims favor. But this is not the case. We do not take someone's word for it that they have good taste. We test the claim. We expect that they will be able to craft and defend articulations not only of their judgments but also how they reached these judgments. We expect they will wield evidence fluidly; we expect they will know relevant contextual information; we expect they will offer insights that had not yet occurred to us but in which we see merit.

While the development of taste in many artforms can be met with skepticism over whether the development of so-called finer taste is true artistic progress or merely the efforts of acculturation to what those respected for their quality of taste admire, this sort of skepticism is rarer when it comes to garden appreciation because the proof of the pudding is in the tasting: to create a great garden is to have the knowledge and skills to keep it alive and vibrant. To appreciate a great garden is to be familiar enough with the biological and horticultural science necessary to be able to talk about it in meaningful and informed terms. In short, it is much harder to bluff garden taste than it may be with other artforms. This allows us more security in claiming that there is actual progress, actual forward momentum, in garden appreciation and the developing of taste in judging gardens.

Our point here is not to argue that today's garden designs, as a whole, are better than those of yesterday. The evidence against this claim is easily seen in our citation of great garden exemplars that not only are centuries old, that not only are in some cases designed by those without formal training, but that are particular to specific cultures, places, and times. Our claim is that there is taste growth in the individual gardenist and gardener.

Before leaving this argument, we should point out that while we believe the argument to have merit, we recognize that it is limited when considered in purely concrete "on the ground" terms. While progress in taste is an expected result of exposure, training, practice, reflection, and the rest, the exercise of that taste can be hampered by some of the realities of garden design work. (1) Gardenists and landscape designers frequently work for others—for patrons, for universities, for municipalities, and so forth—and so in those cases their output typically will reflect the taste of their clients. This is not to say that a gardenist will not incorporate their own signature into their design—this is especially the case when a client seeks out a gardenist specifically because the client appreciates that gardenist's designs—but a gardenist working with/for a client typically will create a design that is not exclusively an expression of that gardenist's taste. Gardenists only have the opportunity to fully showcase their own tastes and signatures when they have complete creative control, as would be the case with their own personal (or professional exhibition) gardens or in a garden show like the famous one in Chelsea. (2) The work of many gardenists, especially during their education, focuses more on the development of drawings, diagrams, and schemata rather than the installation of physical gardens. This removes them one level from the expression of their own tastes in the garden itself. (3) Because gardens are naturally dynamic, because garden designs must accommodate the topography and so forth of their sites, and because gardenists share with gardeners the unfolding development of a garden's design through time, the expression of a gardenist's taste is never absolute.

Shared Judgments and Patterns of Agreement

The most common argument for aesthetic realism is that we commonly find agreement among those who advance judgments about the quality of particular works of art. "Everyone finds Ed Wood's 1957 *Plan 9 From Outer Space* to be lacking as a film." "Everyone finds Orson Welles's 1941 *Citizen Kane* to be a brilliant film." "Everyone finds Wolfgang Mozart's compositions to be better than those of Antonio Salieri." "The greatest American architect was Frank Lloyd Wright, and the greatest American ballet choreographer was George Balanchine." These examples are stated hyperbolically for effect. Even if there are those who challenge such universalist and absolutist judgments, the fact that one can make such statements and others find them reasonable—even if they fail in their universalist and absolutist ambitions—suggests that we expect commonality among at least a class of aesthetic judgments, at least those that have had the benefit of much time and much audience attention. Musō Soseki was a sought-after garden designer in his time, and his surviving gardens continue to be held in high esteem by locals and tourists alike. Whether Damien Hirst will be remembered in five hundred years as a great artist may be in question, but there is no question that Leonardo da Vinci is remembered this way.

In addition to the apparent fact that those artworks and artists that have withstood the test of time enjoy near universal acclaim (within their appropriate cultural contexts), we can to a significant degree express why this is. We can explain why artists and their creations are held in esteem, and the cases we articulate, as is the case with all such justifications, will be received in terms of their merits as cases. Such cases can then be used as bases of prediction: if another artist is like the originally-praised artist in the right ways, that artist can expect the same reception and the same eventual reputation. This is to say that our cases for what counts as aesthetic goodness are lawlike or nomological; they have normative force to the degree to which the case has merit. Our judgments do not stand alone as particular one-off judgments, but they are rather part of a pattern of judgments that implies that quality itself is what is recognizable and recognized.

While this is the most popular argument for aesthetic realism, it is the one most commonly assailed as well. It is difficult to point to instances of universally shared beliefs about the quality of particular artists or particular artworks. Even when we mean to pick out such cases—the *Mona Lisa* is frequently invoked—detractors are quick to point out that there are plenty of folks who have claim to aesthetic sensitivity who do not care for the *Mona Lisa* aesthetically, that it is hard to tease out those who champion the work for its aesthetic merits and those who are simply responding to its popularity as an icon, and that the popularity of the *Mona Lisa* does not cross all cultural lines.

When it comes to The Garden, the situation becomes more complicated. Gardenists do not work in a contextual vacuum, and the reality is that fashion, over the course of the life of The Garden, plays a strong role in what is favored. This has the unfortunate consequence that when a particular fashion wanes, so does acclaim for gardens created in that style. A garden's style sticks with it indelibly; categorization of a garden by identity is more present and perennial than perhaps with other artforms. This complication is not only a matter of time but also a matter of place. Gardens are strongly identified with their cultural locations; a Japanese-style garden in the United States will be recognized as a Japanese-style garden, so it is not a question of physical location but rather of cultural identity. And just as with styles from particular timeframes, cultural identity is equally indelible.

On the other hand, that we were able in chapter one to offer iconic examples of garden types, and that those familiar with these gardens may have been able to recognize the aptness of those examples for their types, suggests that within a firm structure of garden categorization, instances of high quality with regard to particular types are recognizable. Blenheim is a better example of a Capability Brown landscape than is Highclere (say, for instance). Central Park is the quintessential example of a Frederick Law Olmsted landscape design. So within their appropriate categorizational confines, great examples stand out and are received by audiences as examples of significant quality—not merely as great examples of their types but as examples of great gardens that happen to be of a certain type.

Let's now turn to the other side of the argument, the side that holds that aesthetic realism in relation to The Garden is untrue, that ultimately there is nothing to be right or wrong about when we develop and advance aesthetic judgments about particular gardens.

Ockham's Razor

While one easily can argue (on consequentialist grounds) that there is much at stake when it comes to the normativity of ethical judgments—judgments that it is wrong to kill, harm, steal, and so forth have very large real-world consequences—one might argue that there is less at stake when it comes to aesthetic judgments. While some people make their livings based on aesthetics and art—through creation, commerce, criticism, and education—and so judgments based on standards of quality are materially important to them, there seems much less that is exigent about ensuring that we get an aesthetic judgment right. Even those whose lives are based on aesthetics, generally speaking, will not have their lives significantly disrupted if we come to believe that aesthetic judgments carry no true normative force and that such judgments are really nothing more than statements of preference. The ice cream industry is faring well, and so the pleasure we may take in aesthetic objects will continue to keep the "aesthetic industry" alive and well.

Ockham's Razor famously holds that we should not multiply entities beyond necessity. If aesthetic judgment is explained sufficiently well as preference advancement, then what is the need for us to add to this that such judgments must also be true statements about the way the world really is? If nothing essential turns on an aesthetic judgment being more than a statement of preference, why add more? When we recommend that a friend visit a garden we found enjoyable, what is wrong with explaining that recommendation as a prediction based on a pattern of enjoyment—that as we enjoyed a garden, and as our friend is known to us as enjoying the gardens we enjoy, we predict they will enjoy the recommended garden as well? How is this significantly different from discovering a new ice cream flavor we like and then recommending it to a friend who is an ice cream lover?

We think some may find this argument compelling. But only consequentialists. As axiological judgment-making does not differ in structure or, for some, in motivation across the various axiological disciplines, it may make no difference that the consequences of ethical judgments are more weighty than those of aesthetic judgments. Those who see matters this way will not find this "Ockham's Razor" argument compelling because the fault that Ockham warns against is not encountered. In fact, a person who sees matters this way may argue that the reverse is true: if we eliminate the normative force of a certain class of axiological judgments—aesthetic over ethical—we actually move to greater complexity and, so, a need to account for why these two kinds of judgment would be different.

When it comes to garden aesthetics, this argument does little to address the first and second arguments "for" above, the fact that we not only engage in the creation of aesthetic judgments about gardens but also argue for the correctness of those judgments using the same rational and evidentiary structures and processes we do for any of our claims, all axiological ones included. Our judgments about gardens may indeed involve appeal to our preferences, but inclusion of appeal to our preferences is not the end of the story. In advancing aesthetic garden claims, we can say why we have the preferences we do; we can justify our having the ones we do; and we can recommend through argumentation to others that they should have them as well. If the case ends with discussion of enjoyment as the goal (or a goal) of such argumentation, so be it; that matter simply turns on what our goals end up being as we work out our theories of what makes one garden aesthetically better than another.

The Absence of Progress

In 1962 Thomas Kuhn published his famous book *The Structure of Scientific Revolutions*, in which he argued against the notion that science, through a continuous process of theory development and testing to explain the world, is making progress toward offering true statements about the way the world really is. It is of course possible to say something similar about gardens: that if there were standards of goodness that are stable and universally accessible—as perhaps should be the case if the realist arguments above have merit—then we should be able to demonstrate that the aesthetic quality of gardens today has progressed from that of the past.

Kuhn talked about how we trade one paradigm for another, based on what our goals contextualized to our interests at a given time in a given place happen to be. One might say the same of gardens. There are many different kinds of gardens in different locations and from different times, and while it is possible to see among the gardens of a particular type some that stand out as having greater quality than others, it is unclear whether it is possible to compare gardens across types—across cultures, societies, eras, and so forth. Which is better, a great Japanese garden or a great French garden? Which is better, a Capability Brown landscape or a Picturesque landscape? Which is better, a Brown landscape or a cottage garden?

A certain Japanese stroll garden may be comparable with other Japanese stroll gardens, and the chances are high that with time and effort arguments will be forthcoming for why Japanese garden X is aesthetically better—possesses more aesthetic virtues, possesses more of the aesthetic virtues prized by canonical definitions of what a Japanese garden should entail, and so forth—than Japanese garden Y. The same can happen with two French gardens. If the Japanese garden in question ranks well among all Japanese gardens, or ranks well against an idealization of a Japanese garden—and if the French garden in question ranks *very, very* well among all French gardens (and so forth)—the chances are high that if we were forced to recommend a visit of one over the other, we will choose the one that exhibits the higher quality within its domain. While this particular scenario may be uncommon, comparison is at the heart of what it means to evaluate gardens meaningfully, and if the scenario above is unusual, comparison in general is not.

What should we say about "garden progress"? It is difficult to maintain that artwork produced in recent centuries is better than work produced further back in time, that, for instance, Post-Impressionism is better than Impressionism. If we claim that The Garden is an artform, it would be odd then to suggest that gardens created more recently are somehow aesthetically superior to, or demonstrate aesthetic progress over, gardens created earlier. The point we made about some of our garden exemplars coming from centuries ago is apt here as well.

We can justify the claim that the science involved in garden creation and maintenance is better now than it was. Through centuries of garden design experimentation, of plant hybridization and cultivar development, of enhanced education and training of gardenists and gardeners, progress may be witnessed. We understand soils, hydrologies, and climates better than we did, and we have a better appreciation of what it means to create our gardens in accord with nature rather than in opposition to nature. We have better techniques for all manner of gardening practices than we did—propagation and pruning techniques, as well as pest management, are more effective, for instance—we certainly have better tools and machinery, and as standards rise, so we invent or discover still better ways to practice. We may even be able to claim that, through scientific progress, those aesthetic aspects of plants and flowers we value are more present and more vibrant. But scientific progress is not aesthetic progress, and even the point that aesthetic features of plants we favor are more in abundance must be relativized to time and place. What we favor today may not be what we favor tomorrow, and what is favored by one population may not match what is favored by others in another place. Nonetheless, gardening has certainly benefited from scientific advancements, and this in turn no doubt has contributed to garden aesthetics.

In the end, "the absence of progress" argument fails to be compelling because, while it may possibly apply to science, it does not apply to aesthetics—or at least to the sort of aesthetic approach advocated throughout this book. The idea that "real" standards of aesthetic goodness should lead to aesthetic progress (of the sort Kuhn discusses in regard to science) is an idea that would only potentially work with a particularly narrow version of formalism, one in which lawlike relations between the presence of objective features and aesthetic merit not only exist but are discoverable, relations that would allow the absolute ranking of aesthetic merit possessed by gardens that would then, in turn, allow us to chart progress. That we have not discovered such relations is evidenced from the history of formalism starting from Aristotle and going up to the time of the subjective turn ushered in by Hume and Kant. During all that time and through the offering of many formulas purporting to connect objective features with universal aesthetic approbation, none has been successful. Where science can be judged against the enhancement of instrumentality, with better science leading to more predictions coming true, the same mechanism is not available in aesthetics. Aesthetic judgments, as we learned from Kant, necessarily rely on a subjective reaction that cannot be codified in an external form. As Kant said, aesthetic judgments are not judgments of logic.[27]

Some may read this to imply the exact opposite of what we wish to claim. If "progress" is impossible in garden aesthetics, does this not imply that antirealism

---

[27] Proceeding from Hume's statement that patterns should be expected, and perhaps also from the fact that no formula had proven successful in correlating aesthetic praise with arrangements of objective properties, Kant introduced a more subjective focus on our uptake of arrangements of properties. Instead of focusing on objective properties, Kant focused on how we experience formal arrangements of objective properties, which he further particularized to individual experiences.

is the case? That without the progress that should be the result of true normative standards, we are left with a Kuhnian style of relativism when it comes to garden aesthetic judgments?

Happily, that is not the case. We compare gardens one with another, and with former versions of themselves, frequently. These comparisons may be justified by offering arguments that maintain the same standards of logic and evidence that all good arguments incorporate. What limits their applicability across all gardens is the incorporation of (1) relevant contextualities—purposes, styles, cultures, historical periods—that make absolutist comparisons difficult if not impossible, and (2) the sort of subjectivity that Kant describes. So while it is true that we cannot track an unfolding progress of garden aesthetics through the history of gardens, this does not imply the absence of normativity; rather it implies the nonreducibility of garden aesthetic judgments to formulas, the sort we apparently stopped searching for by the nineteenth century.

Irreconcilable Differences among Tastes

The strongest argument against aesthetic realism may be the one that holds that even among aesthetic judges who have similarly advanced skills at recognizing aesthetic quality and arguing cases supporting their observations, there can be irreconcilable differences among their judgments, differences that are commonly attributed to differences in taste. A Jens Jensen landscape is very different from a Le Nôtre garden, and some prefer the minimalism and naturalness of the Jensen landscape to the baroque formalism and control of the Le Nôtre garden. Two judges—of similar skill, similar experience, similar aesthetic sensitivity— may end up with different judgments of a particular garden given that they simply prefer different styles or prefer that their gardens possess certain virtues, a preference not shared by the other judge. "Taste progress" as we discussed above does not answer this problem, as the problem focuses on divergent judgments from two individuals each of whom has full claim to well-developed garden taste. The problem here is not one about "good taste versus bad taste" but rather about patterns of preference—this is the fourth of the four definitions of taste we mentioned above.

We might begin exploring the impact of taste by first asking about the extent of the pattern of preferences. While it is possible that every single person on the planet has a pattern of preferences that is unique to that individual, this is unlikely. Perhaps a good illustration of this point is the fact that it is common for one to find others of similar taste with whom to go to movies, galleries, performances, and restaurants. If we seek company for an aesthetic expedition, we find someone whom we predict will enjoy an experience as much as we do.

One success leads to more as we then continue to invite that person on aesthetic expeditions and make mental note of where their aesthetic preferences match ours and where they do not. The ability to make these predictions—given that many turn out to be successful—depends not only on the stability of the patterns of preference of the individual about whom the predictions are made; they also depend on patterns among those preferences. Someone who enjoys one action-adventure film will likely enjoy another; someone who enjoys Thai food may enjoy Laotian food or Vietnamese food; someone who enjoys an opera is likely to enjoy another.

Perhaps an even more forceful example of patterns of taste comes from the existence of algorithms that populate information-technology-based entertainment delivery systems. We need input little information into Pandora, Netflix, YouTube, Spotify, and so forth, for these systems to generate what strikes many of us as accurate predictions of what we might enjoy. These algorithms work by tracking preferences and then predicting that if you enjoy X, someone else who enjoys X also enjoys Y, and so you might as well. These entertainment systems are not bound by the need to circumscribe a number of taste profiles; they can generate infinite numbers of connections. A similar system is collecting our internet activity to target advertisements to certain profiles of people—they work well enough to be unsettling. But they allow us to appreciate that patterns that expand beyond the individual are not only possible but quite existent.

Whether there are stable patterns among preferences, of course, is an empirical question, one appropriately answered only through scientific investigation, but examples abound to suggest that there are fewer than seven billion taste patterns. If there are fewer, perhaps that number is manageable. Suppose for the sake of argument there are 100 taste patterns with regard to gardens. If such trends exist, and if we are already committed to indexing aesthetic garden judgments (in the ways to be discussed below), then indexing an aesthetic judgment to a certain taste profile—say, one of the 100—may not save aesthetic garden realism in its most robust form but it may still allow us to claim that certain aesthetic judgments are right or wrong within the confines of indexing to specific taste profiles.

Do we need more? Do we need for single aesthetic judgments to be correct regardless of context? No, we do not. This may not even be possible in ethics, where the stakes arguably are higher. Indexing even there may be necessary.[28] Since aesthetic garden judgments must be indexed, one more level of indexing the correctness of aesthetic garden judgments does not seem onerous. What this will allow us to do is make sense of the practical application of aesthetic normativity

---

[28] As argued for by Gilbert Harman, in "Moral Relativism Defended," *Philosophical Review* 84:1 (1975), pp. 3–22.

in relation to gardens. This allows us to continue to compare gardens one with another; it allows us to recommend a visit to one garden over another. And all we require here is to make theoretical sense of the practicalities. While the problem of irreconcilable judgments due to differences in taste may defeat our aspirations for claiming that a single aesthetic judgment is right or is wrong, regardless of context, that aspiration likely was too ambitious from the start. What we need to protect is the ability to compare and evaluate, and indexing to taste profiles—again, if the empirical facts bear out that there are such trends—is enough.

In the next chapter, we move beyond considering whether one garden can be compared with another to asking how such a thing may be done. We segue then to discussing "garden criticism" as an analogue to art criticism. We move from there to talking about the value of gardening, with a particular focus on the virtues that gardening promotes, and the ethics of gardening itself—moral obligations for the garden designer, gardener, and garden visitor.

10

Evaluating Gardens and Gardening

While contemporary aesthetics need not always focus so tightly on the delivery of summative, normative aesthetic judgments—while we can talk about the aesthetic characters and aspects of objects and events we encounter in the absence of judging them or comparing them—a book like this would be incomplete without finally saying how one garden may be compared with another. This discussion naturally takes us to talking about "garden criticism." From there we discuss gardening, garden virtues, and the ethics of gardening and visiting gardens.

How Is the Aesthetic Value of One Garden Compared with That of Another?

There are many ways to judge the merits of a garden: breadth of collection, depth of scientific research based within it, quality of horticultural skill, proximity to the idealization of garden type, how well it fulfills its purpose, how well it attracts visitors, reputation of its gardenist, its provenance, how old it is, and so forth. Here we are interested in the aesthetic evaluation of gardens. We might initially divide the conversation into objective and subjective focused approaches. Our approach will be a familiar amalgam of the two.

Objectivity Focus

Objective approaches have best been represented by varieties of formalism. However, as we have said before, formalism is not a view we can embrace.

First, formulas that connect aesthetic properties with objective properties miss the reality that, first, there are many, many different ways that an arrangement of objective properties can lead to the manifestation of an aesthetic property. (This is a point long recognized in the philosophy of mind.) And, in complement, a

The Vestibule, Wethersfield Estate and Gardens, Stanfordville, NY
Evelyn Poehler, 1950s
Photo by Toshi Yano, Spring 2021

The Art and Philosophy of the Garden. David Fenner and Ethan Fenner, Oxford University Press.
© Oxford University Press 2024. DOI: 10.1093/oso/9780197753590.003.0010

certain arrangement of objective properties can prompt attribution of many, many different kinds of aesthetic properties, depending on the physical perspective of the subject, on the particulars of their experiential context, on what they come to the occasion already valuing (that is, what they already value aesthetically), and finally their taste.

Second, formulas miss the necessary addition of the subject to the equation of moving from the nonevaluative to the evaluative. Aesthetic judgments necessarily involve a satisfaction or pleasure that the subject takes in the arrangement of the objective properties (following both Hume and Kant[1]). The value-aspect of aesthetic properties must be added for aesthetic properties to manifest, as we understand such properties to carry connotations of value. This can only be contributed by the subject; the nonevaluative alone cannot lead to the value-bearing. Rules and formulas cannot account for the connection between evaluative and nonevaluative properties due to both complexity and the value-added nature of moving from the latter to the former.

Nonetheless, there are still lessons we can take from formalism.

First, as we have maintained throughout, aesthetic appreciation begins with appreciation of the formal, or perhaps better stated, the perceptual. While the formalist story may end with appreciation of the perceptual—and perhaps also the representative, expressive, and literary aspects of objects being appreciated—our story continues with inclusion of relevant contextual aspects, such as those having to do with a garden's purpose, its cultural and temporal placement, its origins, its relations to other gardens, and so forth. Yet we can agree with the formalist that we begin with what is immediately present to our senses, and we can agree that this focus remains central to aesthetic appreciation.

A second lesson we might take from the history of evaluative formalism is that a focus that is exclusively subjective harmfully divides the subjectivity of experience from the content of that experience. That is, as we offer and defend aesthetic judgments about gardens, if we do not keep the garden itself as the focus and content of our judgment, it would be easy for the judgment to dissolve into simply possessing a focus on us, the subjects, and how we happen to be feeling, thinking, and so forth, on a given day in the presence of that garden. While an

---

[1] This follows Kant's particularist focus when it comes to what he calls Judgments of Taste. Judgments of Taste are the result of the subjective event of the "free play between the understanding and the imagination"—balancing "agreeableness" and reason—where we see in an object a satisfying internal formal coherence that may be described as "purposeless purposiveness." "Purposeless" because we are viewing disinterestedly (so Kant says); "purposive" because of our appreciation of the formal unity of its (perceptual) aesthetic aspects. The purposiveness we witness is based in our appraisal of a particular object at a particular time, and while we would expect everyone appraising the object to witness this same purposiveness, the purposiveness cannot be captured by appealing to an arrangement of objective properties—Kant follows Hume in his focus on the subjectivity of judging rather than focusing on the thing judged—in the same way that the "free play" cannot be captured by appealing to rules of rationality or logic.

aesthetic experience does as a matter of fact include all manner of subjectivities that are individual to us and our particular experiences—are we hungry, distracted, annoyed, preoccupied, sleepy?—these should not enter as data points into our judgments for the simple reason that if our judgments are meant to translate meaningfully to predictions of how others might experience that same garden, we must include as data for those judgments only aspects that are relevant to the experience of another person. The simplest way to do that is to retain focus on the object—the garden—and, in our case, the relevant contextual aspects of that garden that are appreciable across all subjects. This second lesson reminds us that to lose focus on the object is to lose the primary basis we have for building a shared judgment platform.

Third, while no formula connecting objective properties with aesthetic approbation has proven successful—and no formula is likely to, given the many trajectories of modern art and given Kant's defense of the irreducible inclusion of the subject in evaluation—we should not respond with a deaf ear to attempts to find patterns of aesthetic praise connected to objective factors. It is possible that empirical/psychological exploration will reveal a pattern of universal approbation linked to something like the Golden Ratio or Divine Proportion, or perhaps a pattern describing the development of taste. Earlier we mentioned the idea, discussed by Miller, that humans may be "hard-wired" to gravitate toward landscapes that resemble the African savannah—short grass and scattered trees, as we find in many parks—and we hold the door open for other Darwinian approaches toward discovering aesthetic merits in the garden, such as the beauty found in order and symmetry or the discovery of brightly colored fruit. Evolutionary theories about aesthetic appreciation should be given special consideration with The Garden, as it is the most closely linked artform to the natural context (First and Second Nature) that shapes our survival as a species. An appraisal of the aesthetic worth of the garden involves the teasing apart of aesthetically positive and negative aspects of nature and indexing taste profiles of human habitats—all of which is informed in at least some degree by our biology and may provide some insight into our innate aesthetic attractors.

Subjectivity Focus

More subjectively focused approaches to aesthetic evaluation have been the norm since Kant, even in the face of twentieth-century New Criticism and the formalism of the nineteenth century that still exists today in a smaller way. They may best be represented by "production theories." Production theories focus on value as the content of a subjective state occasioned by aesthetic attention. These theories would be a correlate to consequentialist theories in ethics, where the level of rightness of an act is judged by the level of value or goodness that act produces. This is a family of theories rather than a single one, because what that particular subjective state is differs in terms of how aesthetic value is defined.

The most straightforward account of the subjective state brought into existence through attention to a garden is that one feels pleasure in doing so. On some level, this account must be right, if we understand "pleasure" to refer to any positive response that is valued highly enough by the person experiencing a garden to motivate their attention. However, if we think of pleasure in a narrower way—as the sort of feeling, say, immediately attendant to eating chocolate or opening a birthday present—then pleasure per se may not automatically be right as a description of the motivation, for instance, to endure heat and cold, sore knees and sore back, bites from mosquitos and wasps, and so forth, as one cares for their garden.

There are other values that may account for one's motivation to appreciate gardens and to garden, but for many of the accounts that cite such values, these accounts are not "global" in the sense of describing what is common about all subjective states produced by appreciating gardens and gardening. Instead, these values may feature in accounts that explain the case only for some individuals and only for some occasions for those individuals (not to mention "and likely for only certain garden types"). Such values may include the following:

- Some aesthetic attenders may take satisfaction in the orderliness that is typical of high-quality aesthetic objects, with the garden presenting more occasion for experiencing that order given not only its typical complexity but also given that its orderliness is a fleeting state—and so all the more worthy of appreciation when it is present.
- Some may seek new experiences for the sake of the new experience. A theory explaining such a motivation may fall under the category of escapism, where transport to a different world—the world conjured by the garden (physical, virtual, fantastical, or historical in aspects)—is the goal. The value in question here may be simply the difference of the new garden world from the ordinary world occupied by the attender at nonaesthetically focused times.
- Escape, on occasion, can entail catharsis and quietude, and so some may attend to gardens for the ecopsychological benefits that come with focused attention on nature, to burn off negative feelings, and to foster calm and a feeling of being centered.
- Some may seek out the satisfaction that comes with deep and immersive cognitive engagement, perhaps the same satisfaction that comes with solving a puzzle or a mystery—or working out the answer to a tough question. They might experience this as they work out an account of the coherence of the

formal aspects of a garden or as they work out interpretations of the garden as a work of art.

- Some may take satisfaction in experiencing the garden as a product of the vision and expression of an artist, the garden designer and perhaps the gardeners. As works of art stand as artistic expressions, they present to us in sensible terms something whose expression can only truly take the particular form it does. For gardens that are works of art, they may be appreciated as artistic expressions.

- Along these same lines, some may appreciate the technical expertise that goes into the planning, installation, and maintenance of a (great) garden.

- Some may seek out prompts for various associations, perhaps to bring to mind fond memories, or as a context for appreciating objects they associate with gardens, such as family and friends, homelands, parties, relaxation, and so forth.

- Some may value the garden as a place of education, to experience and learn about plants or ecosystems of different regions, local ecology, site history, horticultural technique, or a garden's makers.

- Some may value the ancillary purposes pursued within the contexts of gardens, and so in a sense they appreciate gardens as settings for times spent playing sports, jogging, walking a dog, and so forth. Children especially may appreciate gardens that present places to run, hide, and explore. While the focus of value may not be exclusively the garden but rather the activity pursued in the garden, the garden is appreciated as the constant context for these events.

There likely are other subject-focused accounts we could offer, but this representative set is sufficient to allow us to say that the values that motivate us to appreciate gardens and to garden may be quite plural. To attempt to reduce the subjective or "reception" side of the equation of garden valuation to a single account is likely a mistake. In the end, it is unnecessary for us to catalog all the different ways in which one who attends aesthetically finds reward. Whatever value is produced for the subject is the appropriate and legitimate content of the state sought to be produced if we are consistent in our view that theory follows practice.

More importantly, an exhaustive list of garden values does not matter for our purposes, as we believe that production theories—taken as complete theories for explaining the aesthetic value of gardens—do not afford the object a stable enough place in the calculation of what the attender finds valuable. While likely every respectable production theory will keep the object focal to some degree in the attention of the subject, that object's value is reduced merely to being the content of that attention. One subject can experience an object very differently from another, given what the two subjects may bring as a context to their attention to the object. This erodes the stability of the object as the experiential focus. But, in addition, there is nothing that holds the subject back from intentionally adjusting the phenomenal representation of the object in ways they find of greater reward. In general we tend to believe that modifying works of art is wrong, yet if the value of an object is as the content of subjective consideration, and the subject finds the experience of even greater value if they modify the object in their mind, then not only is there no reason for them not to do this; there is good reason for them to do exactly this.[2] And so the ensuing loss of stability erodes the possibility that a garden in the minds of two subjects is the same garden. The Garden as dynamically open is already burdened with instability; production theories that locate value primarily or even exclusively subjectively erode the stability necessary for comparison and assessment beyond hope.

A Familiar Amalgam

Our suggestion for how garden evaluation should proceed combines elements of both attention to the object's features and the subject's consideration of those features. It follows in the path cut by Monroe Beardsley.[3] Cases for summative or overall judgments are evidenced by attributing aesthetic properties to the garden being evaluated, and aesthetic properties are evidenced by citing objective or base properties—ideally purely descriptive and value-free perceptual properties—of the garden. While it is rare for an appraisal of an aesthetic object to be exclusively positive, if the summative judgment is positive, the majority, likely the vast majority, of cited aesthetic properties will be positive. The same is true in reverse: if the summative judgment is negative, the aesthetic properties cited in a case for support of that judgment will be negative.

---

[2] David Fenner, "Production Theories and Artistic Value," *Contemporary Aesthetics* 3 (2005); and "Why Modifying (Some) Works of Art Is Wrong," *American Philosophical Quarterly* 43:4 (2006), pp. 329–341. Alan Goldman examines a range of objections to what we call "production theories" in his paper "The Experiential Account of Aesthetic Value," *Journal of Aesthetics and Art Criticism* 64:3 (2006), pp. 333–342. There he works to defeat the objections in support of his view: "I have characterized such experience in terms of the simultaneous challenge and engagement of all our mental capacities—perceptual, cognitive, affective, imaginative, even volitional—in appreciation of the relations among aspects and elements of artworks" (p. 334). He does not take up the objection we offer, one that seems particularly apropos to The Garden.

[3] Monroe Beardsley, *Aesthetics: Problems in the Philosophy of Criticism* (Indianapolis, IL: Hackett Publishing Company, 1981; first published in 1958); and *The Aesthetic Point of View*, eds. M. J. Wreen and D. M. Callen (Ithaca, NY: Cornell University Press, 1982). Our view follows closely the view of Alan Goldman as articulated and defended in his paper "Aesthetic Qualities and Aesthetic Value," *Journal of Philosophy* 87:1 (1990), pp. 23–37. Goldman's views on aesthetic value evolved into what we call a production theory, so we do not hold what he wrote in this paper to be his final word on the subject.

The positive and negative values aesthetic properties carry are context dependent; whether their connotations are positive or negative depends on cultural contexts (as an example), but it also depends on the particulars of the case an individual is making about a certain aesthetic object. An aesthetic property in one formal arrangement may be positive, but it may be negative in another. As mentioned earlier, a judge's employment of the aesthetic aspects of a garden in the creation of a case for its overall aesthetic merit are particular to a single judgment made by a single subject. We have covered a plurality of reasons for why this is true, but the most central is recognition that a garden's formal aesthetic properties only manifest from the phenomenal properties that are the contents of an aesthetic appreciator's experience.

The normative force of aesthetic judgments comes from the epistemologically normative force of citing evidence through logical structures of argumentation in support of one's judgment. Indexed to a particular garden experienced by a particular person who describes that garden's aesthetic character using a particular set of formal aesthetic properties and relevant contextualist aspects, the reasons advanced by that judge should be recognized by anyone hearing the case as reasons for why anyone should judge similarly. If the argument fails to be compelling, then (1) there is a fault in the argument (of a logical or evidentiary nature), (2) there is compelling argument for a competing view, (3) the person receiving the argument is incapable of seeing an aesthetic property manifesting from an arrangement of objective properties, or (4) the person receiving the argument is being irrational (or, perhaps more charitably, nonrational).

While it is possible that a case for a positive summative judgment consists only of citation of separate individual aesthetic properties, this would be unusual. When we think about the quality of aesthetic objects, we usually think about arrangements of properties. A collection of elegant lines, if there is no coherence among these elegant lines, does not make for an elegant aesthetic object. Balance of line in the absence of balance of color does not make for an overall balanced painting. In other words, the case for the aesthetic quality of an object is generally not a matter of quantifying numbers of positive aesthetic properties the object possesses; it is normally about how the positive aesthetic aspects of the object fit together into a coherent whole. This is again why we tend to call those aesthetic properties that are based on objective or descriptive properties "formal"—they have to do with the form of the object, and the broader (or more summative) the judgment, the broader the scope of consideration of how the object's properties come together to form that coherent whole.[4] Miller writes:

---

[4] It is ironic that we rely on Kantian aesthetics to support our rejection of strict formalism yet here we must part company with Kant who held, as we saw earlier, that gardens to his mind lack formal coherence and are only arrangements of pretty things. That is, while we can agree with Kant that aesthetic objects that lack formal coherence are not very aesthetically valuable, we disagree with his view that gardens are incapable of formal coherence.

Gardens exhibit organic form par excellence. And they do it on two levels, that of the individual plant and that of the work of art (the garden as a whole). If they did this only on the level of individual plants, of course, there would be no art, no semblance, no Significant Form. Organic form, however, is achieved on the level of the garden as a whole. This fact provides an important key to understand the peculiar power gardens have to convince. Their illusion is always convincing since it is always based on organic form (organic form being precisely the form exhibited by organisms such as plants). One reason gardens can be successful at revealing the perceived order between macro- and micro-cosm is that they carry evidence of their successful integration of the larger world into the garden in their very existence.[5]

Berthier writes:

The more one contemplates this remarkable garden [the dry garden at Ryōan-ji], and especially the interrelations among the fifteen rocks and the five groups, the more profoundly right the arrangement appears.[6]

This description captures beautifully the formal aesthetic merit of a garden that is the focus of repeated encounters and repeated occasion for reflective contemplation. This is the sort of aesthetic summative judgment we have in mind. Berthier does not talk in terms that focus on his response or reception of the garden but rather about the properties of the garden itself. But in complement the presence of the profound rightness of the arrangement is not a property of the garden in the absence of appreciation of it; the rightness of the arrangement only comes through experience of it by someone capable of discerning the rightness.

Developing the Case

A full aesthetic case will include citation of a garden's formal aesthetic properties, its dynamic aesthetic properties, and its relevant contextual properties, including how the garden engages a visitor emotionally, cognitively, imaginatively, associationally, and so forth. The creation of such cases is the means by which the garden enthusiast articulates the basis for the joy they experience not only to others but to themselves as well. A case for a summative judgment takes thought, and most garden visitors do not compose their arguments until they are

[5] Mara Miller, *The Garden as an Art* (New York, NY: State University of New York Press, 1993), p. 51.

[6] Francois Berthier, *Reading Zen in the Rocks: The Japanese Dry Landscape Garden*, ed. and trans. Graham Parkes (Chicago, IL: University of Chicago Press, 2000), p. 137.

at a point of reflection, likely after they have taken in the whole garden and after their immediate enjoyment of the garden is complete and done. This is not a rule, merely a commonality.

Judgments about the value of gardens begin with understanding a particular garden within its appropriate classifications and with consideration of its expressed goals. In an important sense, the former—understanding a garden within its proper classifications—is the latter, except implicitly rather than as expressed. In other words, a pleasure garden possesses goals that are common to all pleasure gardens, as an Olmsted civic garden possesses goals that are common to all Olmsted-style civic gardens, as a botanical garden—anywhere in the world, at any time—possesses goals in common with others of that classification. So judging a garden within its proper classifications—and "classifications" must be plural here, as many gardens are appropriately considered under more than one categorization—is to understand in broad ways its goals. In addition, a garden's designer or its patron brings to their particular garden a set of goals unique to that garden—Central Park has a different goal than the campus of the University of California at Berkeley, despite that they are both Olmsted designs—and so judging a garden well is to take those particular goals into account.

While correct consideration of a garden's goals is central to appraisal of it, equally important is appraisal of the balance of design and nature in the garden. If a garden contains living elements, as we argue all do (either in each of their garden rooms or in most), then judging the extent to which the design and the maintenance of that garden is done in ways that balance the health of plants and the larger ecosystem is important. A garden that cannot be distinguished from First Nature is hardly a garden at all—because it lacks evidence of its design as a garden—and a garden that does not cooperate with nature as cocreator very quickly becomes a "once-garden" if it ever was a garden to begin with.[7]

Approaching aesthetic evaluation of a garden might best be pursued as an organized endeavor, beginning with organizing one's thoughts about the manifested aesthetic properties of that garden around the structure of "garden form" we introduced earlier, and then continuing with inclusion of consideration of the contextual matters that seem relevant to the aesthetic character of that garden. We introduced eleven taxa of garden form; consideration of the aesthetic character of each of these facilitates a structured appraisal of how the aesthetic properties manifesting in each relate within that category, and then how the aesthetic characters of each category relate together. Coherence within each applicable category is important just as coherence among the characters of

---

[7] This is not to say that a garden complex cannot include garden "rooms" that are meant to look undesigned and purely "natural," like wildflower meadows or forests.

each category is important. The greater the coherence, the greater the unity, the greater the formal excellence of the garden.

In crafting a case for the value of a garden, we might next consider the representational aspects of the garden, if the garden has such aspects (within the experience of the appreciator of that garden). Representational aspects may include human or animal representations in sculpture or topiary, representations of individual landscapes as found in certain Chinese and Japanese gardens, mythological representations as found in gardens that depict paradise or other religious themes, ecological representations that may be found in certain botanic or native plant gardens, historical representations of how certain properties may have looked at a certain time, and so forth. The fact that a garden has such aspects may enhance the aesthetic quality of the garden, because these properties are not as common in gardens as in other artforms, so the achievement of representation in and of itself might be aesthetically relevant. But perhaps more important is how the representation is achieved; that is, the quality of the representation itself, of how the properties of the garden function to represent, is of chief aesthetic importance. (A similar suggestion could be made about a garden's expressive properties, but that may involve reference to the artist, a topic we will take up below.)

Next, we might consider what we earlier termed the garden's "genetic factors," matters having to do with the placement of the garden, with the time period in which the garden was created and existed, with the purpose of the garden (intended and incidental), with the patronage of the garden, with its plan or design, and with issues having to do with its siting. Then we might consider the garden's "lifespan factors": how the garden evolved over its lifespan, how its provenance impacted it, and whether it was open to the public or private. Next we might consider the relationships into which the garden may be seen to have entered—relationships with the land, with the seasons, with how it is classified by history and style, with other gardens, and with other works of art. And then we might consider how the garden has been received by critics, by audiences in general, and by individuals as they identify with the garden, or various parts of it, and as they associate the garden with aspects of their own lives and with general themes. (Genetic, lifespan, relationship, and reception factors are discussed fully in Chapter 6.) Penultimately, we might consider what meaning the garden may hold, in all the ways a garden may have meaning for one experiencing it.

Finally, in considering the aesthetic character of a garden, we might explore the roles of those who have been responsible for shaping the garden into the garden it is. Cooper and Hunt, each on more than one occasion, draw our attention to the importance of the gardeners and the practice of gardening that is an ongoing and integral part of the garden. The aesthetic significance of this impact rests on the ongoing character of their efforts. This can be understood as part of the dynamic and ongoing process of the garden's creation, but it can also be appreciated for its own sake, as a part of the garden in the way that the movement of a dancer is part of a dance and as the playing of an instrument is part of the symphony. Just as one may appreciate the skill, stamina, physical ability, and unique style of a dancer or musician, one may appreciate the technical finesse of the gardener as they coax plants to life and draw out garden scenes. Gardening practice is unique as an aesthetic consideration because unlike the case with performances, gardening is not the garden. It is also unique in the sense that, unlike the case with performances, a garden is an autographic art, existing merely in one instantiation, albeit an instantiation that occupies a lifespan. The inclusion of consideration of those responsible for designing and maintaining a garden is doubly important in evaluating a garden because one of the first marks of a garden's success is how well the cocreators (gardenist, gardener, nature) act harmoniously with one another to produce the desired effect.

The artifactual "half" of a garden, as we have mentioned, is an artistic expression, borne out of an aesthetic vision, of an original gardenist or team of gardenists. That expression cannot be reduced to a description of it; it cannot be replicated in another aesthetic or art form; and it cannot be replicated at another site and retain its original identity. It is therefore autographic and unique in character. To think of aesthetic judgments in terms that focus, even as an amalgam, on a garden as an object and on the experiences of those who appreciate that garden may be to ignore that some degree of value is derived from the garden's character as a unique artistic expression. Appreciation of the artistry itself can directly involve mention of the gardenist and their intentions, but more basically, it is common for appreciators of gardens to mention their respect for the gardenist as an aesthetically relevant matter in their appraisal of a garden.

Garden Criticism

The relationship between garden evaluation and garden criticism may be obvious; the latter clearly relies on the former. Criticism that merely offers information about an aesthetic or art object in the absence of advancing views on its quality seems less than it might be. Offering factual information rather than an evaluative viewpoint is what "garden history"—as a species of "art history"—does. The addition of making and offering judgments is what changes history to criticism, and it happens as early as when we first begin using (value-informed) aesthetic terms in description of gardens.

The argument for encouraging the practice of garden criticism begins here:

- If The Garden is an artform, and

- If a garden can be described—historically, taxonomically, contextually—as richly as any other work of art, and
- If a garden can possess formal aesthetic properties, and
- If a garden can be interpreted to have meaning (ideally) as robustly as any other work of art, and
- If a garden can be evaluated aesthetically as deeply as any other work of art,

Then everything is in place for garden criticism to be a useful activity. Art criticism at its best contains these features:

- nonevaluative description of the work of art, citing its objective features;
- information about the work of art, essentially the "art historical" part of a critical art review, where relevant classificatory, genetic, and relational matters are discussed;
- interpretation of the work; and
- evaluation of the work.

Many gardens can be discussed in all these terms and profitably so. Such discussions not only can enhance how such a garden is regarded but can enliven the connections an appreciator develops with that garden, making their experience of it—and future experiences of it—all the richer and more rewarding.

Garden criticism would not only be a badge of honor for those gardens qualifying as works of art and would enhance experiences of those gardens, but it would bring attention to gardens that would encourage visits and appreciation. In some sense, this already takes place through the many books and websites that describe gardens and include enticing photographs of them. But while praise is always generous in such books, the praise tends to be regularly superficial and uniform. Serious evaluation, much less serious interpretative efforts, are largely missing. True garden criticism would focus attention on gardens in ways that are longer lasting and deeper. The taxonomical approach we took to defining The Garden could be honed for the sake of better and sharper categorization, something we take to be necessary in understanding a garden well enough to interpret and evaluate it. The anthropology of garden design and style evolution could be studied for the sake of seeing new connections and tracing the evolution of a garden style through time and cultures. The relations that exist between garden styles we may take to be disparate could be revealed. If gardens-and-gardening has the place in human history that it clearly does, serious garden criticism could bring that more to light. None of this is meant to suggest that those involved in landscape design and horticultural education do not already engage in many of these efforts, but the systematization of those efforts as they would focus on the aesthetics of gardens has not taken on the status of the sort of disciplinary foci that are commonly found across the academy. Discussion of gardens has heretofore been more focused on science and on history—important components of the conversation, to be sure—but theoretical approaches, especially focused on the aesthetic aspects of garden, have not received the attention they might.

Description of the Garden

A work of garden criticism begins with a description, and perhaps there is no greater challenge for the garden critic than attempting a useful description of a garden, particularly a substantial one, particularly when that description must then serve as a basis for describing the aesthetic character and aesthetic worth of that garden. The task is further complicated because the describer can only describe the results of their own experience, how they experienced the garden utilizing their own "mobile frame," in what seasons and under what climactic conditions, and so forth. But we do not take this problem to be an impediment to the practice of garden criticism but rather simply a challenge, albeit a significant one. What this means is that works of garden criticism likely will be significantly longer than works of criticism dealing with other artforms. This is again not an impediment but merely a challenge. It might be useful, as was the case with interpretation and evaluation, to organize one's description into elements that fall within the "garden form" categories. This not only would provide structure to the description, but it would facilitate ensuring that no important element is overlooked. There will be more elements to a garden than one per eleven categories, of course, and some of the categories will not be relevant to each garden.

In describing a garden's phenomenal and objective properties, and then describing that garden's formal and dynamic aesthetic properties, the garden critic's first challenge is to establish the object, the garden, which the critic means to describe. This can only be done through repetition of direct experiential connection. The Acquaintance Principle[8] is relevant here: such a position holds that only through direct or firsthand experience of an aesthetic object can a critic or describer of the object know that object aesthetically. While this principle has been the subject of discussion in the aesthetic literature, with much of that discussion focused on whether direct acquaintance with a work of conceptual art is necessary to be able to discuss the (presumed) aesthetic character of that work, absent this discussion the Acquaintance Principle is widely accepted as

---

[8] Malcolm Budd explores the Acquaintance Principle in his "The Acquaintance Principle," *British Journal of Aesthetics* 43:4 (2003), pp. 386–392. He attributes the origin of the Principle to Richard Wollheim in *Art and Its Objects* (Cambridge, UK: Cambridge University Press, 1980).

intuitively important. Those who attempt to describe a thing but who work only from descriptions of others have only a derivative perspective.

In the case of garden description, repeated contact is just as necessary as first-hand experience. The greater the repetition in as many different conditions as practical—different seasons, different climatic conditions, different times of the day, ideally over years—the more worthy will be the critic's claim that they are describing the dynamic garden as an object stable enough to support evaluation and interpretation. Since different experiences in different circumstances will yield different descriptions, the critic must also be adept at describing the garden as a plurality of objects—perhaps, for instance, the garden in the spring, the garden in the fall, the garden right after a redesign has taken place, and so forth. The critic must recognize that the stability of the garden object must be indexed to an experience or a set of experiences under similar circumstances. This indexing must be accounted for in the critic's description while at the same time balanced with the need to have a single stable envisage of the garden for the purposes of evaluation and interpretation. The balancing of indexing over repetition and establishing a stable enough object for in-depth discussion of these more summative activities is a special challenge for the garden critic, and mastery of this task would be the basis of enhanced respect.

Many gardens may occupy several places in a taxonomy of garden types. While correct classification of a garden's type is important, mention of the type or types under which a garden properly fits will constitute a discussion in and of itself as the classification of a given garden is explored and evidenced. In general, "more is more" in the sense that any classification that seems reasonable to be applied to a garden likely will serve to illuminate more of its character and its features. This includes matters of purpose, functionality, and utility, as well—again, both planned and incidental.

Part of the reason Cooper rejected factorization of appreciation of a garden into nature-appreciation and art-appreciation is because he thought this ignored what he referred to as a garden's "atmosphere." Synonyms he uses to describe a garden's atmosphere are "mood" and "tone," but perhaps the most revelatory description he offers of atmosphere is "a sense of the place as a whole."[9] What constitutes "the whole" of a garden differs with the perspective of appreciators. "The whole" could be the entire garden, especially if the entire garden can be taken in visually in a single field; "the whole" could be a garden room or some large visual swath of the garden scaled down to the appreciator's perspective; or "the whole" could refer to a certain conceptualization the appreciator has of a garden which may not fit into a single visual field but which could fit within the

---

[9] David E. Cooper, *A Philosophy of Gardens* (Oxford, UK: Oxford University Press, 2006), pp. 48–51.

imagination of the appreciator. When we think of Kew, we imagine the whole of it, although Kew is far too large and complex to fit within a single visual field. With all this perspectival caveat, it is still common to think of gardens as wholes; this certainly fits how we refer to them by single names, many times tied to places. Capturing a garden's atmosphere comes with the added challenge that associations particular to us as individuals may come into play. But for the seasoned garden critic, that person will be able to tease apart the personal from the general, and as they do this they may focus on certain expressive qualities of the garden, again taken as a whole, that they believe would readily be experienced by most attuned visitors. If a garden is calm and serene—if a garden is exciting and vibrant—if a garden is magical—if it is staid and academic—articulating a garden's atmosphere is an interesting aspect of describing it, or, depending on the characterization, of articulating its expressive qualities.

The Place of the Gardenist

In addition to establishing the garden object, the critic should—where possible—also establish the garden's artist. In some cases—when the gardenist in question enjoys the stature of gardenists like Musō Soseki, Brown, and Le Nôtre—the task will be relatively easy. But this may be more the exception than the rule, as some gardens worthy of critical attention may not have gardenists that are identifiable singly or as members of a set. The placement of the fifteen stones at the dry garden at Ryōan-ji was completed by someone whose name has been lost to history, and many contemporary gardens are designed by landscape designers who work within the constraints of a commission and often under the constraints of the firm in which they are employed. In addition, it is unclear that there is not a plurality of gardenists over time involved in the design and redesign, for gardens that are old, in part if not in whole, not to mention the many gardeners who make adjustments to the design as the plants age, as plants die and are replaced, as climatic conditions change, and as the purposes and priorities of gardens alter over time. While all of this may be captured in a detailed exploration of the history of a garden, the critic may still be faced with the question of whether the garden is part of an oeuvre or whether it is a stand-alone—a matter involving the importance of getting the authorship of a garden right.

To undertake serious garden criticism would provide a platform for coming to understand a gardenist's oeuvre in ways that transcend the creation of annotated lists of projects. We could better come to appreciate a gardenist's artistic style and how it stems from a unified vision that underlies the range of art expressions that are that gardenist's works. If we can understand film directors as auteurs, we can do the same with those gardenists who rise to that level. And again, it is not as if this work is not done in limited fashion now. The creation of a list of great gardenists could easily be achieved by any qualified student of landscape architecture, not to mention their teachers. But the full cases for why these artists are deserving of their reputations is uncommon; too often we treat gardenists as we do film directors and think that those who do the big-budget work are those most worthy of attention. But do we *know* they are the ones most worthy of attention? Serious garden criticism would reveal this. It would broaden the applicant pool and allow us to argue over inclusion on the list of gardenist-artists, and such argument would encourage deeper respect for those who are more likely to be remembered deep into the future, as we remember Musō Soseki, Capability Brown, André Le Nôtre, and the like.

The Garden Critic

Garden criticism is only possible if there are garden critics whose views are worthy to be listened to or read. As is the case with every other style of art criticism, the worthiness of the views do not entail that those views must be persuasive. The arguments and evidence advanced by the critic must be high quality, but so long as the critic is offering the reader something they did not have before—some depth of analysis, some new insight, a new interpretation, a new perspective—and what they are offering is taken by the reader to be something of value to their experience of a garden, then that is sufficient for respect to be accorded. (This does not contradict what we said earlier about the evidentiary and argumentative nature of cases backing aesthetic judgments; a good case can be recognized as a good case without being persuasive.) David Hume offered a vision for an "ideal critic":

> [A] true judge in the finer arts is observed . . . to be so rare a character: Strong sense, united to delicate sentiment, improved by practice, perfected by comparison, and cleared of all prejudice, can alone entitle critics to this valuable character; and the joint verdict of such, wherever they are to be found, is the true standard of taste and beauty.[10]

Rory Stuart, in his book *What Are Gardens For?*, devotes two chapters to garden criticism and the garden critic.[11] In the United Kingdom, he says, there are two guides—*The Yellow Book*, published by the National Garden Scheme, and *The*

---

[10] David Hume, "Of the Standard of Taste," *Four Dissertations* (New York, NY: Garland Press, 1970).

[11] Rory Stuart, *What Are Gardens For?* (London, UK: Franklin Lincoln Limited, 2012).

Good Gardens Book—useful to those interested in visiting British gardens. He writes, "the *Good Gardens Book* awards one or two stars to British gardens it considers of exceptional merit, *although its criteria are never made very clear.*"[12] The last emphasis is ours, and little may speak more to the need for principled garden criticism than that one phrase taken in context. He continues by writing, "In the UK the absence of criticism (not in the sense of carping fault-finding but in the sense of intelligent evaluation and sensitive assessment) seems all too common in garden writing, too."[13]

Stuart summarizes Noël Carroll's conception of the critic—from Carroll's book *On Criticism*:[14] "The critic must do four things before coming to a judgment of the work's value: describe, classify, contextualize, and interpret."[15] This list is similar to ours, and both are similar to the views of Terry Barrett as expressed in his *Criticizing Art: Understanding the Contemporary*.[16] Stuart constructs his view of the qualifications of a bona fide garden critic thusly:

> Like the critics of other arts, garden critics must know their subject; they will need to know something of the history of the country they are visiting, something of the social context in which the garden was created, something of the history of taste that guided the designer. . . . [17] [The garden critic must answer] the question of how (if at all) the garden relates to the house . . . and then of how the garden relates to the landscape that surrounds it. . . . Second, the critic will look at the layout of the whole garden and consider the proportions of the spaces into which it is divided. . . .[18] If theme and variation are essential in any work of art—the stability of fulfilled anticipation working hand in hand with the drama of surprise—suspense is also something the garden critic will look for in a good garden. . . .[19] If water is used in the garden, the critic will ask if it is used to its full advantage. . . . The critic will also want to consider the use of colour in the garden. . . .[20] Then the garden critic may consider the quality of . . . the hard landscaping. . . .[21] The critic should also be aware of the problems and advantages of the site. . . . In the end the garden critic, having tried to understand the garden—its context, its historical period, the character of its maker— will assess its quality overall but pay attention also to particular failures and

---

[12] Stuart, p. 41.
[13] Stuart, p. 43.
[14] Noël Carroll, *On Criticism* (New York, NY: Routledge, 2009).
[15] Stuart, *What Are Gardens For?*, p. 42.
[16] Terry Barrett, *Criticizing Art: Understanding the Contemporary* (New York, NY: McGraw-Hill, 2000).
[17] Stuart, *What Are Gardens For?*, p. 47.
[18] Stuart, p. 48.
[19] Stuart, p. 49.
[20] Stuart, p. 51.
[21] Stuart, p. 52.

successes. . . .[22] To sum up, gardens can be "read" like other works of art. . . . With this knowledge and a sharpened critical intelligence, the garden visitor will understand all gardens better, his or her own included, and thus gain more from the experience of visiting them.[23]

After offering three garden reviews, Stuart writes:

> What have I been trying to do in looking at these three gardens in a creatively critical way? In general terms taking the gardens seriously; trying to respond to the particular character of each one, and then assessing what the makers were attempting to do, and where they have been successful and where they have failed. I have tried to respond to the special atmosphere of each of the three gardens, and to judge them without too much of the prejudice of individual taste.[24]

These excerpts offer a variety of lessons. First, they delineate in function-focused terms the activities in which garden critics should be engaged. Second, Stuart focuses on the contextual matters that are relevant to understanding and then to assessing a garden, and then he moves to describing the garden's formal elements. From there he expands to talk about the mindset and approach of the garden critic in carrying out their tasks, echoing in some degree the description Hume offered of the critic, in particular focusing on impartiality, preparation, and sensitivity. Third, he echoes Cooper's focus on the importance of garden atmosphere. Fourth, had he invested even more depth in delineating the formal elements of gardens, it seems he would have proceeded along the path we took in attempting to describe what we call "garden form." Fifth, and perhaps most importantly, Stuart takes seriously the task of articulating what it means to take gardens seriously; he is not demure in his attempt but states this as his goal and further states why this goal is worthy: that through access to more information of a serious critical nature the garden visitor will not only come to understand more and more deeply but also "gain more from the experience" of visiting gardens. This last point mirrors our interest in garden criticism: to facilitate garden visitors gaining more from their experiences of gardens, however that value might be defined for them, but with a particular focus on how they understand the elements of a garden cohering into a unified whole. This work is not exclusively about feeling nor is it exclusively about thinking. While the work of synthesizing the parts into a whole, and then reflectively analyzing those

---

[22] Stuart, pp. 53–54.
[23] Stuart, p. 57.
[24] Stuart, p. 74.

relationships, is cognitive work, engagement with a garden offers opportunities for both cognitive and emotional connection.

Following is our list of what makes a highly qualified garden critic. It is idealized, and so it should not be taken as a set of necessary qualifications, but the closer the garden critic comes to this ideal, the more valuable will be their criticism.

1. The qualified garden critic must understand the history and social-scientific aspects of gardens. They must not only understand appropriate categorization of each garden they encounter, but they must understand how a garden may fit a plurality of categories. They must know how categories of gardens relate to one another. The garden critic must appreciate the context of the garden they aim to criticize including the resources that are allotted to its care.

2. The garden critic must know enough natural science—biology, chemistry, climatology, hydrology, ecology, edaphology, and so forth—to be able to speak authoritatively about the natural aspects of gardens and their sites.

3. The garden critic must know enough horticultural and plant-taxonomical science to speak authoritatively about the plants within a garden. They must also understand the conditions under which plants in the gardens they criticize flourish, how they are cared for, their life cycles, and to what diseases, pests, and problems they are prone. They must be able to discern between intentional acts of the gardener/gardenist and aspects that are purely products of nature. They must understand gardening practice and be able to envision the garden over seasons and years. They must also know enough agricultural science and ecological science to relate the living population of gardens to Second Nature and First Nature living populations.

4. The garden critic must know enough "garden-art history" and landscape architectural history to speak authoritatively about the artistic aspects of gardens and about major gardenists—and be able to place the garden in focus within the influences of other gardens. They must understand enough art history to locate garden-art history within the context of the art world.

5. The garden critic must be able to establish a garden object as a focus of criticism. They must be able to understand, address, and articulate the challenges of fixing a garden object in light of the openly dynamic nature of gardens.

6. The garden critic must understand enough aesthetics to be able to discuss authoritatively the formal aesthetic features of gardens. They must

be able to see and relate the relevant aesthetic contextual matters to their assessments of gardens. They must have enough command of both matters to interpret gardens.

7. The garden critic must be highly sensitive and highly attentive to all material and virtual aspects of a garden. They must be able calmly to reflect on and analyze their own garden experiences. They must understand their own experiential reactions to gardens within a context of experiencing rich and deep connections to the garden under consideration. They must be eloquent enough to capture all this in words.

8. The garden critic must be impartial, able to advance arguments and evidence worthily, and able to speak to the impact of taste on garden preferences and judgments. They must be able to capture the common normative sense of advancing aesthetic judgments about gardens.

9. The garden critic must be strong enough to articulate garden failures and stand by their judgments. They must be able to explain clearly why a failure is a failure, just as they must be able to explain clearly why a success is a success.

10. The garden critic must be highly practiced at garden criticism and must have firsthand experience of a very wide range of gardens, visited ideally repeatedly in different circumstances.

What Is the Value of Gardening?

If The Garden is an artform—which is to say, if some gardens are works of art, a claim for which this book has offered support—then arguably there is no artform that has more potential for teaching ethical lessons, fostering the development of virtues and ethical attitudes, and encouraging ethical actions than The Garden. This is largely because The Garden presents an opportunity for long-term, perhaps lifelong, aesthetic participation. While one can certainly take a "museum approach" to a garden—using Cooper's and Dewey's words—the literature focused on the relationship between The Garden and personal morality primarily focuses on what is gained through gardening, through the participatory activity of getting one's hands dirty.[25]

This should come as no surprise. First, gardening as participatory aesthetics virtually doubles our aesthetic acquaintance and intimacy with gardens. Taking the museum approach likely does not account for a full half of how we interact

---

[25] Even if one does not actually get one's hands dirty—say, because the garden in which they spend time is not theirs or they are not responsible for its maintenance—it is the intimacy between the garden and the garden-appreciator, over that investment of time, that seems to account for the moral effects.

with and appreciate gardens. Second, anyone who has taught ethics, especially environmental ethics, knows that it is rare to move students toward the development of deeper ethical attitudes and actions by the didactic means available through reading books and listening to lectures. To be sure, we commonly move students toward greater insight in developing consistency in their thoughts about ethics. But making them "morally better" more often than not requires them to discover a level of moral meaningfulness that comes through experiences in real-world settings. In some environmental ethical classrooms, students are moved to greater regard for nonhuman animals and toward vegetarianism through introduction to the utilitarian arguments of Peter Singer.[26] Aldo Leopold[27] and Rachel Carson[28] move students as well. But for many students, attitudes and beliefs that make them more active as stewards of the land come from outdoor experiences over the course of their lives, particularly during their formative years. Gardening is certainly on the list of such experiences. Feeling the soil, discovering worms as we dig, rescuing a discarded plant, or choosing from a nursery a living thing—knowing that it will only remain living with care—planting it, watering it (and watering it, and watering it), fertilizing it, pruning it—these activities engage us morally as ongoing and constant caregivers, and through this engagement we typically grow as moral creatures.

No other artform engages us quite like this. While much has been written over the last few decades on the relationship of aesthetics and ethics,[29] the standard model employed for how art impacts the ethical character of an individual focuses largely on (1) identifications and associations the reader (viewer, hearer) makes with characters and themes depicted in the work of art and (2) through the lessons of moral consequences and respect for moral principles that are illustrated through the narrative arc of the work. This model requires the subject to make connections and associations between their own life and what takes place

[26] Peter Singer, *Animal Liberation* (New York, NY: HarperCollins, 1975).

[27] Aldo Leopold, *A Sand County Almanac* (Oxford, UK: Oxford University Press, 1949).

[28] Rachel Carson, *Silent Spring* (New York, NY: Houghton Mifflin, 1962).

[29] As a selection, see Marcia Muelder Eaton, *Aesthetics and the Good Life* (Cranbury, NJ: Associated University Presses, 1989); "Aesthetics: The Mother of Ethics?" *Journal of Aesthetics and Art Criticism* 55:4 (Fall 1997), pp. 355–364; "Integrating the Aesthetic and the Moral," *Philosophical Studies* 67:3 (1992), pp. 219–240. Alan Goldman, "Aesthetic versus Moral Evaluations," *Philosophy and Phenomenological Research* 50:4 (1990), pp. 715–730. Martha Nussbaum, "Exactly and Responsibly: A Defense of Ethical Criticism," *Philosophy and Literature* 22:2 (1998), pp. 343–365. Jerrold Levinson (ed.), *Aesthetics and Ethics: Essays at the Intersection* (Cambridge, UK: Cambridge University Press, 1998).Berys Gaut, "The Ethical Criticism of Art," in Jerrold Levinson (ed.), *Aesthetics and Ethics: Essays at the Intersection* (Cambridge, UK: Cambridge University Press, 1998), pp. 182–203. Noël Carroll, "Moderate Moralism," *British Journal of Aesthetics* 36:3 (1996), pp. 223–238; "Art, Narrative, and Moral Understanding" in Jerrold Levinson (ed.), *Aesthetics and Ethics: Essays at the Intersection* (Cambridge, UK: Cambridge University Press, 1998), pp. 126–160; "Art and Ethical Criticism: An Overview of Recent Directions of Research," *Ethics* 110:2 (2000), pp. 350–387; and "Art and the Moral Realm" in Peter Kivy (ed.), *The Blackwell Guide to Aesthetics* (Malden, MA: Blackwell Publishing, 2004), pp. 126–151.

in the work of art (novel, film, opera, dance—obvious narrative works—and also paintings, sculptures, and so forth). These connections require ethical attention on the part of the subject—the subject must be sensitive to the ethical dimensions of the work—but they also require an empathetic response to what is experienced. That is, one must care about what is being depicted and how that might apply to one's own life. An academic response is not typically going to result in moral growth.

Gardening does not allow for an academic response. Appreciating a garden from a museum perspective may, and for those whose connection to gardens is limited to that approach, then to the extent to which there is a relationship between ethics and the various canonical artforms, that relationship holds for appreciation of a garden as well. But, again, this is a fraction of the normal interaction humans have with gardens, and gardening simply is too participatory to allow for responses that do not endemically include an empathetic response. Unlike the case with living things found in First Nature, living things found in Second Nature and Third Nature require interaction with humans, with those who purpose the planting and ensure that the items planted continue to thrive. Our moral response to gardens is not only about caring in the sense that the moral lessons available are lessons we care about; our moral response is care itself: we care for the garden and its inhabitants, both living and nonliving. We who garden are therefore two steps removed from a purely academic or intellectualized response when it comes to gardens.

Virtue Development

Michael Pollan explores the development of an environmental ethic that is not based on the idea of wilderness but rather on the idea of The Garden.[30] He writes:

> The gardener in nature is that most artificial of creatures, a civilized human being: in control of his appetites, solicitous of nature, self-conscious and responsible, mindful of the past and the future, and at ease with the fundamental ambiguity of his predicament—which is that though he lives in nature, he is no longer strictly *of* nature.[31]

Pollan's gardener is a person who possesses a collection of traits as they inform his character from the outset. Others add to this by describing the impact gardening

---

[30] Michael Pollan, *Second Nature: A Gardener's Education* (New York, NY: Dell, 1991), p. 179 and pp. 190–196.
[31] Pollan, p. 196.

has on the development of such traits. Angela Kallhoff and Maria Schörgenhumer, in "The Virtues of Gardening: A Relational Account of Environmental Virtues,"[32] begin to sketch the virtues that gardening encourages.[33] They write:

> The gardener is a person who is trained in cultivating plants and who knows that action needs to be related to skills that rest on knowledge and experience; and he or she is aware of the limited capacities to regulate nature. In particular, the preconditions for being a competent gardener are not a position of distance, of wonder, or of appreciation of values in the garden. Instead, a gardener needs to endure the otherness that nature provides and simultaneously needs to get involved in nature throughout the process of cultivation in the garden. . . .[34] Important to the notion of gardening . . . (1) that it is practiced on a small scale, in direct, personal contact and interaction with nature (embodied relation), which involves (2) learning about and learning from nature, and (3) goes beyond purely instrumental purposes, with regard to more than just the mere utility of plants. . . .[35] David Cooper proposes an array of virtues that can be learned by growing food in the garden: **care and respect, discipline** and **self-mastery** through adapting one's life to what the garden needs and provides, **humility**, and **hope**.[36]

To the list they attribute to Cooper, Kallhoff and Schörgenhumer add **being disposed to wonder, wisdom, patience, long term perspective, prudence,** and **tenacity**.[37] Each of these takes only the least amount of reflection to appreciate how gardening is particularly adept at fostering them. In complement, Isis Brook writes:

> I look at those gardening practices that, as an incidental side effect of their purpose, increase our **patience, humility, respect for reality, caring for others,** and **open-heartedness**. . . . I argue that they are brought together in a unique way in the relationship between garden and gardener.[38]

---

[32] Angela Kallhoff and Maria Schörgenhumer, "The Virtues of Gardening: A Relational Account of Environmental Virtues," *Environmental Ethics* 39:2 (2017), pp. 193–210.

[33] In their article, Kallhoff and Schörgenhumer note that a list of about 170 environmental virtues can be found in Louke van Wensween's book *Dirty Virtues: The Emergence of Environmental Virtue Ethics* (Amherst, NY: *Prometheus Books*, 2000), pp. 163–167.

[34] Kallhoff and Schörgenhumer, "The Virtues of Gardening," p. 194.

[35] Kallhoff and Schörgenhumer, p. 198.

[36] Kallhoff and Schörgenhumer, p. 200.

[37] Kallhoff and Schörgenhumer, pp. 193 and 204ff.

[38] Isis Brook, "The Virtues of Gardening," in Dan O'Brien (ed.), *Gardening—Philosophy for Everyone: Cultivating Wisdom* (Chichester, UK: Wiley-Blackwell, 2010), pp. 13–25. The quote comes from page 13.

Damon Young adds:

> For Voltaire, the garden did not symbolize monkish quietism—quite the contrary . . . it was a bold metaphor for **compassion, responsibility**, and **pragmatism**—a call to improve his immediate surroundings.[39]

What do all these virtues, dispositions, and attitudes have in common? Cooper writes:

> [C]ertain garden-practices necessarily induce virtues since, when properly or "seriously" engaged in, the engagement is an understanding one, imbued with an appreciation of what is being done: and it can only be this if, at the same time, it "invites" or "brings on" the exercise of virtues.[40] . . . [S]ubmission to the discipline of caring for [a] plant and [a] garden imposes a structure and a pattern on life. . . .[41] To engage in a project with the understanding that its outcome is only partly in one's hands, but without any trust or confidence in the co-operation of the world—of "grace"—would be futile. Hope, therefore, is a virtue induced by the same gardening-practices as humility is. . . . Hope is not simply induced by this or that gardening-practice . . . but pervades the very ethos of gardening.[42] [Each virtue induced by gardening-practices] belongs to the wider economy of virtue that Iris Murdoch called "unselfing," a process of detachment from absorption in what peculiarly concerns one's own interest and ambition. . . . The Buddha for one argued this: each virtue contributes in its own way to a transformation from a "conditioned" state of a person in the grip of "the conceit 'I am'" to an "unconditioned" state liberated from that conceit. . . .[43] I want to suggest that the style of meditation—"reverie," as I'll call it—to which the garden is especially hospitable is not at all the kind that focuses on one's self.[44]

Here Cooper offers the case for the necessary exercise of virtues as one gardens, and he characterizes that exercise in a very specific way: as a movement away from focus on oneself to a focus on the whole of nature, of which the gardener is only a part, understood as the ground from which grows recognition that one is only a part and a confidence that nature will, as a "grace," provide.

Gardening takes one outside oneself. This happens in almost every aspect of gardening. The original garden plan requires one to think of creating something

---

[39] Damon Young, *Philosophy in the Garden* (Minneapolis, MN: Scribe Publications, 2020), p. 169.
[40] Cooper, *A Philosophy of Gardens*, p. 93.
[41] Cooper, p. 95.
[42] Cooper, p. 96.
[43] Cooper, p. 97.
[44] Cooper, p. 83.

beyond oneself; reliant on processes one does not control; perhaps inclusive of trees the planter will never see in maturity; that will require resources of time, effort, and money; and that will almost certainly belong to someone else eventually as one moves on to other projects, sells one's house, or dies. We envision how the garden will look to others, how the plants will fill in over the years, what kinds of services it provides to the ecosystem. And maintaining a garden requires regular and ongoing care that does not respect one's moods, one's tolerance of weather conditions, or how many scratches from branches, thorns, and in response to mosquito bites one sustains.

Visiting a garden takes one outside oneself as well, as the pleasures a garden visit occasions are only substantial or lasting when a garden is permitted to envelop the visitor and attention is directed to the many, many stimuli a garden provides. Even when we are enjoying activities where the garden simply provides a venue—having a meal, having a chat, playing a game, or just sitting in the sun— the naturalness of the garden and its continuity with the rest of nature (humans included) moves us from an exclusive focus on self to appreciation of the greater context of which one is only a small part. It must or there is no explanation for why we prefer the garden and its physicality to the creature comforts of soft chairs in an air conditioned or heated room of our own exclusive design.

Rene Descartes's ruminations proverbially took place in a chair in a room; had they taken place in a garden the hyper focus on the nature of self that dominated those ruminations might have been different. Who knows? It is hard to imagine the results that Kant and Hegel reached about metaphysics, art, and gardens would have been as they were had they done their work in gardens. Philosophy is an inwardly focused sport, and so it is likely no surprise that some of the most enduring modern philosophy in the West seems to come from places where one is forced to spend the bulk of the year indoors. None of these speculations should be taken to suggest that gardens cannot be venues for deep thought par excellence. Rather the suggestion is that being out in the nature of the garden will influence the direction and contents of those thoughts, orienting them in line with a recognition of that larger context. It is likely no accident that Epicurus developed his pleasure-focused thoughts about the constitution of a good life in a garden, that countless monks found understanding through working and contemplation in cloister gardens, and that the Buddha reached his spiritual insights while sitting under a tree.[45]

---

[45] It is a shame that trees do not figure more positively in Christian scripture—the Tree of Life (from which Adam and Eve are exiled) seems dwarfed by the Tree of the Knowledge of Good and Evil, Jonah and the withered shade tree, the fig tree that Jesus curses, the tree upon which Judas killed himself, and the tree upon which Jesus was crucified. No doubt a scholar of Christian scripture could find more balance than we present here.

The virtues that are fostered by gardening—at least the representative list compiled above—are of this outward facing sort, where concern for the self is not central in one's thoughts. While virtues are states that are constitutive of a person's character, and so all virtues necessarily point back to the person who possesses them, the contents of the virtues that gardening promotes are focused on care of "the other." To go back to a point made above, a response to gardening does not allow for an academic response; it requires an attitude of empathy and engagement.

Hope

Miller writes:

> Every garden is an attempt at the reconciliation of the oppositions which constrain our existence; the act of creating a garden, however limited it may be, is not only an assertion of control over our physical surroundings but a symbolic refusal of the terms under which life has been presented to us and an insistence on determining the terms of our existence. As such it is always an act of hope. . . .[46] Every garden moves with an almost unnerving equability back and forth between the "is" of everyday reality and the "ought to be" of the ideal.[47]

While the following claim may not apply to the creation of all works of art, the creation of all aesthetic objects—insofar as they are purposed to serve as foci of aesthetic appreciation and so incorporate or otherwise evoke aesthetically positive features—are efforts directed at making the world a better place. In the case of most artifactual aesthetic objects, there is a point at which the object is finished and the creator of that object—be it a work of art or a decoration—is ready to have the object experienced and admired. Gardeners occasionally bring the garden into states in which they are eager to have it visited and admired. This may come at a time when flowers are most in bloom, when new spring plantings have been installed to replace those killed by the winter, or when fresh mulch has been added to the beds and the edges of those beds are crisply cut. Every seasoned gardener knows these states are fleeting and is reconciled (in some measure) to that reality. We know there will be another time, not too distant in the future, when another opportunity to take pride in the state of our garden will present itself. And just as importantly we generally know what we can do in terms of gardening practices to encourage that state into manifestation. We live

---

[46] Miller, *The Garden as an Art*, p. 25.
[47] Miller, p. 130.

in hope that our efforts, in cooperation with the processes of nature, will produce those states where the world includes that much more beauty as presented by our garden. While the creation of these beauteous states may not be the cause of goodness on a global scale, it is a contribution to it, no matter how modest the garden. But unlike the case with other artifactual aesthetic objects, we do this on a continual basis. Our investment does not wane; it cannot, or at least it cannot for very long, or our garden will not have the capacity for contributing to beauty. The constancy of our gardening means that we are always invested in efforts at creating good.

This observation seems entailed in what Miller writes, but her central point seems to be a deeper one about the mindset of the gardener. Through investing the attention, planning, care and so forth that gardens require, the gardener works from the perspective that our efforts will be useful, that they will result in making at least our corner of the world a little brighter, and that we help create something of value. The hope that is inculcated by gardening practice extends to the most basic of human wants—that the future will be hospitable for continued life. Society grows great, the proverb goes, when old men plant trees the shade of which they will never know; planting trees, like so many other garden practices, is an act that demonstrates hope for future generations of one's family, the continued prosperity of the society, and the basic hospitable conditions that allow trees to flourish. In an era of environmental degradation and climatic uncertainty, bringing fragile seedlings into the world and carrying the weight of responsibility for their care can seem like an uphill battle. But we persist in these tasks because of the hope we have for the future of our garden, our species, and our planet.

A gardener works, as Miller says, from an attitude of hope. We labor in local optimism that our efforts will bear fruit (metaphorically and sometimes literally) and from the more global optimism that the world really is such that it provides, that it produces good, that it sustains us not merely biologically but spiritually, intellectually, and aesthetically as well. The gardener's time spent in the garden—despite sunburn, mosquitos, wasps, blisters, and the rest—is time spent in a worthwhile fashion. It must be seen this way for gardeners working in a context of Third Nature—that is, working beyond simple biological necessity—to continue our efforts month after month, season after season, year after year. There is no requirement for anyone to garden—to engage in agriculture perhaps, but not to garden, especially when this requires so much investment of time, talent, and treasure. The benefits that come from gardening are like the benefits that come from other aesthetic investments; while they may be largely intangible, they nonetheless are highly significant and highly valuable in the constitution of one's time commitments, one's character, and one's life.

Connection

Roger Paden writes:

> [A]rchitecture aesthetically appropriates and humanizes nature, understood in terms of space and time: it orients us to space, time, and geography, thereby providing us with a place in the world. But because early gardens, unlike primitive huts, always contained plants, landscape architecture differs from architecture in that it aesthetically appropriates and humanizes nature, understood in terms of life. Landscape architecture can orient us to the biological world by offering an interpretation (1) of the living world around us; (2) of our own nature; (3) and of our place among other living things. It can help humanize our relationship with living nature, help orient us to that nature, and help us dwell in it. This is its ethical function.[48]

Gardening allows us to reconnect intimately with the natural world, to be in touch with the ultimate context of our own biological survival and flourishing, and to reestablish rhythms of life in sync with those of that context. Gardening is an outwardly focused activity, and one of the most important skills of the gardener is close observation of how the outside world changes day to day—how the plants grow, what sorts of animals visit, how much rainfall there has been this week, what anomalies are present. Those who garden come to appreciate the interconnectedness and interdependence between ourselves and other life, our common evolutionary ancestors, the equal aptitude to life shared by all living species, and the process of nature of which we all are part.

Some of the living things that gardens can assist in our orientation toward are other people, that is, other gardeners, those with whom we either share the maintenance of a particular garden or those who, like us, create and nurture their own gardens and with whom we can then share a kind of kinship of common purpose and common investment. Nathan Nun, in discussing themes of societal and social alienation, writes:

> [C]ommunity garden projects, grassroots efforts to grow food on public and private plots of land in an urban setting, offer a promising example of an environment in which what [Herbert] Marcuse calls the "new sensibility" is fostered and plays itself out through practical-aesthetic activity, activity that positively overcomes various forms of alienation. . . . Gardeners see their activities in the

---

[48] Roger Paden, "The Ethical Function of Landscape Architecture," *Environmental Philosophy* 15:2 (2018), pp. 139–158. The quote comes from page 149.

gardens as "part of a social agreement that builds cohesion within the group and further strengthens a common aesthetic judgment of the garden."[49]

The connections we make with the earth are made with other human beings through gardening. We share in common our love of gardens and gardening, our interests in demonstrating ongoing care of living things and of learning all about them for the sake of their flourishing, for the sake of the knowledge itself, and our purposiveness in gardening. Beyond our connections with those with whom we garden, we can appreciate the connections we have with all humans who ultimately must rely on efforts of farming and foraging that at least are substantively like gardening for our survival—and not just with humans but with all animals. Granted that this moves us into the realms of Second Nature and First Nature, but it is through engagement with others in Third Nature that many of us come to appreciate these more basic connections.

Continuing Creativity

In reviewing Cooper's book, *A Philosophy of Gardens*, Ronald Moore, making a point about gardening but also contributing to a point about the undilutability of appreciating gardens as wholes, writes:

> The fundamental idea is that properly involved apprehension of an aesthetic object must be comprehensive in taking stock of the whole of that object. In the case of gardens, Cooper points out, it is not enough to experience wholeness as the totality of objects of delectation. True garden wholeness must be apprehended as an integration of natural materials with the purposeful activity of gardening that undergirds the display of materials.[50]

The creative act of bringing a garden into being does not end with the garden designer or the garden artist. "Creation" of a garden may be thought of in two ways. One way is to think about the very first installation of a garden, the manifestation of the gardenist's (s') design, the date we set as the start of a particular garden. The other way is the more philosophically robust, and there the garden is always in a state of creation. This is entailed by The Garden's constant dynamism—the "nature side" of the ongoing creativity—but it is also a consequence of that fact that gardeners are always undertaking activities that not only maintain a garden

[49] Nathan Nun, "Practical Aesthetics: Community Gardens and the New Sensibility," *Radical Philosophy Review* 16:2 (2013), pp. 663–677. Quotes come from pages 664 and 672.

[50] Ronald Moore, "A Review of David E. Cooper's *A Philosophy of Gardens*," *Journal of Aesthetic Education* 41:3 (2007), p. 122.

in some state relative to its initial design and keep it coherent with its expressed purpose but also make changes that truly amount to alterations of the original design. Earlier, we identified gardening practices as the following:

1. Designing
2. Preparing
3. Planting, Placing (as in rocks), Building, and Ornamenting
4. Irrigation
5. Tending to the Ongoing Health of the Living Elements (including finding and treating for pests, staking, cabling and bracing trees, air spading, adding compost and fertilizer, and so forth)
6. Propagating and Grafting
7. Weeding and Pressure Washing
8. Pruning, Cut-Backs, Shaping, and Deadheading
9. Editing and Curating (including transplanting and removing desirable plants that do not belong or are too thick)
10. Replacing and Resurfacing
11. Redesigning

All but perhaps the first are handled by gardeners and are attended to continuously. As so many of these activities require the decision-making and creativity definitive of creation, it is right to think of gardeners as being coequal partners in the creation—the ongoing, never-ceasing creation—of a garden.

The Garden may not be unique in this unusual sense of ongoing creation, but no artform involves this sort of thing more than The Garden. Conductors and musicians interpret scores, directors and actors interpret scripts, artistic directors and dancers interpret the work of choreographers, but in all these forms we have guidelines that to one degree or another limit the amount of divergence from scores and scripts allowed while maintaining identity conditions. Dance is the exception, as we mentioned, but even with dance, the dancers are not making up the moves they execute on the spot out of whole cloth. Improvisational jazz is perhaps the closest musical example of what takes place in The Garden. If we think of nonscored jazz as a partnership of musicians following general themes in tandem with one another, acting out their own creative vision and reacting to the creative vision of their bandmates, this may be close to illustrating the work of gardeners as we cooperate with one another, certainly with nature, and in line with original designs and current purposes as we work our gardens.

Those who are engaged with gardens—gardenists, garden visitors, garden enthusiasts, garden owners, and especially gardeners—demonstrate by our investments of time, talent, and treasure our commitments not merely to our gardens, which is obvious, but to our efforts to be engaged in aesthetic creation.

Simply put, gardens and gardening function as indicators—as barometers—of one's commitment to and investment in their own personal aesthetic sensibility.

As Aristotle taught us with respect to moral virtues, a pattern repeated can turn into a character trait, and for some children who grow up with gardens, their love of gardens, begun early and reinforced positively and often, becomes lifelong. We believe this pattern is not contained, that it is not limited to simply the love of gardens, but that it forms a basis for appreciation of living things in general and, at the same time, for beauty. E. O. Wilson describes this phenomenon in his book *Biophilia*.[51]

Garden Ethics

This final section in our consideration of gardens and value focuses on how one should act as a gardenist, a gardener, and a garden appreciator, with each role possessing for the most part different obligations—though of course a single person could easily fit into any of these three roles, perhaps even simultaneously. Garden ethics of all stripes are motivated by achievement of idealized states of gardens, either of The Garden or of a particular garden style.

That there are such things as "garden ethics" stems from one of the criteria we included in Chapter 2 as a necessary aspect of all gardens: all gardens are rule governed. Our job now is to explicate those rules in their proper spheres; that is, in terms that specify to whom they apply and what obligations are attendant to them. There certainly will be many different codes of ethics, depending on the garden type. The following suggestions are meant to apply to all gardens, no matter how grand or modest.

Gardenist Ethics

1. A gardenist should plan a garden in concert with nature, being respectful that every element of the planned garden is designed to ensure that the desired living elements thrive. This will include ensuring that gardeners have appropriate access to do their work; that topography, edaphology, climate, and appropriate scientific aspects are all taken into account; that plants are chosen that are appropriate to the location and their location within the garden.

2. A gardenist should ensure that a garden is designed effectively to meet its purpose and, if it is meant to illustrate a particular garden style, that it

---

[51] Edward O. Wilson, *Biophilia* (Cambridge, MA: Harvard University Press, 1986).

effectively represents that style to the degree to which the particular style demands (Persian gardens are typically more prescriptive than American pleasure gardens, for instance).

3. A gardenist should ensure that a garden is designed in concert with the environment—natural, built, and human—surrounding and contextualizing it. This includes being careful to avoid introduction of invasive plants and to avoid creating gardens that, in their installation or their maintenance, are morally problematic.

4. A gardenist should design a garden keeping in mind how it will be experienced by visitors.

Gardener Ethics

1. A gardener should ensure every element of the garden is in good order, in particular ensuring that the living elements thrive. This will include attending to the soil, to irrigation and moisture retention, to fertilizer (if the garden requires fertilizer or compost), finding and treating for pests, weeding, deadheading, pruning, shaping, editing, grafting, staking, cabling and bracing trees, air spading, replacing, resurfacing, pressure washing hardscapes, and so forth.

2. A gardener should work in cooperation with nature. This includes giving due consideration to aspects of the garden that may seem counterintuitive to garden maintenance, such as leaving dead stems in winter for the sake of beneficial insects, allowing dead trees to decompose and contribute to soil biota, restricting indiscriminate pesticide use as much as possible, and so forth.

3. A gardener should seek in their work to maintain the original garden design and purpose of the garden to the extent feasible. When redesign is called for, a gardener should endeavor to pursue redesign deliberately and in optimization of the garden's identity and purpose.

4. A gardener should understand all aspects of the garden they tend well enough to provide answers to questions from those who visit the garden.

5. A gardener should work with diligence, patience, and consciousness of the reality that the work is constant, ongoing, and subject to the uncontrollable effects of nature.

Garden Visitor Ethics

1. A garden visitor should never disturb or harm a garden in any way. This includes remaining on the designated paths, touching plants only when such activity is sanctioned, and certainly not picking flowers, taking cuttings, or pulling up plants. Normal wear and tear are expected, but a visitor should strive to disturb as little as possible, including the experiences of other visitors and those who tend the garden.

2. A garden visitor should be properly prepared to visit the garden by wearing appropriate clothing, being in a receptive state of mind, understanding what they might expect to experience in the season of their visit, and being prepared to seek out those elements of the garden likely to provide an optimally valuable experience of the garden.

3. Garden visitors should educate themselves about a garden, its purpose, and its population, ideally before visiting. They should (unobtrusively) ask questions of gardeners to learn more about the garden.

4. A garden visitor should consider whether through resource donation of time, talent, and/or treasure they can support the garden.

Conclusion

In the introduction, we raised several questions we promised to answer in due course. Below we briefly review the questions and the answers we offered as a way to summarize the ground we covered.

What is "a garden"? When does a garden come into being?

This was, of course, the focus of both the first and second chapters, and while it took some time to come finally to our ten-point account, we think thorough discussion of why we took the teleologically organized taxonomic approach we did was important. Our account was that a place is a garden when it possesses these ten properties:

1. Spatially Three Dimensional
2. Geographically Bound
3. Human Scaled and Perspectivally Bound
4. Dynamic and Time Dependent
5. Nature Interactive and Ecologically Bound
6. Unbounded in Time and Lifespan
7. Designed
8. Human Interactive and Requiring Maintenance
9. Rule Governed
10. Aesthetically Engaging

What is it "to garden"? What are gardening practices?

It was in the discussion leading to our definitional account that we discussed gardening practices, our list of which was:

1. Designing

Southern African Collection, University of California Botanical Garden at Berkeley, Berkeley, CA
Multiple gardenists starting 1928, redesigned 2001
Photo by Eric Hupperts, Spring 2020

2. Preparing
3. Planting, Placing (as in stones), Building, and Ornamenting
4. Irrigation
5. Tending to the Ongoing Health of the Living Elements (including finding and treating for pests, staking, cabling and bracing trees, air spading, adding compost and fertilizer, and so forth)
6. Propagating and Grafting
7. Weeding and Pressure Washing
8. Pruning, Cut-Backs, Shaping, and Deadheading
9. Editing and Curating (including transplanting and removing desirable plants that do not belong or are too thick)
10. Replacing and Resurfacing
11. Redesigning

In Chapter 10, we amplified our discussion of gardening in exploration of what motivates us to do it and what virtues and values it fosters.

Is a garden "an object"? If so, what kind?

We discussed in Chapter 5 what it can mean to call a garden an "object" both in view of the nature-dynamic character of all gardens and in view of the challenge that we need a stable set of properties attached to a stable object for the sake of comparison, evaluation, and interpretation. In the end, we settled on a two-fold answer: (1) we can think of a garden as a four-dimensional "real" place, but that conception is endemically and inescapably openly dynamic, and (2) we can think of an experience-focused, phenomenal-capture of a garden that is the bearer of those properties and aspects we can use as evidence for making normative and meaning claims about that garden.

When does a garden cease being a garden or change into a different garden?

A garden ceases being a garden when, absent regular maintenance, it reverts to being First Nature (or something similar to First Nature). In the second chapter we talked about how gardens are site-specific, and that a garden moved from one location to another only retains the name and principles of the original garden.

What is it for a garden to be an aesthetic object?

This conversation, largely the focus of Chapter 5, was an extension of our discussion about the objecthood of the garden.

What is it to be an aesthetic property (feature, aspect, etc.) of a garden?

Ultimately, we developed a schema to capture our thoughts on how the various properties of gardens relate to one another and how they build upon one another in movement toward interpretative and evaluative claims. A reduced version of that schema, focused on the aesthetic character of gardens, follows.

Objective Physical Properties

Item or Property	Examples	Derivation
Site	*Terroir*, topography, climate, hydrology, soils	These items are chosen because of (1) practicalities concerning available land, patronage, resources, etc., and (2) the anticipated phenomenal or elemental properties of the forthcoming phenomenal garden.
Materials	Plants, soil, water, stone, mulch, hardscape, hidden ha-ha walls and irrigation pipes	
Hidden Nature	Unperceived root structures, unperceived shoots of tree limbs above the canopy	These are real aspects of the physical garden that only enter experiences of the garden imaginatively.

Dynamic Process Properties

Item or Property	Examples	Derivation
Nature-based processes of creation	Weather, seasonal change, climate change, ecological activity, plant growth/competition	These are processes that make up nature's contribution to the creation of garden at any given moment
Human-based processes of creation	Arranging, planting, shaping, pruning, thinning, deadheading, watering, fertilizing, building	These are gardening practices, responsible for human contribution to the garden at any given moment

Phenomenal/Experiential Properties

Item or Property	Examples	Derivation
Garden Elements	The pink of the rose, the thorniness and height of its stems, the bright yellow of variegated schefflera against the deep purple of a carefully shaped loropetalum hedge, the scent of calamondin blossoms, the sound of a fountain of water	These properties are derived from garden appreciators experiencing states of and perspectives on the four-dimensional, dynamic garden.
Borrowed Scenery	Mountains, skyscrapers, visible trees and architecture outside the garden, sensible sounds and smells outside the garden	

Aesthetic Properties

Item or Property	Examples	Derivation
Formal Aesthetic Properties	Balance, charm, elegance, grace, harmony, peacefulness, grandness, leanness	These properties arise from interpretation of the perceived properties of the phenomenal garden, typically based on the form of the phenomenal garden but occasionally based on individual phenomenal properties
Contextual Properties	Genetic, lifespan, reception, relationship, and scientific factors— "The New York Botanical Garden is sited on 250 acres in the Bronx." "Jennie Butchart transformed a former limestone quarry into the Butchart gardens we know today." "This garden is made up entirely of invasive plants." "This species only flowers once in its lifetime."	Contextual properties have to do with (1) genetic aspects of the garden (seen at the top of this table); (2) external relations the garden bears to other objects, themes, and persons—historically and currently; (3) responses of audience members, both expressed and not; and (4) those processes discovered through science

Item or Property	Examples	Derivation
Summative Aesthetic Properties	"This garden is beautiful." "This garden is more beautiful than that garden."	These properties are derived from and evidenced by the formal aesthetic properties of the phenomenal garden and from the "garden form" exhibited by the garden

What is it to have an aesthetic experience of a garden? How does this relate to other aesthetic experiences? To experiences commonly? What is to be an aesthetically relevant contextual property of a garden or the experience of a garden?

These questions were the focus of Chapters 5 and 6. Our view is that aesthetic experiences of gardens, while having aspects distinct to the appreciation of gardens, are not different in kind from other aesthetic experiences, as evidenced by the fact that we use the same terminology and modes of speaking about our aesthetic experiences of gardens as of other aesthetic objects. On the second question, our view is that it is even more common in consideration of gardens than with other aesthetic or art forms to bring relevant contextualities into our discussions and our claims. We divided relevant contextual aspects into four kinds—genetic aspects, lifespan aspects, reception aspects, and relationship aspects—and we included science as an important part of this discussion.

Can a garden be a work of art? Is The Garden an artform? A craft? A subsidiary or subset of another artform? If The Garden is an artform unto itself, how does it relate to other artforms or art kinds? If The Garden is an artform, what is it to be a "garden artist"?

This, of course, was the focus of Chapters 3 and 4. In the end, despite the many differences between gardens as works of art and other works of art that we reviewed, our view is that some gardens are indeed works of art and can be regarded just as fully from that perspective as any other artworks.

What is it to be a member of a garden's audience? What is it to be an appreciator of a garden?

This question was largely addressed in our conversation about a garden's objecthood.

Can a garden be "faked" or "forged"?

This question was the topic of the last section of Chapter 2. We believe that while an exact copy of a garden is impossible—due to the particulars of the siting—gardens can be "modestly" copied and garden styles can be forged.

How do/can we come to know (see, hear, sense) a particular garden as an object?

Again, this was the topic of our conversations about garden objecthood in Chapter 5. For those who might have thought those conversations lengthy, the number of salient questions those discussions addressed are evident in this list.

Can a garden be a conveyor of knowledge or meaning? If so, how? How can a garden be "read" or interpreted? Can The Garden possess a meaning? If so, how do we come to know it?

This was the focus of Chapters 7 and 8. We believe that all gardens that qualify as works of art are interpretable and that other gardens may be, as well. Our recommendation—although this is but one path to follow in interpreting gardens—is to focus first on articulating an understanding of the coherence of the various aspects of the form of a garden, of which we listed eleven:

1. Path
2. Structure, Hardscape, or Hard Components
3. Artifactual Components
4. Space
5. Water
6. Light
7. Topography
8. Setting
9. Ecology
10. Plant Palette
11. Plant Character

We did not focus, as Cooper did, on whether The Garden as an aesthetic form possesses a meaning. Our inclination is to say that a suitable answer cannot be constructed due to the variety of subjective value responses gardens occasion—and due to the variety of purposes gardens are employed to serve. But, of course, Cooper may have done exactly what we are inclined to say cannot be done. The proof is only available by reading his book.

Should gardens exist? If so, under what conditions or constraints (if any)? Is there any kind of garden that should not exist?

While it is our position that all gardens, as gardens, are good, there are purposes to which gardens may be put and motivations to create gardens that may be morally suspect. There may be ethical questions of resource allocation. In those cases, calculations must be made as to whether the good of the garden, along with the value attendant to its employment for a potential plurality of good purposes, is overridden by the morally negative dimension(s). On a rare occasion, a garden should not exist. This was discussed in Chapter 9.

Are there particular ways with which gardens should be regarded, valued, or engaged? Is there any special perspective one should take to value a garden correctly?

We believe the aesthetic enjoyment of a garden is consonant with the enjoyment of any other aesthetic object, with the one caveat being that aesthetic appreciation of a garden is typically more complex and may be spread over more time than appreciation of simpler aesthetic forms. No special attitude—especially not one of disinterest—should be employed. On the other hand, a perspective informed by the relevant science is likely to make the experience all the richer—which is a claim we might make about the relevant contextual aspects of any art object or many aesthetic objects. This was discussed primarily in Chapters 5 and 6.

Can one garden be better than another? Can a garden be evaluated? Aesthetically evaluated? Artistically evaluated?

Yes, comparison of one garden with another is entirely possible. It is common practice when building an agenda of gardens one wishes to visit on a trip or over the course of a lifetime. The argument for this answer was given in Chapter 9, and the "how" question was addressed in Chapter 10.

How do we manage the subjective in our appreciation or evaluation of a garden? Should personal associations and identifications play a role?

Our approach to garden evaluation is both object-focused and subject-focused, and so there is indeed a role for "the subjective" in the evaluation of a garden. While aesthetic experiences of a garden typically will involve associations of a personal sort—and these typically will not be included in a serious evaluation of a garden—some subject-focused aspects involved in garden appreciation transcend personal boundaries and are relevant to an evaluation.

What is the role of taste in the evaluation of gardens?
The role of cultural context? The role of function?

We discussed, in Chapter 9, the role of taste in our consideration of the question of whether the value of one garden can be compared with another; while we take the plurality of preference patterns to be a matter of fact, we do not believe this eliminates the possibility of making meaningful normative claims about gardens. Gardens, at least as much as any other "cultural product" that is appreciated aesthetically, are situated within cultures, and many times within national identities and within timespans when creation of their sort was popular or particularly meaningful. We argued that inclusion of consideration of the function or purposes of a garden is important to any serious evaluation of a garden.

How are the values associated with gardens related to other values?
Virtues? Vices? Religious values?

While religious values might best be addressed in classifying a garden and among relevant contextual aspects of gardens, the fact that gardens foster in garden enthusiasts and particularly in gardeners a wide range of virtues is undeniable. While gardening, a gardener is engaged in a worthy endeavor. This was discussed in Chapter 10.

Can or should gardens be thought about in moral/ethical terms?

Yes—"can" and "should." Some gardens involve moral dimensions—both positive and negative—and they are relevant to how those gardens are appreciated and how they are evaluated. This was discussed in Chapter 9.

These questions, posed in the introduction, created a frame within which to conceive a comprehensive approach to considering garden aesthetics. Each was addressed in this book—plus several others that were raised along the way.

The significance of a study like this one rests in three places.

- Aesthetic appreciation is one of the primary ways, if not the primary way, we approach gardens. It motivates us to visit and spend time in them, to work in and care for them, and to think about them and their places in our lives. A study like this one provides an occasion for having a conversation about what aesthetic appreciation of gardens entails and to raise issues for continued conversation and, frankly, debate.

- Such a conversation as the one this book occasions is not common in the philosophical or even in the "garden-theoretical" literature. Given the ubiquity of gardens in our contemporary lives and in our history going back at the very least four thousand years, the paucity of these sorts of conversations in the contemporary literature seems odd.

- While some branches of philosophical inquiry—such as metaphysics—might not appropriately be held to a standard of practical usefulness, this is not the case with axiological disciplines like value theory, ethics, and aesthetics. Indeed, one common way to critique an ethical theory is to raise issues about its applicability for solving actual problems. Aesthetics that does not deliver the means by which aesthetic experiences can be enhanced in both quantity and quality is less than it might be. This study is meant to deliver exactly that with regard to the aesthetic appreciation of gardens. In other words, considering the various topics raised in this study is meant, ultimately, to have a practical positive impact on our appreciation of gardens.

Bibliography

Addison, Joseph, "On the Pleasures of the Imagination," Nos. 411–421, in Robert Allen (ed.), *Selections from The Tatler and The Spectator* (New York, NY: Holt, Rinehart and Winston, 1957).

Alison, Archibald, *Essays on the Nature and Principles of Taste* (Boston, MA: Cummings & Hilliard, 1812).

Anstey, Peter R., & Stephen A. Harris, "Locke and Botany," *Studies in History and Philosophy of Science Part C: Studies in History and Philosophy of Biological and Biomedical Sciences* 37:2 (2006), pp. 151–171.

Aristotle, *The Poetics* (Richard Janko, trans.) (Indianapolis, IN: Hackett Publishing Company, 1987).

Arnheim, Rudolph, *Film as Art* (Berkeley, CA: University of California Press, 2006).

Barrett, Terry, *Criticizing Art: Understanding the Contemporary* (New York, NY: McGraw-Hill, 2000).

Barwell, Ismay, and John Powell, "Gardens, Music, and Time," in Dan O'Brien (ed.), *Gardening—Philosophy for Everyone: Cultivating Wisdom* (Chichester, UK: Wiley-Blackwell, 2010), pp. 135–147.

Bazin, Andre, *What Is Cinema?* (Berkeley, CA: University of California Press, 1967).

Beardsley, Monroe C., *The Aesthetic Point of View* (M. J. Wreen and D. M. Callen, eds.) (Ithaca, NY: Cornell University Press, 1982).

Beardsley, Monroe C., *Aesthetics: Problems in the Philosophy of Criticism* (Indianapolis, IN: Hackett, 1981).

Beardsley, Monroe C., "What Is an Aesthetic Quality?" *Theoria* 39 (1973), pp. 50–70.

Bell, Clive, *Art* (London, UK: Chatto and Windus, 1914).

Berleant, Arnold, *The Aesthetics of Environment* (Philadelphia, PA: Temple University Press, 1992).

Berleant, Arnold, "Reconsidering Scenic Beauty," *Environmental Values* 19:3 (2010), pp. 335–350.

Berthier, Francois, *Reading Zen in the Rocks: The Japanese Dry Landscape Garden* (Graham Parkes, trans. and ed.) (Chicago, IL: University of Chicago Press, 2000).

Booth, Wayne, *The Rhetoric of Fiction* (Chicago, IL: University of Chicago Press, 1983).

Bourassa, Steven, *The Aesthetics of Landscape* (London, UK: Belhaven Press, 1991).

Brady, Emily, "Imagination and the Aesthetic Appreciation of Nature," *Journal of Aesthetics and Art Criticism* 56:2 (1998), pp. 139–147.

Brady, Emily, Isis Brook, and Jonathan Prior, *Between Nature and Culture: The Aesthetics of Modified Environments* (Lanham, MD: Rowman and Littlefield, 2018).

Brook, Isis, "Dancing with Time: The Garden as Art," *British Journal of Aesthetics* 60:2 (2020), pp. 231–234.

Brook, Isis, "The Importance of Nature, Green Spaces, and Gardens in Human Well-Being," *Ethics, Place and Environment* 13:2 (2010), pp. 295–312.

Brook, Isis, "Making Here Like There: Place Attachment, Displacement, and the Urge to Garden," *Ethics, Place, and Environment* 6:3 (2003), pp. 227–236.

Brook, Isis, "The Virtues of Gardening," in Dan O'Brien (ed.), *Gardening—Philosophy for Everyone: Cultivating Wisdom* (Chichester, UK: Wiley-Blackwell, 2010), pp. 13–25.

Brook, Isis, "Wildness in the English Garden Tradition: A Reassessment of the Picturesque from Environmental Philosophy," *Ethics and the Environment* 13:1 (2008), pp. 105–119.

Brook, Isis, and Emily Brady, "Topiary: Ethics and Aesthetics," *Ethics and the Environment* 8:1 (2003), pp. 128–141.

Brown, Kendall, and David Cobb, *Quiet Beauty: The Japanese Gardens of North America* (Rutland, VT: Tuttle, 2013).

Brown, Kendall, and David Cobb, *Visionary Landscapes: Japanese Garden Design in North American* (Rutland, VT: Tuttle, 2017).

Brown, Kendall, and Melba Levick, *Japanese-Style Gardens of the Pacific West Coast* (New York, NY: Rizzoli, 1999).

Budd, Malcolm, *The Aesthetic Appreciation of Nature: Essays on the Aesthetics of Nature* (Oxford, UK: Clarendon Press, 2002).

Bullough, Edward, "'Psychical Distance' as a Factor in Art and as an Aesthetic Principle," *British Journal of Psychology* 5 (1912), pp. 87–98.

Campbell, Gordon, "Epicurus, the Garden, and the Golden Age," in Dan O'Brien (ed.), *Gardening—Philosophy for Everyone: Cultivating Wisdom* (Chichester, UK: Wiley-Blackwell, 2010), pp. 220–222.

Carlson, Allen, *Aesthetics and the Environment: The Appreciation of Nature, Art and Architecture* (New York, NY: Routledge, 2000).

Carlson, Allen, "Appreciating Art and Appreciating Nature," in Salim Kemal and Ivan Gaskell (eds.), *Landscape, Natural Beauty and the Arts* (Cambridge, UK: Cambridge University Press, 1995), pp. 199–227.

Carlson, Allen, "Appreciation and the Natural Environment," *Journal of Aesthetics and Art Criticism* 37:3 (1979), pp. 267–275.

Carlson, Allen, "Formal Qualities in the Natural Environment," *Journal of Aesthetic Education* 13:3 (1979), pp. 99–114.

Carlson, Allen, "Nature, Aesthetic Judgment, and Objectivity," *Journal of Aesthetics and Art Criticism* 40:1 (1981), pp. 15–27.

Carlson, Allen, "Nature and Positive Aesthetics," *Environmental Ethics* 6:1 (1984), pp. 5–34.

Carlson, Allen, "On the Possibility of Quantifying Scenic Beauty," *Landscape Planning* 4 (1977), pp. 131–172.

Carlson, Allen, "The Requirements for an Adequate Aesthetics of Nature," *Environmental Philosophy* 4:1 (2007), pp. 1–12.

Carroll, Noël, "Art and Ethical Criticism: An Overview of Recent Directions of Research," *Ethics* 110:2 (2000), pp. 350–387.

Carroll, Noël, "Art and the Domain of the Aesthetic," *British Journal of Aesthetics* 40:2 (2000), pp. 191–208.

Carroll, Noël, "Art Appreciation," *Journal of Aesthetic Education* 50:4 (2016), pp. 1–14.

Carroll, Noël, "Art, Narrative, and Moral Understanding" in Jerrold Levinson (ed.), *Aesthetics and Ethics: Essays at the Intersection* (Cambridge, UK: Cambridge University Press, 1998), pp. 126–160.

Carroll, Noël, "Historical Narratives and the Philosophy of Art," *Journal of Aesthetics and Art Criticism* 51:3 (1993), pp. 313–326.

Carroll, Noël, "Moderate Moralism," *British Journal of Aesthetics* 36:3 (1996), pp. 223–238.

Carroll, Noël, "On Being Moved by Nature: Between Religion and Natural History," in S. Kemal and I. Gaskell (eds.), *Landscape, Natural Beauty and the Arts* (Cambridge, UK: Cambridge University Press, 1993), pp. 244–266.

Carroll, Noël, *On Criticism* (New York, NY: Routledge, 2009).

Carson, Rachel, *Silent Spring* (New York, NY: Houghton Mifflin, 1962).

Chisholm, Linda, *The History of Landscape Design in 100 Gardens* (Portland, OR: Timber Press, 2018).

Classens, Michael, "The Nature of Urban Gardens: Toward a Political Ecology of Urban Agriculture," *Agriculture and Human Values* 32:2 (2015), pp. 229–239.

Coffin, David R., "The English Garden: Meditation and Memorial," *Journal of Aesthetics and Art Criticism* 55:3 (1997), pp. 459–463.

Cooper, David E., "In Praise of Gardens," *British Journal of Aesthetics* 43:2 (2003), pp. 101–113.

Cooper, David E., *A Philosophy of Gardens* (Oxford, UK: Oxford University Press, 2006).

Cotton, Anne, "Gardener of Souls: Philosophical Education in Plato's Phaedrus," in Dan O'Brien (ed.), *Gardening—Philosophy for Everyone: Cultivating Wisdom* (Chichester, UK: Wiley-Blackwell, 2010), pp. 232–244.

Crawford, Donald, "Scenery and the Aesthetics of Nature," in Arnold Berleant and Allen Carlson (eds.), *The Aesthetics of Natural Environments* (Peterborough, Canada: Broadview Press, 2004), pp. 253–268.

Daniel, Terry, "Whither Scenic Beauty? Visual Landscape Quality Assessment in the 21st Century," *Landscape and Urban Planning* 54:1 (2001), pp. 276–281.

Danto, Arthur, *The Philosophical Disenfranchisement of Art* (New York, NY: Columbia University Press, 1986).

Danto, Arthur, *Transfiguration of the Commonplace* (Cambridge, MA: Harvard University Press, 1981).

Davies, Stephen, "Is Architecture Art?" in Michael H. Mitias (ed.), *Philosophy and Architecture* (Amsterdam, NL: Rodopi, 1984), pp. 31–47.

Davies, Stephen, "John Cage's 4'33": Is It Music?," *Australasian Journal of Philosophy* 75:4 (1997), pp. 448–462.

Davies, Stephen, *Philosophical Perspectives on Art* (Oxford, UK: Oxford University Press, 2007).

Day, Jo, "Plants, Prayers, and Power: The Story of the First Mediterranean Gardens," in Dan O'Brien (ed.), *Gardening—Philosophy for Everyone: Cultivating Wisdom* (Chichester, UK: Wiley-Blackwell, 2010), pp. 65–78.

Dewey, John, *Art as Experience* (New York, NY: Perigee, 1934).

Dickie, George, *Art and the Aesthetic: An Institutional Analysis* (Ithaca, NY: Cornell University Press, 1974).

Dickie, George, "Bullough and the Concept of Psychical Distance," *Philosophy and Phenomenological Research* 22 (1961), pp. 233–238.

Dickie, George, *Evaluating Art* (Philadelphia, PA: Temple University Press, 1988).

Dickie, George, "The Myth of the Aesthetic Attitude," *American Philosophical Quarterly* 1:1 (1964), pp. 56–65.

Dickie, George, "The New Institutional Theory of Art," in G. Dickie, R. Sclafani, and R. Roblin (eds.), *Aesthetics: A Critical Anthology*, 2nd ed. (New York, NY: St. Martin's Press, 1989), 196–205.

Dickie, George, "Psychical Distance: In a Fog at Sea," *British Journal of Aesthetics* 13 (1973), pp. 17–29.

Dickie, George, "'Stolnitz' Attitude: Taste and Perception," *Journal of Aesthetics and Art Criticism* 43 (1984), pp. 195–204.

Dickie, George, "Taste and Attitude: The Origin of the Aesthetic," *Theoria* 39 (1973), pp. 153–170.

Eaton, Marcia Muelder, *Aesthetics and the Good Life* (Cranbury, NJ: Associated University Presses, 1989).

Eaton, Marcia Muelder, "Dangerous Beauties," *Philosophic Exchange* 30 (1999–2000), pp. 35–52.

Eaton, Marcia Muelder, "Fact and Fiction in the Aesthetic Appreciation of Nature," *Journal of Aesthetics and Art Criticism* 56:2 (1998), pp. 149–156.

Eaton, Marcia Muelder, "Integrating the Aesthetic and the Moral," *Philosophical Studies* 67:3 (1992), pp. 219–240.

Eaton, Marcia Muelder, "Kantian and Contextual Beauty," *Journal of Aesthetics and Art Criticism* 57:1 (1999), pp. 11–15.

Eaton, Marcia Muelder, "A Strange Kind of Sadness," *Journal of Aesthetics and Art Criticism* 41:1 (1982), pp. 51–63.

Endersby, Jim, "A Garden Enclosed: Botanical Barter in Sydney, 1818–39," *British Journal for the History of Science* 33:3 (2000), pp. 313–334.

Endersby, Jim, "A Visit to Biotopia: Genre, Genetics and Gardening in the Early Twentieth Century," *British Journal for the History of Science* 51:3 (2018), pp. 423–455.

Evans, Susan, "The Garden of the Aztec Philosopher-King," in Dan O'Brien (ed.), *Gardening—Philosophy for Everyone: Cultivating Wisdom* (Chichester, UK: Wiley-Blackwell, 2010), pp. 207–219.

Fenner, David, "Aesthetic Appreciation in the Artworld and in the Natural World," *Environmental Values* 12:1 (2003), pp. 3–28.

Fenner, David, "The Aesthetic Impact of the Garden of Eden," *Contemporary Aesthetics* 20 (2022).

Fenner, David, "Environmental Aesthetics and the Dynamic Object," *Ethics and the Environment* 11:1 (2006), pp. 1–19.

Fenner, David, "Gardens and Plasticity," *Contemporary Aesthetics* 21 (2023).

Ferrari, G. R. F., "The Meaninglessness of Gardens," *Journal of Aesthetics and Art Criticism* 68:1 (2010), pp. 33–45.

Fish, Stanley, *Is There a Text in This Class?* (Cambridge, MA: Harvard University Press, 1980).

Francis, Mark, and Randolph T. Hester Jr. (eds.), *The Meaning of Gardens: Idea, Place, and Action* (Cambridge, MA: MIT Press, 1990).

Frydryczak, Beata, "Landscape Garden as a Paradigmatic Model of Relationships between Human and Nature," *Dialogue and Universalism* 24:4 (2014), pp. 103–114.

Gaskell, Ivan, and Salim Kemal (eds.), *Landscape, Beauty, and the Arts* (Cambridge: Cambridge University Press, 1993).

Gaut, Berys, "The Ethical Criticism of Art," in Jerrold Levinson (ed.), *Aesthetics and Ethics: Essays at the Intersection* (Cambridge, UK: Cambridge University Press, 1998), pp. 182–203.

Gillette, Jane, "Can Gardens Mean?" *Landscape Journal* 24:1 (2005), pp. 85–97.

Goldman, Alan H., "Aesthetic Qualities and Aesthetic Value," *Journal of Philosophy* 87:1 (1990), pp. 23–37.

Goldman, Alan H., *Aesthetic Value* (Boulder, CO: Westview Press, 1995).

Goldman, Alan H., "Aesthetic versus Moral Evaluations," *Philosophy and Phenomenological Research* 50:4 (1990), pp. 715–730.

Goldman, Alan H., "The Experiential Account of Aesthetic Value," *Journal of Aesthetics and Art Criticism* 64:3 (2006), pp. 333–342.

Goldman, Alan H., "Interpreting Art and Literature," *Journal of Aesthetics and Art Criticism* 48:3 (1990), pp. 205–214.

Goldman, Alan H., *Philosophy and the Novel* (Oxford, UK: Oxford University Press, 2013).

Goodman, Nelson, *Languages of Art: An Approach to a Theory of Symbols* (Indianapolis, IN: Hackett Publishing Company, 1976).

Graham, Gordon, "Art and Architecture," *British Journal of Aesthetics* 29:3 (1989), pp. 248–249.

Hall, Matthew, "Escaping Eden: Plant Ethics in a Gardener's World," in Dan O'Brien (ed.), *Gardening—Philosophy for Everyone: Cultivating Wisdom* (Chichester, UK: Wiley-Blackwell, 2010), pp. 38–47.

Hargrove, Eugene, *Foundations of Environmental Ethics* (Englewood Cliffs, NJ: Prentice Hall, 1989).

Harrison, Robert Pogue, *Gardens: An Essay on the Human Condition* (Chicago, IL: University of Chicago Press, 2008).

Hegel, G. W. F., *Aesthetics: Lectures on Fine Art* (Oxford, UK: Oxford University Press, 1975).

Hepburn, Ronald, "Contemporary Aesthetics and the Neglect of Natural Beauty," in Bernard Williams and Alan Montefiore (eds.), *British Analytical Philosophy* (London, UK: Routledge & Kegan Paul, 1966), pp. 43–62.

Hepburn, Ronald, *The Reach of the Aesthetic: Collected Essays on Art and Nature* (New York, NY: Routledge, 2019).

Hepburn, Ronald, "Trivial and Serious in Aesthetic Appreciation of Nature" in Salim Kemal and Ivan Gaskell (eds.), *Landscape, Natural Beauty, and the Arts* (Cambridge, UK: Cambridge University Press, 1995), pp. 65–80.

Hernández, Jo Farb, "Peter's Garden: Case Study of a Spanish Art Environment," *Environment, Space, Place* 6:1 (2014), pp. 97–124.

Herrington, Susan, "Gardens Can Mean," *Landscape Journal* 26:2 (2007), pp. 302–317.

Hettinger, Ned, "Evaluating Positive Aesthetics," *Journal of Aesthetic Education* 51:3 (2017), pp. 26–41.

Heyd, Thomas, "Thinking Through Botanic Gardens," *Environmental Values* 15:2 (2006), pp. 197–212.

Hipple, Walter J., Jr., "Review of *Jacques Boyceau and the French Formal Garden* by Franklin Hamilton Hazelhurst," *Journal of Aesthetics and Art Criticism* 26:4 (1968), pp. 548–549.

Hirsch, E. D., *Validity in Interpretation* (New Haven, CT: Yale University Press, 1967).

Hume, David, "Of the Standard of Taste," *Four Dissertations* (New York, NY: Garland Press, 1970).

Hunt, John Dixon, "Gardens: Historical Overview," in Michael Kelly (ed.), *Encyclopedia of Aesthetics* (Oxford, UK: Oxford University Press, 1998), pp. 271–274.

Hunt, John Dixon, *Greater Perfections: The Practice of Garden Theory* (London, UK: Thames and Hudson, 2000).

Hunt, John Dixon, "Greater Perfections: The Practice of Garden Theory," *Journal of Aesthetics and Art Criticism* 59:3 (2001), pp. 341–343.

Kallhoff, Angela, and Maria Schörgenhumer, "The Virtues of Gardening: A Relational Account of Environmental Virtues," *Environmental Ethics* 39:2 (2017), pp. 193–210.

Kant, Immanuel, *Critique of Judgment* (Indianapolis, IL: Hackett, 1987).

Katz, Eric, "Geoengineering, Restoration, and the Construction of Nature," *Environmental Ethics* 37:4 (2015), pp. 485–498.

Kivy, Peter, *The Seventh Sense: Francis Hutcheson and Eighteenth-Century British Aesthetics* (Oxford, UK: Clarendon Press, 2003).

Langer, Susanne, *Feeling and Form* (New York, NY: Charles Scribner's Sons, 1953).

Leddy, Thomas, "The Garden as an Art," *International Studies in Philosophy* 28:4 (1996), pp. 126–127.

Leddy, Thomas, "Gardens in an Expanded Field," *British Journal of Aesthetics* 28:4 (1988), pp. 327–340.

Lee, Michael G., *The German "Mittelweg": Garden Theory and Philosophy in the Time of Kant* (New York, NY: Routledge, 2013).

Leopold, Aldo, *A Sand County Almanac* (Oxford, UK: Oxford University Press, 1949).

Levinson, Jerrold (ed.), *Aesthetics and Ethics: Essays at the Intersection* (Cambridge, UK: Cambridge University Press, 1998).

Levinson, Jerrold, "Defining Art Historically," *British Journal of Aesthetics* 19 (1979), pp. 232–250.

Levinson, Jerrold, "Extending Art Historically," *Journal of Aesthetics and Art Criticism* 51:3 (1993), pp. 411–423.

Levinson, Jerrold, "The Irreducible Historicality of the Concept of Art," *British Journal of Aesthetics* 42:2 (2002), pp. 367–379.

Levinson, Jerrold, "Refining Art Historically," *Journal of Aesthetics and Art Criticism* 47:1 (1989), pp. 21–33.

Levinson, Jerrold, and Philip Alperson, "What Is a Temporal Art?," *Midwest Studies in Philosophy* 16 (1991), pp. 439–450.

Liu, Yu, "The Metaphysics of Disinterestedness: the Chinese Gardening Style and Shaftesbury's New Aesthetics," *European Legacy* 9:2 (2004), pp. 195–212.

Liu, Yu, "Transplanting a Different Gardening Style into England: Matteo Ripa and His Visit to London in 1724," *Diogenes* 55:2 (2008), pp. 83–96.

Loftin, Robert, "Psychical Distance and the Aesthetic Appreciation of Wilderness," *International Journal of Applied Philosophy* 3 (1986), pp. 15–19.

MacDonald, Eric, "Hortus Incantans: Gardening as an Art of Enchantment," in Dan O'Brien (ed.), *Gardening—Philosophy for Everyone: Cultivating Wisdom* (Chichester, UK: Wiley-Blackwell, 2010), pp. 121–134.

Miller, Mara, *The Garden as an Art* (New York, NY: State University of New York Press, 1993).

Miller, Mara, "The Garden as an Art," *Journal of Aesthetics and Art Criticism* 52:4 (1994), pp. 480–482.

Miller, Mara, "The Garden as Significant Form," *Journal of Speculative Philosophy* 2:4 (1988), pp. 267–287.

Miller, Mara, "Gardens as Art," in Michael Kelly (ed), *Encyclopedia of Aesthetics* (Oxford, UK: Oxford University Press, 1998), pp. 274–280.

Miller, Mara, "Gardens as Works of Art: The Problem of Uniqueness," *British Journal of Aesthetics* 26:3 (1986), pp. 252–256.

Miller, Mara, "Time and Temporality in the Garden," in Dan O'Brien (ed.), *Gardening—Philosophy for Everyone: Cultivating Wisdom* (Chichester, UK: Wiley-Blackwell, 2010), pp. 178–191.

Mitias, Michael H. (ed.), *Philosophy and Architecture* (Amsterdam, NL: Rodopi, 1984).

Mitias, Michael H., *What Makes an Experience Aesthetic?* (Amsterdam, NL: Rodopi, 1988).

Moore, Charles W., William J. Mitchell, and William Turnbull Jr., *The Poetics of Gardens* (Cambridge, MA: MIT Press, 1988).

Moore, Ronald, "The Framing Paradox," *Ethics, Place and Environment* 9:3 (2006), pp. 249–267.

Moore, Ronald, "A Review of David E. Cooper's A Philosophy of Gardens," *Journal of Aesthetic Education* 41:3 (2007), p. 122.

Moss, Michael, "Brussels Sprouts and Empire: Putting Down Roots," in Dan O'Brien (ed.), *Gardening—Philosophy for Everyone: Cultivating Wisdom* (Chichester, UK: Wiley-Blackwell, 2010), pp. 79–92.

Nun, Nathan, "Practical Aesthetics: Community Gardens and the New Sensibility," *Radical Philosophy Review* 16:2 (2013), pp. 663–677.

Nussbaum, Martha, "Exactly and Responsibly: A Defense of Ethical Criticism," *Philosophy and Literature* 22:2 (1998), pp. 343–365.

O'Brien, Dan (ed.), *Gardening—Philosophy for Everyone: Cultivating Wisdom* (Chichester, UK: Wiley-Blackwell, 2010).

Olin, Laurie, "Form, Meaning, and Expression in Landscape Architecture," *Landscape Journal* 7:2 (1988), pp. 149–168.

Orenstein, Gloria Feman, "The Greening of Gaia: Ecofeminist Artists Revisit the Garden," *Ethics and the Environment* 8:1 (2003), pp. 102–111.

Paden, Roger, "A Defense of the Picturesque," *Environmental Philosophy* 10:2 (2013), pp. 1–21.

Paden, Roger, "The Ethical Function of Landscape Architecture," *Environmental Philosophy* 15:2 (2018), pp. 139–158.

Paden, Roger, "Picturesque Landscape Painting and Environmental Aesthetics," *Journal of Aesthetic Education* 49:2 (2015), pp. 39–61.

Parkes, Graham (ed.), *Reading Zen in the Rocks: The Japanese Dry Landscape Garden* (Chicago, IL: University of Chicago Press, 2000).

Parsons, Glenn, "Nature Appreciation, Science, and Positive Aesthetics," *British Journal of Aesthetics* 42:3 (2002), pp. 279–295.

Parsons, Glenn, "Theory, Observation, and the Role of Scientific Understanding in the Aesthetic Appreciation of Nature," *Canadian Journal of Philosophy* 36:2 (2006), pp. 165–186.

Parsons, Glenn, and Allen Carlson, "New Formalism and the Aesthetic Appreciation of Nature," *Journal of Aesthetics and Art Criticism* 62:4 (2004), pp. 363–376.

Parsons, Russ, and Terry Daniel, "Good Looking: In Defense of Scenic Landscape Aesthetics," *Landscape and Urban Planning* 60:1 (2002), pp. 43–56.

Pollan, Michael, *Second Nature: A Gardener's Education* (New York, NY: Dell, 1991).

Ralston, Shane, "A Deweyan Defense of Guerrilla Gardening," *The Pluralist* 7:3 (2012), pp. 57–70.

Ray, Meghan, "Cultivating the Soul: The Ethics of Gardening in Ancient Greece and Rome," in Dan O'Brien (ed.), *Gardening—Philosophy for Everyone: Cultivating Wisdom* (Chichester, UK: Wiley-Blackwell, 2010), pp. 26–37.

Robinson, Jenefer, "Style and Personality in the Literary Work," *Philosophical Review* 94 (1985), pp. 227–248.

Rogers, Elizabeth Barlow, *Landscape Design: A Cultural and Architectural History* (New York, NY: Abrams, 2001).

Rolston, Holmes, "Does Aesthetic Appreciation of Nature Need to Be Science Based?" *British Journal of Aesthetics* 35:4 (1995), pp. 374–386.

Rolston, Holmes, *Philosophy Gone Wild* (Buffalo, NY: Prometheus Books, 1986).

Ross, Stephanie, "*Ut Hortis Poesis*—Gardening and Her Sister Arts in Eighteenth-Century England," *British Journal of Aesthetics* 25:1 (1985), pp. 17–32.

Ross, Stephanie, *What Gardens Mean* (Chicago, IL: University of Chicago Press, 1998).

Saito, Yuriko, "The Aesthetics of Unscenic Nature," *Journal of Aesthetics and Art Criticism* 56:2 (1998), pp. 101–11.

Saito, Yuriko, "Appreciating Nature on Its Own Terms," *Environmental Ethics* 20:2 (1998), pp. 135–149.

Salwa, Mateusz, "The Garden as a Performance," *Estetika* 51:1 (2014), pp. 42–61.

Santayana, George, *The Sense of Beauty* (New York, NY: Scribner's, 1936).

Schmidt, Dennis J., "Klee's Gardens," *Research in Phenomenology* 43:3 (2013), pp. 394–404.

Schopenhauer, Arthur, *The World as Will and Idea* (London, UK: Routledge & Kegan Paul, 1896).

Scruton, Roger, *The Aesthetics of Architecture* (London, UK: Methuen, 1979).

Scruton, Roger, *Art and Imagination* (London, UK: Methuen, 1974).

Shadidi, Mohammadsharif, Mohamad Reza Bemanian, Nina Almasifar, and Hanie Okhovat, "A Study on Cultural and Environmental Basics at Formal Elements of Persian Gardens (before & after Islam)," *Asian Culture and History* 2:2 (2010), pp. 133–147.

Shaftesbury, Anthony, *Characteristics of Men, Manners, Opinions, Times* (New York, NY: Bobbs-Merrill, 1964).

Shapiro, Gary, "The Pragmatic Picturesque: The Philosophy of Central Park," in Dan O'Brien (ed.), *Gardening—Philosophy for Everyone: Cultivating Wisdom* (Chichester, UK: Wiley-Blackwell, 2010), pp. 148–160.

Shelley, James, "The Problem of Non-Perceptual Art," *British Journal of Aesthetics* 43:4 (2003), pp. 363–378.

Sibley, Frank, "Aesthetic Concepts," *Philosophical Review* 68:4 (1959), pp. 421–450.

Silvers, Anita, "The Story of Art Is the Test of Time," *Journal of Aesthetics and Art Criticism* 49:3 (1991), pp. 211–224.

Singer, Peter, *Animal Liberation* (New York, NY: HarperCollins, 1975).

Sontag, Susan, "Against Interpretation," in *Against Interpretation* (New York, NY: Farrar, Straus and Giroux, 1966), pp. 3–14.

Stewart, Robert, and Roderick Nicholls, "Virtual Worlds, Travel, and the Picturesque Garden," *Philosophy and Geography* 5:1 (2002), pp. 83–99.

Stolnitz, Jerome, "The Aesthetic Attitude' in the Rise of Modern Aesthetics," *Journal of Aesthetics and Art Criticism* 36 (1978), pp. 409–422.

Stolnitz, Jerome, *Aesthetics and Philosophy of Art Criticism* (New York, NY: Houghton Mifflin, 1960).

Stolnitz, Jerome, "The Artistic Values in Aesthetic Experience," *Journal of Aesthetics and Art Criticism* 32 (1973), pp. 5–15.

Stolnitz, Jerome, "On the Origins of 'Aesthetic Disinterestedness,'" *Journal of Aesthetics and Art Criticism* 20 (1961), pp. 131–144.

Stolnitz, Jerome, "On the Significance of Lord Shaftesbury in Modern Aesthetic Theory," *Philosophical Quarterly* 11 (1961), pp. 97–113.

Stolnitz, Jerome, "A Third Note on Eighteenth-Century 'Disinterestedness,'" *Journal of Aesthetics and Art Criticism* 22 (1963), pp. 69–70.

Stuart, Rory, *What Are Gardens For?* (London, UK: Franklin Lincoln Limited, 2012).

Townsend, Dabney, "Archibald Alison: Aesthetic Experience and Emotion," *British Journal of Aesthetics* 28:2 (1988), pp. 132–144.

Townsend, Dabney, "From Shaftesbury to Kant: The Development of the Concept of Aesthetic Experience," *Journal of the History of Ideas* 48 (1987), pp. 287–305.

Treib, Marc (ed.), *Meaning in Landscape Architecture and Gardens* (New York, NY: Routledge, 2011).

Treib, Marc, "Must Landscapes Mean? Approaches to Significance in Recent Landscape Architecture," *Landscape Journal* 14:1 (1995), pp. 47–62.

Turner, Tom, *Garden History: Philosophy and Design 2000 BC to 2000 AD* (New York, NY: Spon Press, 2005).

Van Tonder, Gert Jakobus, "Visual Geometry of Classical Japanese Gardens," *Axiomathes* 32 (2022), pp. 841–868.

Van Tonder, Gert Jakobus, and Michael J. Lyons, "Visual Perception in Japanese Rock Garden Design," *Axiomathes* 15:3 (2005), pp. 353–371.

Van Wensween, Louke, *Dirty Virtues: The Emergence of Environmental Virtue Ethics* (Amherst, NY: *Prometheus Books*, 2000).

Walton, Kendall L., "Categories of Art," *Philosophical Review* 79 (1970), pp. 334–367.

Wear, Andrew, "The Garden as a Fine Art," *British Journal of Aesthetics* 20:4 (1980), p. 376.

Wilson, Edward O., *Biophilia* (Cambridge, MA: Harvard University Press, 1986).

Wimsatt, William K., and Monroe C. Beardsley, "The Intentional Fallacy," *Sewanee Review* 54 (1946), pp. 3–23.

Wimsatt, William, and Cleanth Brooks, "Art for Art's Sake," in their *Literary Criticism: A Short History* (New York, NY: Alfred A. Knopf, 1969), pp. 475–498.

Young, Damon, *Philosophy in the Garden* (Minneapolis, MN: Scribe Publications, 2020).

Zangwill, Nick, "Feasible Aesthetic Formalism," *Nous* 33:4 (1999), pp. 610–629.

Zangwill, Nick, "In Defence of Moderate Aesthetic Formalism," *Philosophical Quarterly* 50 (2000), pp. 476–493.

Zemach, Eddy, "Aesthetic Properties, Aesthetic Laws, and Aesthetic Principles," *Journal of Aesthetics and Art Criticism* 46:1 (1987), pp. 67–73.

Zemach, Eddy, "Emotion and Fictional Beings," *Journal of Aesthetics and Art Criticism* 54:1 (1996), pp. 41–48.

Zemach, Eddy, "Real Beauty," *Midwest Studies in Philosophy* 16 (1991), pp. 249–265.

Zemach, Eddy, *Real Beauty* (University Park, PA: Pennsylvania University Press, 1997).

Index

For the benefit of digital users, indexed terms that span two pages (e.g., 52–53) may, on occasion, appear on only one of those pages.